NEW CONCEPTS AND TECHNOLOGIES IN PARALLEL INFORMATION PROCESSING

NATO ADVANCED STUDY INSTITUTES SERIES

Proceedings of the Advanced Study Institute Programme, which aims at the dissemination of advanced knowledge and the formation of contacts among scientists from different countries.

The series is published by an international board of publishers in conjunction with NATO Scientific Affairs Division

A	Life Sciences	Plenum Publishing Corporation
B	Physics	London and New York
C	Mathematical and Physical Sciences	D. Reidel Publishing Company Dordrecht and Boston
D	Behavioural and Social Sciences	Sijthoff International Publishing Company Leyden
E	Applied Sciences	Noordhoff International Publishing Leyden

Series E: Applied Sciences
Volume 9. New Concepts and Technologies in Parallel Information Processing

NEW CONCEPTS AND TECHNOLOGIES IN PARALLEL INFORMATION PROCESSING

edited by

E. R. CAIANIELLO

Director of the Laboratory
of Cybernetics, CNR
Arco Felice, Italy

NOORDHOFF – LEYDEN – 1975

Proceedings of the NATO Advanced Study Institute
on New Concepts and Technologies in Parallel Information Processing,
held at Capri, Italy, June 17-30, 1973

ISBN-13: 978-94-010-1919-4 e-ISBN-13: 978-94-010-1917-0
DOI: 10.1007/978-94-010-1917-0

FOREWORD to the Proceedings of the NATO ADVANCED STUDY INSTITUTE on "New Concepts and Technologies in Parallel Information Processing".

The field of Parallel Information Processing is undergoing a phase of exponential growth; interdisciplinary by its very nature, it comprises a wide range of researches, from sophisticated solid state technologies to abstract mathematics. The point of view of the engineer, of the biologist, of the physicist and of the abstract thinker are each indispensable; it is to be emphasized, though, that the more each contribution shows to be a concrete part of a well established field of investigation, the more it acquires value. Parallel Information Processing is in fact a very wide term, which may easily be rephrased as "Artificial Intelligence", "General Systems Theory", "Cybernetics" or some such wording; these denominations are in fact more or less synonymous, the use of one or the other depending more upon local tradition and personal inclination than on the nature of the contributions made by scientists: whatever the name, what really stands ahead of us is a new paradigm in science, that calls for a global philosophy and methodology. Discussions about generalities turn out quite often, however, to be regrettably void of content; maturity is achieved only when global aims and perspectives form, so to say, the background and the motivation: scientific contributions should stay strictly technical, in the field of which each is master. An interdisciplinary endeavour, with ambitions so great as those which are declaredly shared by all who take part or interests in Parallel Information Processing, can make sound scientific sense only if two conditions are fulfilled: the first is that a deep sense of humility be felt by everybody concerned, as he faces a task certainly bigger not only than his personal lifetime, but his generation; the second, that the general perspective be conceived, as I just said, as the design of a mosaic, of which each scientist must be aware but to which he can at most contribute one or a few tassels.

It was the general consensus of many of the leaders in this field, with whom I have been long discussing the matter, that the time was just right to organize a gathering of scientists all interested in the subject, each either distinguished for notable contributions already made, or promising for interest and alertness. The International School of which the Proceedings are reported here was the result of these consultations, and purported to be at the same time a School in the true sense of the word, to teach and stimulate brilliant young ones, a Symposium in which new ideas and contributions should be presented and a Forum for frank discussions and criticism. According to the tradition fruitfully established in many previous ventures of this sort, with which I was involved in the past, once a substantial group of main lecturers and an appropriate site were chosen, the detailed organization was left completely spontaneous; daily lectures and discussions were arranged by common agreement, in discussions chaired mostly by Professor A. C. Scott and S. Winograd (to whom I am indebted also in many other ways). The reader of these proceedings will notice, it is hoped, unity through disparity; each lecture was felt by the audience to be a piece of a larger picture the ordering given ranging from solid state physics to abstract models or theories. Young scientists were also given their chance, with results not to be regretted. I feel I express the general feeling of the participants, as it was many times kindly conveyed to me, by stating that this A. S. I. was a happy and refreshing compromise between the narrowly specialized or overly generalizing customary meetings. Although the special atmosphere of Capri, an ideal spot for living together and communing in meditation, cannot certainly be recaptured here and has to stay in the memory of those who honoured us with their presence, I do very much hope that this Volume may nevertheless be of use, in bringing to a wider audience the essential scientific content of this School.

My warmest thanks go to NATO (Advanced Study Institute Programme) and to the Gruppo Nazionale di Cibernetica e Biofisica of the Italian National Research Council, whose generous support rendered this School possible.

Eduardo R. Caianiello

TABLE OF CONTENTS

ELEMENTS OF SUPERCONDUCTIVE JOSEPHSON TUNNELING TECHNOLOGY

Antonio Barone

Laboratorio di Cibernetica del CNR -Arco Felice-Naples
and
Istituto di Fisica Universita di Napoli, Naples-Italy

INTRODUCTION

Parallelism in electronic system organization can provide increased computer performance. Particular problems can in fact be more easily solved by "clever" systems which employ parallel processing combined with a little imagination in the design, depending on the particular task to be performed. However, the basic advantage over sequential procedures is that when all the elements are independently and simultaneously processed higher speeds of operation can be achieved. In practice significant improvements toward parallel processing imply the development of suitable large-scale integration technologies.

The investigation of superconductive tunneling junctions not only yields important results in the general aspects of superconductivity but also leads to a wide variety of interesting applications in different fields. Josephson tunnel junctions[1], as high speed and low power computing elements, have been considered in the last few years by several investigators from both theoretical and experimental points of view.

The main purpose of the present lecture is to give the physical background of superconductive tunneling to provide the basis for the more specific lectures which will follow*.

*The reader is referred to the lectures by R.D. Parmentier and by A.C. Scott in this volume.

The state of the art of Josephson tunneling structures as possible devices to be applied in the area of information processing and storage is also briefly outlined.

2. - BASIC THEORY

The superconducting state is a macroscopic quantum state. This is described by the association of a single wave function Ψ to a macroscopic number of electrons which are assumed to "condense" in the same quantum state. There is no contraddiction with Pauli principle since the theory deals with particles which are infact pairs of electrons (Cooper pairs). The paring results from an attractive phonon exchange interaction which is larger than the Coulomb repulsion. The macroscopic wave function is of the form $\Psi = \rho^{\frac{1}{2}} e^{i\phi}$, where ϕ is the phase common to all the particles and $\rho = |\Psi|^2$ represents in this picture the actual density of the particles.

Since the number of pairs N and the phase ϕ are conjugate variables [2] there is an uncertainly relation $\Delta N \Delta \phi \sim 2\pi$. Therefore, within an isolated superconductor, N will be fixed and, consequently, the absolute phase ϕ undefined. On the contrary, if we consider a system of two "weakly coupled" superconductors a transmission of pairs can occur and the number of pairs in each superconductor will change determining the relative quantum phase. Josephson theory is concerned with the behaviour of systems of two superconductors coupled by any link which is able to "transmit" phase coherence [1] across the boundary.

We shall confine our attention to the tunneling structures neglecting other types of weak links. The next section is devoted to the tunneling of single particles. Josephson effect is discussed in section 2b.

2a) ELECTRON TUNNELING

The superconductive tunneling has been introduced with the experiments performed by Giaever [3]. The tunneling structure consists of two metal films separated by a thin dielectric barrier (Fig.1). A great deal of information can be obtained by measuring the dependence of the tunneling current on the voltage applied across the junction. A semi-phenomenological theory which agrees very well with the original experimental results has been discussed by Giaever and Megerle [4]. The basic assumption of this approach is that the quasi-particles[*] can be regarded as fermions. The tunneling current $i_{1\to 2}$ from metal 1 to metal 2 is supposed to be proportional to the density of filled states in metal 1, to the density of empty states in metal 2 and to the tunneling probability. All quantities referred to a given energy (specular tunneling).

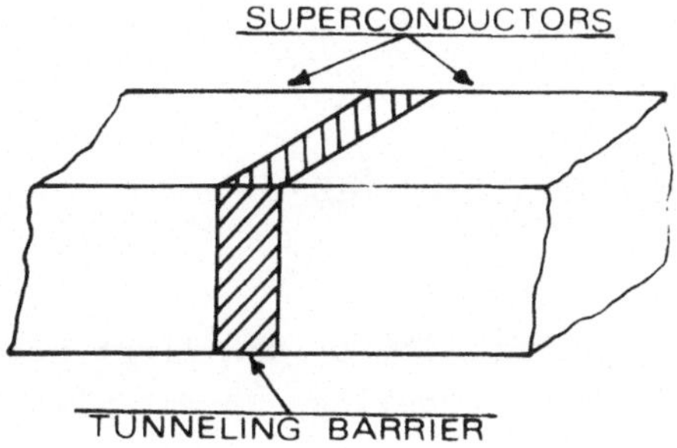

Fig. 1

Schematic tunnel junction structure

[*] The term quasi-particles indicates the elementary excitations in the superconductive state. In spite of a lack of rigor it is often convenient to identify quasi-particles with electrons. In particular the superconductive tunneling can be regarded in the framework of the more familiar electron hole semiconductor picture.

Thus, integrating over all the energies, we have

$$i_{1\to 2} = \frac{2\pi}{\hbar}\int_{-\infty}^{\infty} |M|^2 n_1 F_1 n_2 (1-F_2)\, dE$$

where F_1 and F_2 are the Fermi factors $F = (1+e^{\beta E})^{-1}$ and n_1, n_2 are the densities of states in the two metals respectively. M is the matrix element connecting states of same energy between the two metals which is assumed to be energy independent. Following same arguments we obtain the expression for the tunneling current from metal 1 to metal 2

$$i_{2\to 1} = \frac{2\pi}{\hbar}\int_{-\infty}^{\infty} |M|^2 n_1 n_2 F_2 (1-F_1)\, dE$$

where the same tunneling probability of the first process has been assumed. Thus, the actual tunneling current will be given by

$$I = i_{1\to 2} - i_{2\to 1} = \frac{2\pi}{\hbar} |M|^2 \int_{-\infty}^{\infty} n_1 n_2 (F_1 - F_2)\, dE$$

For an applied voltage V across the junction the Fermi energy levels will be relatively shifted by the energy eV so that

$$I = \frac{2\pi}{\hbar} |M|^2 \int_{-\infty}^{\infty} n_1(E)\, n_2(E+eV) \left[F_1(E) - F_2(E+eV)\right] dE \qquad (1)$$

The densities of states for the normal metals are supposed to be constant, in the limit of low temperatures and small voltages, equation 1 can therefore be integrated. The tunneling current I_{NN} between normal metals is then

$$I_{NN} = \text{constant} \int_{-\infty}^{\infty} \left[F(E) - F(E+V)\right] dE$$

which gives

$$I_{NN} = GV$$

The constant G can be interpreted as the conductance of the normal junction which, therefore, exhibits an ohmic behaviour.

For the superconductor - superconductor tunneling it is assumed that the matrix element M is unchanged. The densities of states are now referred to superconductors so that we can use the expressions given by the B.C.S. theory

$n_s = n_n |E| / |E^2-\Delta^2|^{\frac{1}{2}}$ for $|E| \geq \Delta$ and $n_s = 0$ for $|E| < \Delta$, where Δ is the energy gap. Therefore the tunneling current I_{ss} between two superconductors is obtained specifying equation 1 as

$$I_{ss} = \text{const} \int_{-\infty}^{\infty} \frac{|E|}{\sqrt{E^2-\Delta_1^2}} \frac{|E+eV|}{\sqrt{(E+eV)^2 - \Delta_2^2}} |F(E) - F(E+eV)| \, dE$$

The numerical calculation of this integral leads to a logarithmic singularity for I_{ss} at a voltage $V = \frac{|\Delta_1 + \Delta_2|}{e}$ and a finite discontinuity at $V = \pm \frac{1}{e} |\Delta_2 - \Delta_1|$ [5]. The complete voltage-current characteristic obtained in this way is sketched in fig. 2a. Fig. 2b shows a typical V - I characteristics of a superconductive tunnel junction made in our laboratory.

In spite of the simplicity of the model just outlined, it does predict almost all the essential features of the phenomenon. Various Authors[6-8] gave important contributions to the theory of the tunneling justifing " a posteriori " the more or less strong assumptions in the original semiphenomenological model. The Hamiltonian formulation for the electron tunneling has been given by Cohen et al.[7]. In this approach the system is described by

$$H = H_o + H_T$$

2a)

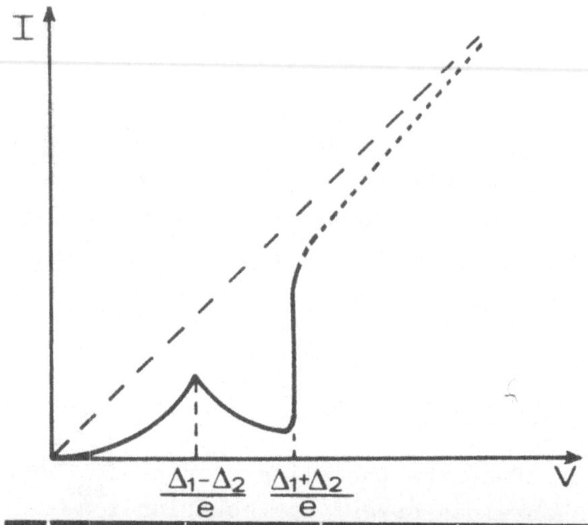

2b)

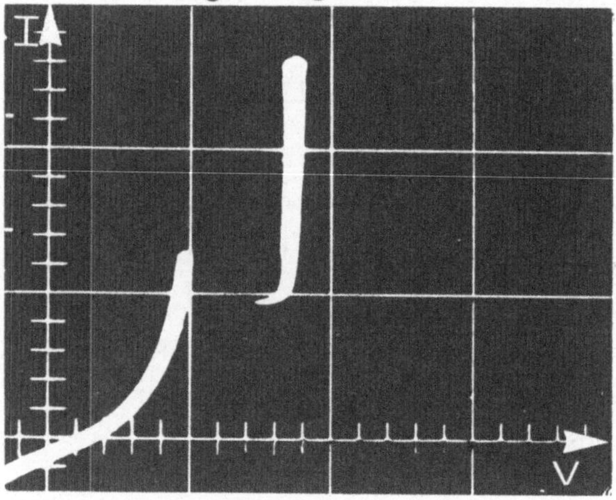

Fig. 2

2a) Sketch of the theoretical V-I characteristic of junction with different superconductors.

2b) Observed V-I characteristic for a Sn - Sn_xO_y - Pb junction.

where H_o is the Hamiltonian of the two non-interacting superconductors and

$$H_T = \sum_{K,K',\sigma} T_{KK'} (C^+_{K\sigma} d_{K'\sigma} + C^+_{K'\sigma} d_{K\sigma}) + H.C.$$

is the tunneling interaction Hamiltonian. $T_{KK'}$ is the tunneling matrix element connecting single particle state K on one side of the barrier and K' on the other side*. In this picture the "passage" of an electron through the barrier is described by the destruction of an electron on one side (state K) and the simultaneous creation of an electron on the other side (state K'). Electron interactions inside the barrier are not taken in to account. An extensive discussion of the argument of the present section can be found in the review article by Douglass and Falicov[9].

2b) JOSEPHSON TUNNELING

The Josephson equations can be obtained in different ways. We will follow the simple derivation given by Scalapino[10] and based on the model of Ferrel and Prange[11] (analogy with the tight binding approximation).

Let us consider a system of two superconductors S_1 and S_2 which possess N_1 and N_2 electrons respectively. The two ground states will be described by the macroscopic wave functions $\Psi^{(1)}_o(N_1)$ and $\Psi^{(2)}_o(N_2)$. The ground state of the total system (i.e. two isolated superconductors) is given by

$$\Psi_o = A\Psi^{(1)}_o(N_1)\ \Psi^{(2)}_o(N_2) \text{ with energy } E_o = E^{(1)}_o(N_1)+E^{(2)}_o(N_2)$$

---

* The time reversal symmetry of H_T implies $T_{K,K',\sigma} = T_{-K-K'\sigma} = T_{KK'}$

A represents the antisymmetrization operator. Consider now all the states generated by transferring different numbers of electron pairs from S_1 to S_2. The generic state Ψ_n corresponding to the transfer of n pairs (2n electrons) will be:

$$\Psi_n = A\,\Psi_o^{(1)}(N_1 - 2n)\,\Psi_o^{(2)}(N_2 + 2n) \text{ with energy given by}$$

$$E_n = E_o^{(1)}(N_1 - 2n) + E_o^{(2)}(N_2 + 2n) + \frac{Q^2}{2C}$$

where $Q = 2nC$ and C is the relative capacitance of the two superconductors. By indicating with μ_1 and μ_2 the chemical potentials of S_1 and S_2, we have

$$E_n = E_o + \frac{Q^2}{2C} + 2n\,(\mu_2 - \mu_1).$$

If $\mu_1 = \mu_2$, neglecting the electrostatic term, is

$$E_n = E_o$$

Therefore all states Ψ_n will be degenerate in energy with Ψ_o. When the two superconductors are not anymore isolated but separated just by a microscopic distance (in practice a dielectric barrier about 20 Å thick) the electron pairs can flow from one superconductor to the other by means of tunneling effect. Since H_T transfers only single electron, whereas all the Ψ_n's differ for pairs ($\langle \Psi_m | H_T | \Psi_n \rangle = 0$), in order to remove the degenerancy we have to consider the second order tunneling interaction Hamiltonian $H_T^{(2)} = H_T\,(E_o - H_o)^{-1}\,H_T$. The matrix elements

$$\langle \Psi_m | H_T (E_o - H_o)^{-1} H_T | \Psi_n \rangle \qquad \text{connect state } n = m \text{ and } n = m \pm 1.$$

Following the degenerate-state perturbation theory we have to diagonalize the matrix $\varepsilon_{ij} = \langle m | H_T^{(2)} | n \rangle$ to obtain the energy splitting produced by the interaction. The resulting eigenstates[10] will be given by the following linear combination of the Ψ_n's

$$\Psi_\phi = \sum_n e^{in\phi}\,\Psi_n \qquad (3)$$

with the corresponding eigenvalues

$$E_\phi = E_o + \frac{<\Psi\phi|H_T^{(2)}|\Psi\phi>}{<\Psi\phi|\Psi\phi>}$$

from 3 and with appropriate normalization

$$E_\phi = E_o + \sum_n \sum_m <e^{im\phi}\Psi_m|H^{(2)}|e^{in\phi}\Psi_n>.$$

since n = m or n = m ± 1, we have

$$E_\phi = E_o + (-E_1) + (e^{i\phi} - e^{-i\phi})\,(-\tfrac{1}{2}E_1) \quad \text{thus}$$

$$E_\phi = E_o - E_1 - E_1\cos\phi$$

$(-E_1)$ and $(-\frac{1}{2}E_1)$ are the values of the matrix elements for n = m and n = m ± 1 respectively (for their explicit values see reference 10). If we consider a phase change $\Delta\phi$ of the pairs in one superconductor relative to the phase of the other will be

$\Psi_n \rightarrow e^{in\Delta\phi}\,\Psi_n$ and the coherent superposition $\Psi\phi$ will transform as $\Psi_\phi \rightarrow \Psi_{\phi+\Delta\phi}$.

Therefore ϕ can be interpreted as the phase difference between the two superconductors[1,10]

Since n and ϕ are conjugate variables[1,2], $[n,\phi] = -i$, the Hamilton equations will give

$$\begin{cases} i\hbar \dfrac{\partial n}{\partial t} = i\,\dfrac{\partial E}{\partial \phi} \\[2ex] \hbar\,\dfrac{\partial \phi}{\partial t} = -\dfrac{\partial E}{\partial n} \end{cases} \Longrightarrow \begin{cases} \dfrac{i\hbar}{2e}\,I = i\,E_1 \sin\phi \\[2ex] \hbar\,\dfrac{\partial \phi}{\partial t} = 2eV \end{cases}$$

which can be written

$$I = I_o \sin\phi \qquad \text{and} \qquad \frac{\partial\phi}{\partial t} = \frac{2e}{\hbar} V \tag{4}$$

where $I = \frac{2e}{\hbar} E_1$. These are the Josephson equations. We can see that for V = 0 the phase difference ϕ is a constant which can be not zero, so that a finite current (d.c.) of maximum value I can flow through the barrier even with a zero-voltage across the junction (see fig. 3). Qualitatively this can be regarded as an "extension" of the superconductive properties over the all structure including the barrier.

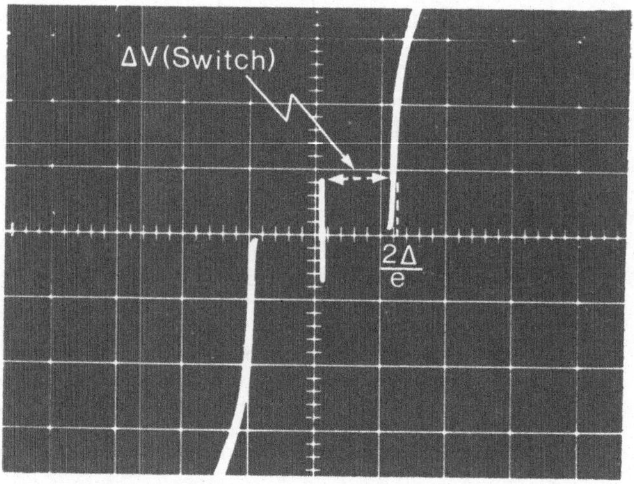

Fig. 3

V-I characteristic of a Sn- Sn_xO_y - S_n Josephson junction at T = 1.99 K;Hor. 1 mV/div Vert. 50 mA/div.

For this reason the term "weak superconductivity" has been coined for such weakly coupled structures. If V = const $\neq 0$ by integration we have $\phi = \phi_o + \frac{2e}{h} Vt$ and, consequently there is an alternating current $I = I_o \sin(\phi_o + \frac{2e}{h} Vt)$ with a frequency of about 484 MHz /μV.

It is easy to show that ϕ has also a spatial dependence given by [10]

$$\nabla\phi = \frac{2e}{hc}(\lambda_1+\lambda_2+d)\ \underline{B} \times \underline{n} \qquad 5)$$

where $\underline{B}$ is the magnetic flux density, λ_1 and λ_2 are the London penetration depths and d is the barrier thickness.
Equations 4 and 5 describe almost completely the behaviour of the Josephson junctions. These can be combined with the Maxwell equations giving

$$\frac{\partial^2\phi}{\partial x^2} + \frac{\partial^2\phi}{\partial y^2} - \frac{1}{v^2}\frac{\partial^2\phi}{\partial t^2} = \frac{1}{\lambda_J}\sin\phi \qquad 6)$$

where $v = \left|\frac{c^2}{4\pi(\lambda_1+\lambda_2+d)C}\right|^{\frac{1}{2}}$ and $\lambda_J = \left|\frac{h^2 c}{8\pi J_o e(\lambda_1+\lambda_2+d)}\right|^{\frac{1}{2}}$

λ_J is called Josephson penetration depth. For "large" junctions (dimension larger than λ_J) the conduction of the tunneling current (d.c.) occurs only within a distance λ_J from the edges of the junction[11]. This is due to the effect of the self magnetic field associated with the currents flowing in the junction[12,13]

Equation 6 can be normalized as $\frac{\partial^2\phi}{\partial x^2} + \frac{\partial^2\phi}{\partial y^2} - \frac{\partial^2\phi}{\partial t^2} = \sin\phi$.

This is the socalled "sine-Gordon equation"[14] which will be discussed in one of the next lectures*.

---

* The reader is referred to the lecture by A.C. Scott in this volume.

It is possible to show (see for example ref. 10) that the Josephson current (d.c.) is periodic in the applied magnetic field. For some geometrical conditions (direction of magnetic field, size and symmetry of the junction) this dependence describes a Fraunhofer diffraction pattern. The periodicity (magnetic field interval ΔH between two minimus) is given by $\Delta H = \frac{\Phi o}{A}$, where $\Phi o = \frac{hc}{2e}$ is the flux quantum ($\sim 10^{-7}$ Gauss cm.) and A is the transverse area threaded by the magnetic flux.

3. - JOSEPHSON JUNCTION DEVICES

Josephson tunnel junctions have been considered in the design of high performance logic circuits[17,20]. Various aspects of the superconductive Josephson tunneling technology, applied to ultra high speed computer systems, have been extensively discussed in an excellent article by Anacker[21]. In that reference a complete investigation of the feasibility of a general purpose computer based on such technology is given and potential performances summarized. The author carefully concludes in terms of "desirability of the technology" and "encouraging technological results". To which extent is, today, the situation changed? Certainly a straight answer is difficult to give. It is perhaps more correct simply to mention some of the further results obtained in the last few years and than try to draw possible conclusions.

Since the description of various devices will be the subject of other lectures I shall outline only the general features of Josephson tunnel components making few general considerations about some particular aspects as operation speed and power dissipation.

3a) SWITCHING PROPERTIES

Ultra fast operation of Josephson tunneling devices has been widely demonstrated. The first concrete experimental results are due to Matisoo[17] who discussed a two state logic element based on a Josephson junction. He called this component "the tunneling cryotron" since it is reminiscent of the in-line cryotron but, it does not involve problems connected with superconductive normal phase transition. The two states correspond to the zero-voltage Josephson state and the quasi-particles tunneling state at $V = 2\Delta/e$ (at typical operation temperatures $2\Delta/e$ is $\sim$1.13mV for tin and $\sim$2.77mV for lead). The switching from one state to the other (see fig. 3) in the tunneling cryotron is accomplished by a magnetic field. A transition time t_s less than 800 psec. was measured by Matisoo (this was the time resolution limit of his electronic apparatus). Two devices in a flip-flop configuration have been also considered by the same Author. In a current steering mode of operation, current transfer time of 2 nsec. was measured and a value less than 200 psec. was calculated for an optimized geometry.

More recentely other very interesting results have been obtained by Zappe and Greebe[22] working on "large" (see section 2b) junctions (area about $10^4 \mu m^2$). They have measured a switching time of 60 psec. with a power dissipation of few microWatts in continuous operation. A switching time value of 38 psec. has been achieved by Jutzi et al[23] using a junction of only 3.8 μm^2. In this case the power involved was about 100 n Watts leading to the value of the power delay-time product of 10^{-18} Watts-sec. Furthermore Anderson, et al[24] have discussed a new shift register employing large Josephson junctions in which fluxons play the role of bits of information. The switching of the position can occur in a time given by the Josephson plasma oscillation period which is $\sim$10 psec. A power dissipation of $\sim$1 μ Watt per single shift (energy 10^{-18} Joule/shift) has been found[25].

An interesting objective toward parallelism in electronic systems would be the possibility of realize simultaneous optical inputs. Light sensitivity in superconductive tunnel junctions has been first observed by Giaever[26]. Recent results have extended previous investigations also to different materials[27-29].
In fig. 4 is reported the V-I characteristic of a light-sensitive Giaever type junction made in our laboratory. The storage-like response to the input of light is clearly displayed. The photosensitive junctions can be examined with the prospect of making components which can be optically controlled. Scott[30] has discussed a tunnel junction oscillators array in the context of the data processing and dynamic storage of information[31]. On this basis, a two dimensional array employing superconductive tunnel junctions has been experimentally investigated[32]. The switching of this array from one configuration to another was accomplished by an external magnetic field perturbation. The application of light-sensitive elements to such array could, for instance, provide a parallel optical input.

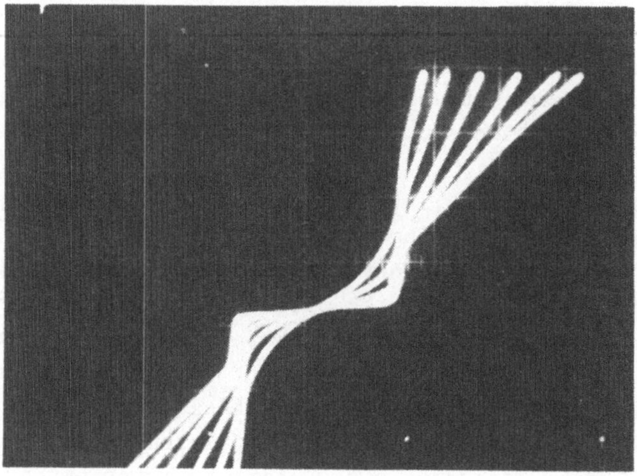

Fig. 4

V-I characteristics of a light-sensitive Pb-CdS-Pb tunnel junction for different exposure time at T = 4.2 K; Hor. 2mV/div. Vert. 0,2 mA/div.

3b) POWER DISSIPATION

When problems connected with information processing and storage are to be considered, it becomes necessary to investigate the properties of more complex systems such as large arrays of electronic components[20]. A larger number of elements in the array with the requirement of fast transmission and processing implies making devices as small as possible leading to a further increase of the packing density in the system.

Many problems[33] arise in reducing the size of electronic components, one is the heat dissipation. Although this is a general problem for any component it becomes of paramount importance in computing circuits since they operate in an extremely non-linear regime. In fact the intrinsic nonlinearity of operation such as digitalization, storage of binary information etc. require relatively large signals so that a significant amount of power can be involved in the activity on an electronic array. Since for Josephson junctions the basic power level is several orders of magnitude smaller than for transistor switches it can be concluded that power dissipation should not be a limiting factor for this technology. In addition interconnections made by superconducting lines should avoid losses due to propagating signals between elements of the array.

Although the signals involved in the operation of Josephson tunneling devices could appear very low, it should be remembered that at helium temperatures the thermal energy is $kT \lesssim 10^{-4}$ eV so that even signals of a millivolt can be regarded as "large". The difficulty of driving external circuits by such signals, i.e. interface problem between superconducting-semiconducting technologies should deserve further consideration.

It is worth while also to mention the problem of the cryogenic effort which is required by the superconducting technology. An estimate of the refrigeration necessary for the continuous operation of a memory module with a capacity of 30 M bit is given in reference [21]. A power dissipation of $\sim$ 110 m Watts is calculated considering a parallel access per cycle of 256 bits with a cycle time being 15 or 50 nsec. (since two different arrays are considered). However the most significant contribution to the total power dissipation $\sim$ 800 m Watts, comes from the heat transmission between the nitrogen external dewar (77 K) and the helium bath (3.6 K operation temperature) via electric transmission lines connections (in the particular design 200 only) and through the vacuum jacket of the helium cryostat. This leads to a value of $\lesssim$ 1 Watt of refrigeration capacity which is well within the capabilities of present day closed cycle helium refrigeration systems.

CONCLUSIONS

In spite of the required low temperature environment a reasonable conclusion which can be draw from this discussion is that superconductive junctions devices based on Josephson tunneling could complete as very favorable technology as far as concernes higher speed operation requirements. As we have seen, no limitation due to power dissipation problems should be encountered. Furthermore, even if Josephson structures have, in the past, been very delicate, recent improvements of the technology[32,34,35] concerning both reproducibility and reliability legitimate cautious optimisme for future applications of such structures in the area of the information processing and storage.

REFERENCES

1) Josephson, B.D. - Possible new effects in superconductive tunneling - Phys. Letts. 1, 251, (1962); Weakly coupled superconductors in Superconductivity, Vol. n° 1, Parks R.D. Ed., Dekker Inc. New York 1969.

2) Anderson, P.W., - Special effects in superconductivity in Lectures on the many body problems, Vol. n° 2 Caianiello E.R. Ed. Academic Press New York 1964.

3) Giaever, I. - Electron tunneling between two superconductors, Phys. Rev. Letts. 5, 464, 1960.

4) Giaever, I. and Megerle K., - Study of superconductors by electron tunneling, Phys. Rev. 22, 1101, 1961.

5) Nicol, J., Shapiro S. and Smith P.H. - Direct measurements of the superconducting energy gap, Phys. Rev. Letts. 5; 461 1960.

6) Bardeen, J. - Tunneling from a many particle point of view Phys. Rev. Letts. 6, 57, 1961.

7) Cohen, M.H., Falicov L.M. and Phillips J.C. - Superconductive tunneling, Phys. Rev. Letts. 8, 316, 1962.

8) Prange, R.E. - Tunneling from a many particle point of view; Phys. Rev. 131, 1083, 1963.

9) Douglass, D.H. and Falicov L.M. - The Superconducting energy gap, in Progress in low temperature physics Vol. n° 4, Gorter C.J. Ed. North-Holland, Amsterdam 1964 see also Meservey R. and Schwartz B.B., Equilibrium properties, in Superconductivity, Vol. n° 1, Parks R.D. Ed., Dekker Inc. New York 1969.

10) Scalapino, D.J. - The theory of Josephson tunneling, in Tunneling phenomena in solids Burstein E. and Lundquist S. Eds., Plenum Press, New York 1969, pag. 477.

11) Ferrel, R.A. and Prange R.E. - Self field limiting of Josephson tunneling of superconducting electron pairs, Phys. Rev. Letts. 10, 479 1963.

12) Schroen, W. and Pritchard J.P. - Maximum tunneling supercurrents through Josephson barriers, J. Appl. Phys.40, 2118, 1969.

13) Johnson, W.J. and Barone A. - Effect of junction geometry on maximum zero-voltage Josephson current J. Appl. Phys. 41 2958, 1970.

14) Barone, A., Esposito F.,Magee C. and Scott A.C. - Theory and application of the sine-Gordon equation, Rivista del Nuovo Cimento 1, 227, 1971.

15) Rowell, J.M., - U. S. Patent n° 3, 281, 602, 1966.

16) Matisoo, J. - Subnanosecond pair-tunneling to single-particle tunneling transitions in Josephson junctions, Appl. Phys. Letts. 9, 167, 1966.

17) Matisoo, J. - The tunneling cryotron - A superconductive logic element based on electron tunneling, Proc. IEEE 55, 172, 1967; Measurements of current transfer time in a tunneling cryotron flip-flop, Proc. IEEE 55, 2052, 1967.

18) Clark, T.D. and Baldwin J.P. - Superconducting memory device using Josephson junctions, Electronics Letts. 3, 178, 1967.

19) Matisoo, J. - Josephson-type superconductive tunnel junctions and applications, IEEE Trans. Magn., MAG - 5 848, 1969.

20) Pritchard, J.P. and Schroen W.H. - Superconductive tunneling device characteristics for array application, IEEE Trans. Magn. MAG - 4, 320, 1968.

21) Anacker, W. - Potential of superconductive Josephson tunneling technology for ultrahigh per-

formance memories and processors, IEEE Trans. Magn. MAG - 5, 968, 1969.

22) Zappe, H.H. and Greebe K.R., - Dynamic behaviour of Josephson tunnel junctions in the subnanosecond range J. Appl. Phys.44, 865, 1973.

23) Jutzi, W., Mohr H.O., Gasser M. and Gschwind H.P. - Josephson junctions with 1 μm dimensions and with picosecond swtching times, Electronics Letts. 8, 589, 1972.

24) Anderson, P.W., Dynes R.C. and Fulton T.A., - Josephson flux quantum shuttles, Bull. Am. Phys.Soc. 16, 399, 1971.

25) Fulton, T.A., Dynes R.C. and Anderson P.W. - The flux shuttle - A Josephson junction shift register emploing single flux quanta, Proc. IEEE 61, 28, 1973.

26) Giaever, I. - Photosensitive tunneling and superconductivity, Phys. Rev. Letts. 20, 1286, 1968.

27) Giaever, I. and Zeller H.R. - Subharmonic structure in superconducting tunneling, Phys. Rev. B 1, 4278, 1970.

28) Rissman, P. - Photosensitivity in superconducting tunnel junctions with a cadmium selenide barrier, J. Appl. Phys. 44, 1893, 1973.

29) Barone, A., Rissman P. and Russo M. - Light sensitivity in superconductive tunnel junctions, International Conference on Detection and Emission of Electromagnetic Radiation with Josephson Junctions. Sept. 3-5, 1973. Perros - Guirec France

30) Scott, A.C. - Tunnel diode arrays for information processing and storage, IEEE Trans. Syst. Man and Cyber. SMC-1, 267, 1971.

31) Caianiello, E.R. - Outline of a theory of though-processes and thinking machines, J. Theor. Biol. 2, 204, 1961.

32) Hoel, L.S., Keller W.H., Nordman J.E. and Scott A.C., - Niobium Superconductive tunnel diode integrated circuit arrays, Solid State Elec. 15, 1167, 1972.

34) Anacker et al. - Dispositifs à jonction de Josephson avec des electrodes en alliage au plomb, Brevet d'invention n°-2, 120, 739 - France, 1972.

35) Pritchard, J.P. and Schroen - Process for preparation of tunneling barriers, U.S. Patent n°-3, 673, 071, 1972.

JOSEPHSON JUNCTION DEVICES FOR LARGE COMPUTERS

R. D. PARMENTIER

Istituto di Fisica
Università di Salerno
Salerno, Italy

INTRODUCTION

The growth of computing capacity, measured in terms of the number of istructions executed per second, has been approximately exponential over the past two decades. This fact derives essentially from the enormous advances in semiconductor device technology that have been made during this period. The key point here has been the steady reduction in the size of circuit elements that has resulted from the perfection of the various processes used in the manufacture of integrated circuits. Since information, in whatever form, is propagated only at finite velocities, a reduction in element size allows both an increase in the basic switching speeds of individual devices and a reduction in the delay times involved in inter-device communication, both of which imply an increase in the operating speed of the system.

In order to continue this process, one must eventually confront the fundamental limitations imposed by the size, speed, reliability, and power dissipation of the individual elements from which the computer is constructed. This is particularly true in the context of parallel computation where an increase in speed is normally obtained at the cost of a proportional or greater increase in the amount of hardware involved. These fundamental factors of size, speed, reliability, and power dissipation are intimately related, and are such that improvements with respect to one factor are frequently obtained only at the cost of aggravating the problems associated with another [1]. For example, reducing device size allows faster system operating speeds, but it also seriously increases the problem of removing generated heat, since the thermal resistance seen by the heat flowing away from a device into a solid varies inversly with the linear dimensions of the device. Moreover, for a given device, the power dissipation increases as the number of switching operations executed per second is increased. Particularly for fast switching circuits, the dynamic power dissipation may be much larger than the quiescent stand-by power. Thus, although the basic power level of a device should decrease when the size of the device is decreased, this reduction tends to be offset by the fact that the smaller device, working in a faster system, executes more operations per second. Furthermore, the fact that device packing densities are increased as the individual devices are made smaller further complicates the heat removal problem.

Although computers based on semiconductor integrated circuits have not yet arrived at the performance limits imposed by fundamental device considerations, it has been estimated that the thermal problems mentioned above will limit further increases in system operating speeds to something like one order of magnitude [1]. Thus, even though predictions of technological limitations often prove unduly pessimistic, it becomes reasonable to consider the possibilities offered by other technologies.

JOSEPHSON TUNNEL JUNCTIONS *

Of the various possible alternative technologies for constructing computer elements, one of the more promising at present is that of the Josephson tunnel junction. The basic structure of a Josephson tunnel junction is simply a sandwich consisting of two films of superconducting metal separated by a thin barrier layer. The two superconductive layers, which typically measure 0.1-1.0 μm in thickness, may be either both of the same metal or of two different metals. The thickness of the barrier layer, which may be an insulator, a semiconductor, or a normal metal, is about 20 Å in the case of an insulator and somewhat thicker for a semiconductor or normal metal. Very often the barrier layer is generated

---

* The reader is referred to the lecture by A. Barone in this volume for a detailed discussion of the physics of these devices.

by oxidizing the surface of the lower metal layer.
The transverse geometry of these structures is defined by the same photoresist techniques as are used for semiconductor integrated circuits. Thus, minimum linear dimensions of the order of 1 μm are presently realizable [2]; with the development of electron beam photoresist techniques, an order of magnitude reduction may be expected [3].

The essential advantage of Josephson junctions over transistors as switching devices are very low power dissipation levels and very fast basic switching speeds. The first of these derives from the fact that the operating voltage level of a Josephson junction switch is several millivolts as opposed to several hundreds of millivolts for a transistor (nonsaturating logic), and that, simultaneously, the operating current density level for the Josephson junction is typically lower than for the transistor. The second depends upon the fact that the switching speed of Josephson junctions is governed by a quantum mechanical tunneling time, which is inherently much faster than the charge carrier transit time that plays the corresponding rôle in determining the switching speed of a transistor.

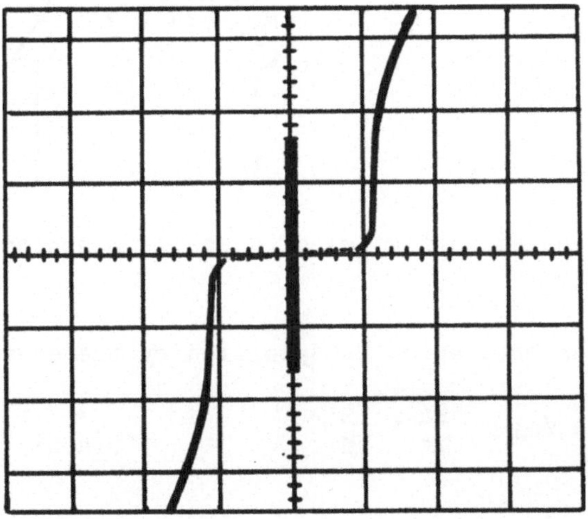

Fig. 1. Typical current-voltage characteristic for a 0.2 x 0.2 mm tin-tin oxide-tin Josephson junction.
Vertical: 2mA/major div.; horizontal: 1 mV/major div.; temperature 1.9 K.

A typical current-voltage characteristic for a small Josephson junction is shown in Fig. 1. The salient features of these devices are that: 1) a dc supercurrent with magnitude less than or equal to a maximum value I_o can tunnel through the barrier at zero voltage; 2) I_o depends strongly on temperature, the normal tunneling resistance of the barrier, the energy gaps of the superconductors, and, in particular, magnetic field; and 3) a voltage V applied across the barrier induces an ac supercurrent through the barrier of frequency 484 MHz/μV. In addition, for $V \neq 0$, there is a component of current due to the tunneling of normal electrons, or so-called quasi-particles.

The electronic behavior of a small Josephson junction can, to a first approximation, be modelled by the equivalent circuit of Fig. 2, in which C is the capacitance of the barrier, R(V) represents the voltage dependent quasi-particle tunneling resistance, and I_j is

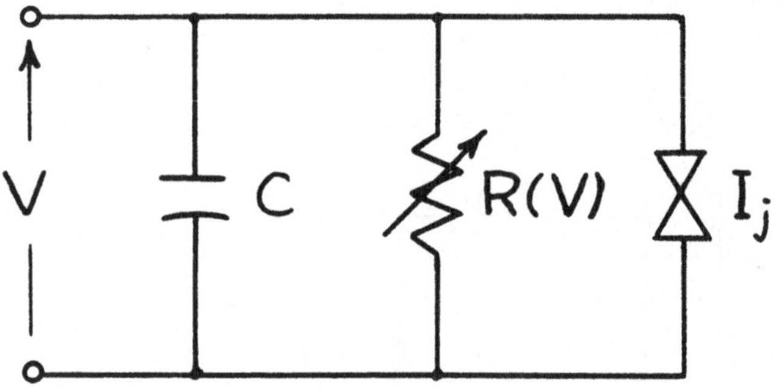

Fig. 2. Equivalent circuit model of a small Josephson junction.

the Josephson current, which is specified by the equations

$$I_j = I_o \sin \phi \; ; \qquad \frac{d\phi}{dt} = \frac{2eV}{\hbar} \qquad (1a,b)$$

in which ϕ is the quantum mechanical phase difference across the barrier, e is the electronic charge, and $\hbar$ is Planck's constant divided by 2π. In Fig. 1, one sees essentially just the quasi-particle current for $V \neq 0$ since the capacitance C effectively shunts the ac Josephson current for all but very small values of the voltage.

THE JOSEPHSON TUNNELING CRYOTRON

The double-valued nature of the current-voltage characteristic of the Josephson junction, as shown in Fig. 1, led Matisoo to propose a logic element based on this device [4]. If a junction is fed by a constant dc current I less than the maximum Josephson current I_o, the junction can rest either in the V=0 or the V>0 state (from Fig. 1 it is evident that the behavior is completely symmetrical for negative currents and voltages). If now I is raised above I_o, or, as is done in practice, I_o is reduced below I, a junction initially in the V=0 state must necessarily switch to the V>0 state. The speed of this switching transient, which is the logic delay in a Josephson switching circuit, has been measured to be of the order of tens of picoseconds [2,5]. In Matisoo's cryotron, this switching is accomplished by constructing above a junction, in close proximity but electrically insulated from it, a superconductor control line, i.e., a metal film having dimensions roughly the same as those of the films comprising the junction. A current pulse applied to this control line generates a magnetic field in the junction, momentarily reducing I_o and thus initiating switching. Matisoo showed that this structure can be built such that a given control current causes switching of a larger current, i.e., the device shows gain, which is of course necessary in order to perform logic operations [4].

Matisoo also demonstrated the operation of a bistable flip-flop based on the tunneling cryotron [6].

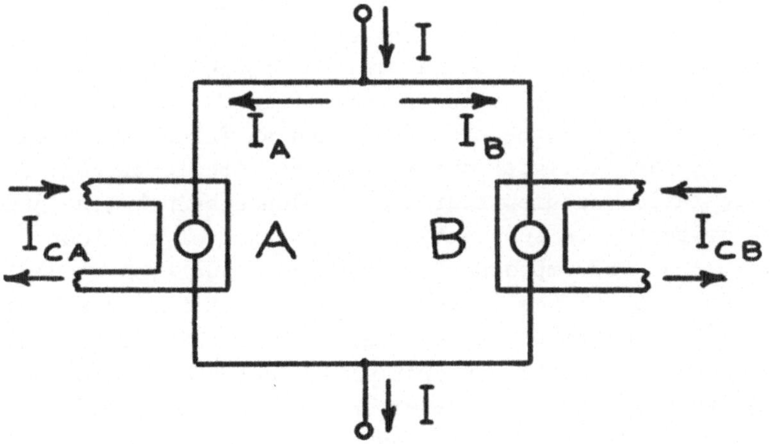

Fig. 3. Schematic diagram of a Josephson tunneling cryotron flip-flop.

The structure of this basic computer element, as indicated schematically in Fig. 3, is a parallel connection of two identical tunneling cryotrons in a superconducting loop fed by a dc current $I < I_o$. The basic mode of operation involves steering the current I between the two legs of the flip-flop by means of suitable control current pulses applied to the two individual cryotrons. The operation may be summarized as follows: Suppose initially that $I_A = I < I_o$ and $I_B = 0$ (both junctions in the V = 0 state). A control current pulse applied to A causes that junction to switch to the V > 0 state. Current is thus diverted from leg A to leg B at the rate

$$\frac{dI}{dt} = \frac{V_A}{L} \qquad (2)$$

where V_A is the istantaneous value of the voltage of junction A and L is the total inductance of the superconducting loop. At the end of the current transfer, $I_B = I$, $I_A = 0$, and both junctions are again in the V = 0 state. Current transfer times of several hundreds of picoseconds have been measured [5].

These values, which represent the cycle time of a Josephson junction memory cell, were obtained for structures that are fairly large in terms of the capabilities of present day photoresist technology. It might be thought, following the line of reasoning regarding speed vs. miniaturization, that reducing circuit dimensions (in order to reduce L) would therefore necessarily yield faster times. This argument, however, is not completely valid. In fact, the operation of the tunneling cryotron flip-flop is rather more complicated than indicated above, for three reason: 1) The dynamics of the flip-flop during current transfer may be roughly modelled by placing a current source I and an inductor L (the loop inductance) in parallel with the equivalent circuit of Fig. 2. This representation assumes that the impedance seen by the junction undergoing switching is substantially just that of the loop inductance, i.e., that the impedance of the second junction is negligible with respect to L. Neglecting for the moment the Josephson current I_j, it is evident that the response of this circuit may be underdamped, overdamped, or critically damped depending on the value of R (which, it must be remembered, is not constant) relative to the critical value R_c given by

$$R_c = \left(\frac{L}{C} \right)^{\frac{1}{2}} \qquad (3)$$

It is generally agreed that an underdamped loop response will lead to unacceptable behavior [5,7,8]. 2) The effects of I_j, however, cannot be neglected. It is easily shown that this element, which is defined by equations (1), is completely equivalent to a Josephson inductance L_j given by

$$L_j = \frac{\hbar\phi}{2eI_o \sin \phi} \quad (4)$$

The effective inductance of the equivalent circuit thus becomes the parallel combination of L, which, depending only on the geometry of the loop, is constant, with L_j, which has a complicated time dependence governed by equations (1b) and (4). Thus, equations such as (2) and (3) are at best only rough, average approximations to the actual situation during current transfer. In particular, it has been demonstrated in a computer simulation using the circuit model described here that variations in the value of L as small as 0.25 % may lead to large differences in the current transfer behavior [9]. 3) Moreover, as L is reduced, the simple equivalent circuit model begins to lose validity inasmuch as it is no longer possible to neglect the second junction in the loop. In this case, quantum interference effects begin to dominate [10], and the dynamics of the flip-flop become considerably more complicated.

It is not yet completely clear what conditions will have to be imposed on the various parameters of the flip-flop in order to obtain reliable operation in the manner described in the simplified summary above. Whatever be the final solution, however, it will undoubtedly involve a stringent control of the flip-flop parameters.

FABRICATION TECHNOLOGY

The requirement of an extremely thin, uniform tunneling barrier, sufficiently stable to withstand the rigors of temperature cycling between room temperature and liquid helium, was for many years the chief stumbling block in the experimental study of Josephson junctions. Junctions made from the soft superconducting metals (tin, lead, indium, etc.), with a thermally grown oxide barrier, typically had lifetimes of only a few days. Moreover, repeatability for these devices was poor, with order of magnitude parameter variations for ostensibly identical junctions being quite common.

This situation was much improved by the adoption of plasma glow discharge oxidation rather than thermal oxidation to generate the barrier layer [11].

The reason for this is thought to be that whereas thermal oxidation is essentially a diffusion controlled process, plasma oxidation proceeds by bombarding the metal surface with energetic atomic oxygen, resulting in a denser oxide structure. Moreover, whereas thermal oxide growth rates depend strongly on the presence of minute traces of impurities, e.g., water vapor, in the oxidizing atmosphere, the thickness of a plasma oxide seems to depend only on relatively easily controlled fabrication parameters such as gas pressure and oxidizing voltages and times. Since tunneling currents depend exponentially on the oxide thickness, a strict control of this parameter is of utmost importance in order to construct systems of identical devices. A refinement of the plasma oxidation technique has been proposed in which the oxide layer is generated by means of a dynamic balance between rf sputter etching and glow discharge oxidation [12]. This process offers the promise of extremely precise control over the oxide layer thickness. In addition, the durability of device based on the soft superconducting metals has been further enhanced by using alloys of these metals rather than the pure metal in order to minimize the mismatch in thermal expansion coefficient between metal and substrate [13].

Niobium has for some time been recognized as a desirable base metal for Josephson junction devices because of its good mechanical properties, relatively tough and non-porous oxide, and high transition temperature. Because of its refractory nature, niobium cannot be vacuum evaporated by means of a simple resistance heated crucible, as can the soft superconducting metals. Films of niobium are deposited either by high power electron beam evaporation or, more commonly, by dc or rf sputtering. This fact renders difficult the construction of niobium-insulator-niobium devices, which would be most desirable to exploit the good mechanical properties of niobium, because the deposition of the top metal film is sufficiently energetic to damage or destroy the insulating barrier layer. For this reason, the top metal layer of niobium based devices is generally a soft superconductor. The hardness of niobium also complicates the photoresist processing by which the geometry of the device or array of devices is defined: one must use either extremely corrosive acid etches or sputter etching through a soft metal overlay. In spite of these and other difficulties, the fabrication technology for niobium based junctions is now reasonably well advanced [12, 14, 15]. Further perfection of this technology will undoubtedly yield highly durable devices.

Josephson junctions with semiconductor barriers have been the subject of some interest because the lower barrier height (in energy) of a semiconductor with respect to that of an insulator should allow the use of a thicker barrier (spatial dimension) for a given tunneling characteristic. This offers the possibility of two advantages: 1) a greater tolerance for slight thickness variations of the barrier, and 2) a reduced junction capacitance, both of which might prove desirable for computer applications. Niobium based junctions with semiconductor barriers having tunneling and lifetime characteristics quite comparable to those of niobium based oxide barrier junctions have been reported [16]; however, it is not yet possible to say that one type of junction is superior to the other.

CONCLUSIONS

In 1969, Anacker proposed that high performance computers could be constructed based entirely on Josephson junction devices, in particular, Matisoo's tunneling cryotron [17]. As one example, he projected the design of a 30 Mbit memory with parallel access of 256 bits per cycle and a maximum cycle time of 50 nsec. Assuming a value of 100 μm for the minimum linear dimension of devices and device spacing, he estimated that this memory could be housed in a volume of one cubic foot and that it would require a continuous refrigeration power of one watt at 3.6 K.

The realization of such performance levels will be determined by the extent to which the problems discussed in this lecture can be overcome. In the Introduction, the factor of size, speed, reliability, and power dissipation were mentioned as the fundamenatal device parameters to be confronted in the expansion of computing capacity. With respect to speed and power dissipation, Josephson junction devices remain decidedly attractive. The size of individual Josephson devices is governed by the same photoresist technology as governs the sizes of others types of computer elements; however, the minimum size for Josephson junction circuits will probably be determined by quantum interference effects and will probably be somewhat larger than the minimum technological limit. The remaining question mark is reliability. Considerable progress has been made in improving basic fabrication technology, but both here and in the construction of circuits and systems much yet remains to be done.

REFERENCES

[1] R. W. Keyes, "Physical problems of small structures in electronics," Proc.IEEE,vol.60, pp.1055-1062, Sept.1972.

[2] W. Jutzi, et al., "Josephson junctions with 1 μm dimensions and with picosecond switching times," Electronics Letters, vol.8, pp. 589-591, 30 Nov. 1972.

[3] A. N. Broers and M. Hatzakis, "Microcircuits by electron beam," Scientific American, vol. 227, pp. 34-44, Nov.1972.

[4] J. Matisoo, "The tunneling cryotron — a superconductive logic element based on electron tunneling" Proc. IEEE, vol. 55, pp. 172-180, Feb. 1967.

[5] H.H. Zappe and K.R. Grebe, "Dynamic behavior of Josephson tunnel junctions in the subnanosecond range," J.Appl.Phys. vol. 44, pp.865-874, Feb. 1973.

[6] J. Matisoo, "Measurement of current transfer time in a tunneling cryotron flip-flop," Proc. IEEE (Letters),vol. 55, pp. 2052-2053, Nov. 1967.

[7] T.A. Fulton, "Punchthrough and the tunneling cryotron," Appl. Phys. Lett., vol. 19, pp. 311-313,1 Nov. 1971.

[8] H.H. Zappe, "Minimum current and related topics in Josephson tunnel junction devices," J. Appl. Phys., vol. 44, pp. 1371-1377, March 1973.

[9] P. Guéret, "Flux-quantization effects in large inductance loops containing Josephson junctions," J. Appl. Phys., vol. 44, pp. 1771-1773, April 1973.

[10] J. Clarke, "Low-frequency applications of superconducting quantum interference devices," Proc. IEEE, vol. 61, pp.8-19, Jan. 1973.

[11] W. Schroen, "Physics of preparation of Josephson barriers" J. Appl. Phys., vol. 39, pp. 2671-2678, May 1968.

[12] J. H. Greiner, "Josephson tunneling barriers by rf sputter etching in an oxygen plasma," J. Appl. Phys., vol. 42, pp. 5151-5155, Nov. 1971.

[13] W. Anacker, et al., "Dispositifs à jonction de Josephson avec des èlectrodes en alliage au plomb" Brevet d'inven-

tion no. 2, 120, 739 (France), 24 July 1972.

[14] L. S. Hoel, et al., "Niobium superconductive tunneling diode integrated circuit arrays," Solid-State Electronics, vol. 15, pp. 1167-1173, Oct. 1972.

[15] P. Rissman and T. Palholmen, "Preparation and lifetest of niobium Josephson junction tunnel diodes and arrays" Solid-State Electronics, (to be published).

[16] W. H. Keller and J. E. Nordamn, "Niobium thin film Josephson junctions using a semiconductor barrier," preprint.

[17] W. Anacker, "Potential of superconductive Josephson tunneling technology for ultrahigh performance memories and processors," IEEE Trans. Magnetics, vol. MAG-5, pp. 968-975, Dec. 1969.

QUASIHARMONIC INFORMATION PROCESSING SYSTEMS

Alwyn C. Scott

Department of Electrical and Computer Engineering
The University of Wisconsin
Madison, Wisconsin

INTRODUCTION

During past years students of information processing systems have inclined toward a "logical" representation of nonlinear effects. In the design of machines this inclination is evidenced by the dominant role of "switching theory" usually based upon Boolean algebra, and in neurophysiology by the somewhat fictitious "all or nothing" concept. Many investigators seem to assume that a deeper consideration of nonlinearities would be too difficult to carry through.

While general nonlinear dynamic problems are more difficult than those which arise in linear theory or in Boolean algebra, they are by no means hopeless. Indeed it should be emphasized that they are _difficult_ to the degree that they are _interesting_, because they present many rich and unexplored possibilities for solutions. Furthermore, a broadly based study of nonlinear dynamics is instructive; for the same effect (from a mathematical point of view) often appears in widely different physical situations. In this lecture and those to follow ("Transmission of information by solitary waves" and "Nonlinear wave problems in neurophysiology") we hope to encourage an optimistic attitude toward such problems by indicating some possibilities for analysis of and relations between man made, cerebral and ecological information processing systems.

THE VIBRATING MEMBRANE

Let us begin by considering the square mechanical membrane indicated in Fig. 1a. The (clamped) boundary conditions at the edges require that any oscillating mode can have only an integral number of half wavelengths in the x-direction ($m\lambda_x/2=A$) and also in the y-direction ($n\lambda_y/2=A$). Thus the wave numbers ($\beta=2\pi/\lambda$) in the two directions are

$$\beta_x = m\pi/A \qquad \text{m, n integers} \qquad \beta_y = n\pi/A \tag{1}$$

and each mode of oscillation corresponds to a point in "β-space" as indicated in Fig. 1b.

In the linear case the modes can be individually excited without interaction and the mode frequency is determined by a <u>dispersion equation</u> of the form $\omega=\omega(\beta_x,\beta_y)$. For real ω the system is lossless and any mode which is excited will remain in the oscillatory state for an indefinite length of time. Thus we can consider the linear, lossless membrame as a <u>distributed</u> memory. A single "bit" of information can be stored by exciting a single mode of oscillation; but notice, from Fig. 1, that this bit of information is localized in β-space while it is distributed in real space. A somewhat similar distributed memory effect takes place in the hologram which may, in turn, be related to memory effects in the brain.[1-6]

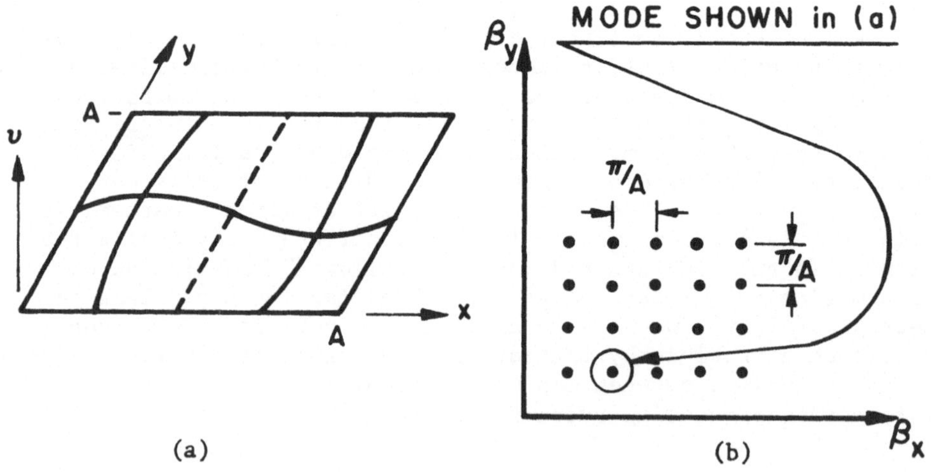

Fig. 1. The vibrating membrane

MULTIMODE STABILITY FOR AN ELECTRONIC ANALOG

The "linear-lossless" assumption of the preceding section is often not satisfied well enough to make information storage practical. The energy in a mode (and the corresponding bit of information) decays exponentially with time and eventually disappears. One way to circumvent this difficulty is to introduce an "active" element which will inject energy into the mode. This is accomplished electronically as indicated in Fig. 2. An active, nonlinear conductive element

$$I(v) = -Gv\left(1 - \frac{4}{3}\frac{v^2}{V_o^2}\right) \qquad (2)$$

is connected across the L-C "tank circuit" of Fig. 2a. Without this additional element (Fig. 2a) the voltage across the capacitor is sinusoidal

$$v = V \cos \omega t \qquad (3)$$

where the amplitude, V, is independent of time. With the additional element and the assumption that the current I(v) is small compared with the current through the capacitor C, we can still use (3) if we suppose that V is a slowly varying function of time. This quasiharmonic approximation was introduced by van der Pol[7] and extensively developed by Kryloff and Bogoliuboff.[8] It is satisfied if we assume $G \ll C\omega$; then the rate of energy input to the tank circuit can be evaluated by averaging the instantaneous power input $(-vI(v))$ over a cycle. This yields

$$\frac{dE}{dt} = \frac{E}{\tau}\left(1 - \frac{E}{E_o}\right)$$

where $\tau \equiv C/G$ and $E_o \equiv 1/2CV_o^2$ is the saturation energy of the oscillator. For $E \ll E_o$, the energy grows exponentially as $\exp(t/\tau)$,

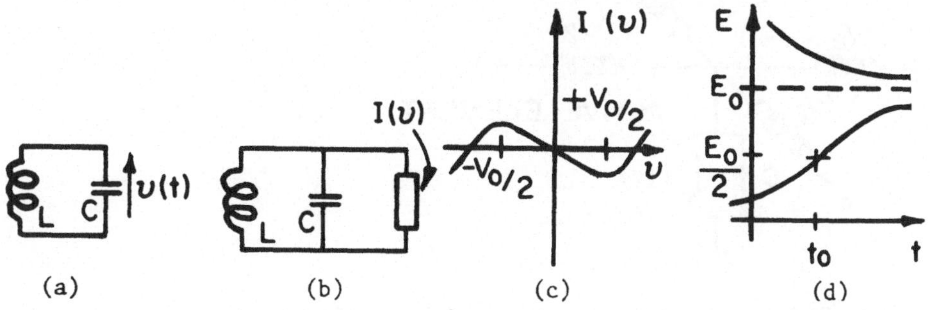

Fig. 2. Growth of energy in a simple nonlinear oscillator.

otherwise it approaches E_o asymptotically. Equation (4) is readily integrated to obtain the logistic function indicated in Fig. 2d.

$$E(t) = \frac{E_o}{1 + \exp\,[-(t-t_o)/\tau]} \tag{5}$$

where t_o is the time at which $E=E_o/2$. This equation was originally discussed in 1844 by Verhulst who was concerned with the population growth of Belgium, and is related to many other growth problems.[9]

We have been interested in applying these basic ideas to the development of an electronic analog for the mechanical membrane shown in Fig. 1.[10-15] The simplest structure which is compatible with superconductive tunneling technology is a two dimensional array of the unit cells shown in Fig. 3. It is of interest to determine the conditions under which several modes can simultaneously and stably oscillate at their saturation levels. Assuming n modes are excited (i.e., have nonzero energy) and no resonant interactions, the time derivatives of the mode energies are given by

$$\begin{aligned}
\frac{dE_1}{dt} &= \frac{E_1}{\tau}\left(1 - 2\,\frac{(\frac{9}{8}E_1 + E_2 + \ldots + E_n)}{E_o}\right)\\
\frac{dE_1}{dt} &= \frac{E_1}{\tau}\left(1 - 2\,\frac{(E_1 + \frac{9}{8}E_2 + \ldots + E_n)}{E_o}\right)\\
&\vdots\\
\frac{dE_n}{dt} &= \frac{E_n}{\tau}\left(1 - 2\,\frac{(E_1 + E_2 + \ldots + \frac{9}{8}E_n)}{E_o}\right).
\end{aligned} \tag{6}$$

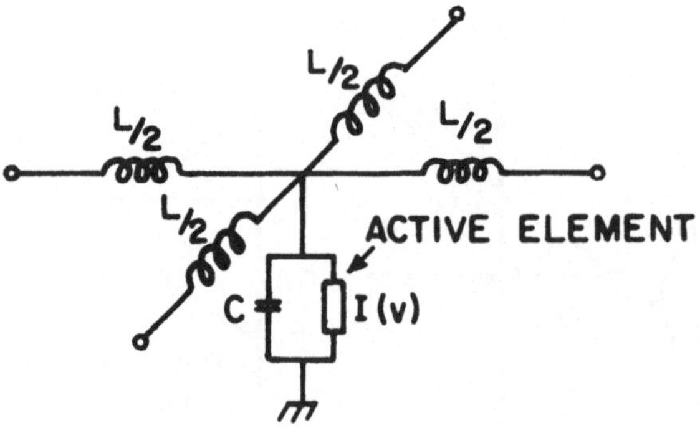

Fig. 3. Unit cell for an electric analog of the mechanical membrane

These have the stationary solution $E_1=E_2=\ldots=E_n=E_o/(2n+\frac{1}{4})$, but we must also be concerned with the stability of this solution. This stability is related to the coefficient matrix K for the quadratic terms.

$$K = \begin{pmatrix} \frac{9}{8} & 1 & \cdots & 1 \\ 1 & \frac{9}{8} & \cdots & 1 \\ \cdot & \cdot & \cdots & \cdot \\ 1 & 1 & \cdots & \frac{9}{8} \end{pmatrix}_{n \times n} \tag{7}$$

Since the diagonal term is greater than unity, it can be shown that small perturbations from the stationary solution decay exponentially with time.[10] For a more general class of systems with M spatial dimensions, rather than 2 as in Fig. 3, the "9/8" in (7) becomes $3^M/2^{M+1}$. For M=0 (6) reduces to (4). For M=1, this number is 3/4 (less than unity) and the stationary solution is not stable.

There are several aspects of this analysis which are not yet satisfactory: (a) Although nonresonant multimode oscillation is often observed, resonant interaction is by no means uncommon. It certainly should not be neglected. (b) The unexcited modes will, according to (6), grow exponentially as $\exp[t/\tau(8n+1)]$.[11] (c) This is a small growth rate for large n which may (or may not) be damped by other loss effects not considered in (6).[11,15] However for the (4x4) sixteen unit cell superconductive array shown in Fig. 4[13] the number of states is at least 9 as indicated in Fig. 5. This is in rough agreement with previous estimates.[11]

FEATURE DETECTION

Another possible application of these arrays is to detect "features" of two dimensional optical patterns as a preprocessing step for some sort of pattern classification. A basic unit cell is indicated in Fig. 6. When the array is illuminated by the rectangular pattern in Fig. 7, the lowest eigenfrequency is

$$\omega_o = \pi\left(\frac{a}{b}+\frac{b}{a}\right) \tag{8}$$

where unit area and wave velocity have been assumed. In this array the oscillator is always linear, but the shape is determined by the optical pattern. Since the eigenfrequencies are determined by the shape, they can give some useful information concerning the shape. Other patterns with the same degree of elongation (such as the ellipse) come close to satisfying (8).[16]

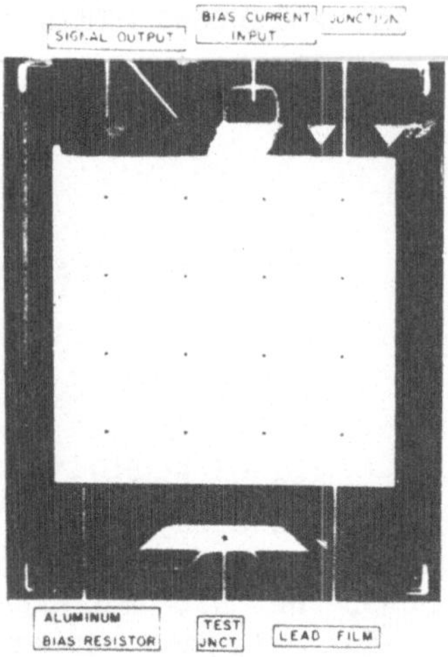

Fig. 4. Nb-NbO-Pb multimode array. Substrate dimensions are 1 x 3/4 in.

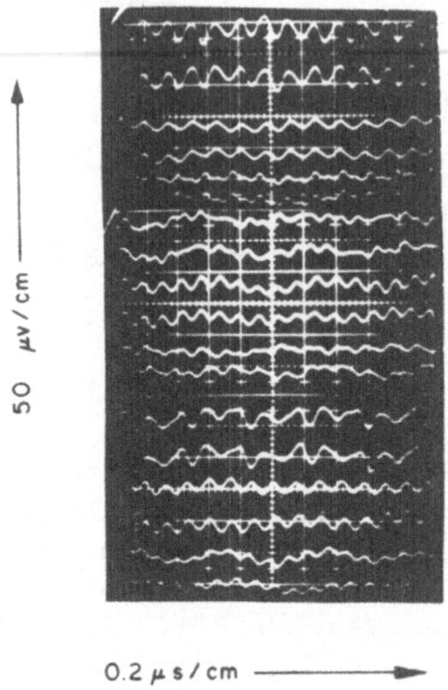

Fig. 5. Multistate oscillations of the array. Two sweeps are presented for each state.

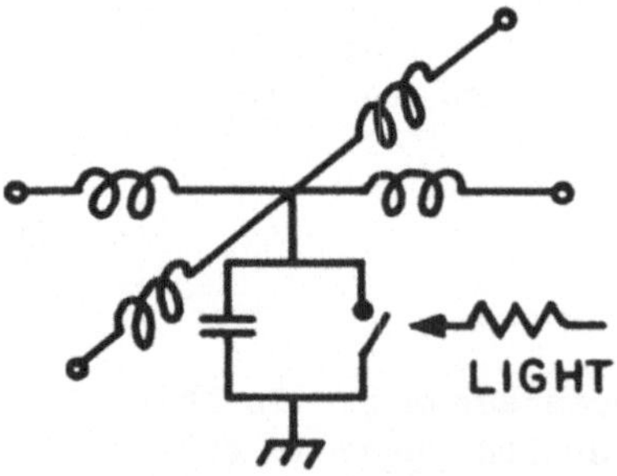

Fig. 6. Unit cell for a light sensitive array.

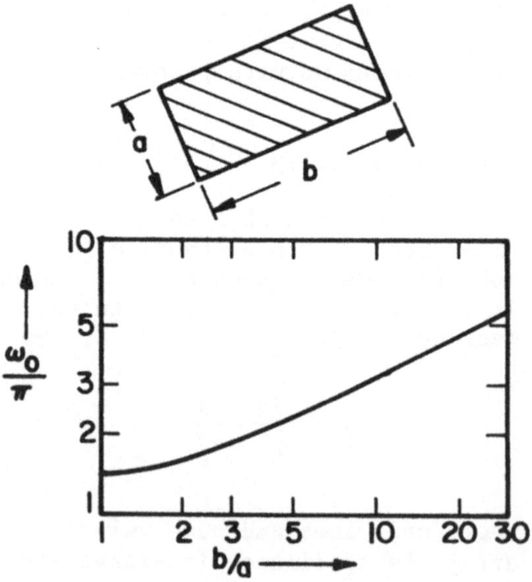

Fig. 7. Lowest eigenfrequency vs. elongation.

HIGH PASS FREQUENCY MEMORY

The unit cells shown in Figs. 3 and 6 are essentially elements of low pass filters. Recently Aumann has investigated in detail the multistate properties of the high pass line oscillator a unit cell of which is shown in Fig. 8.[14] Since the inductors have a common ground, this circuit has the important practical advantage that they can be synthesized from semiconductor integrated circuits. Furthermore, Aumann has shown, both theoretically and experimentally, that the individual modes on such a line oscillator are stable (as long as the number of sections plus one is a prime number). Thus this circuit may be useful as an element in a frequency memory.

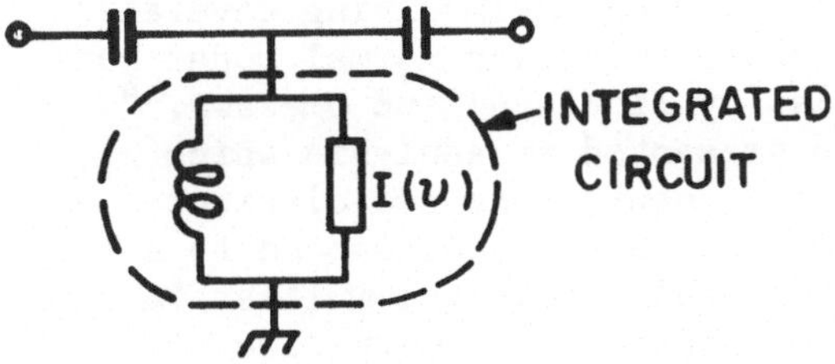

Fig. 8. Unit cell for a high pass line oscillator

RELATED DYNAMICAL SYSTEMS

Interacting species

Just as (4) is related to the population growth of a single species, (6) can represent the interaction of several species. An important consideration is the symmetry properties of the coefficient matrix K for the quadratic terms (7). If K is <u>symmetric</u>, the time dependence of small perturbations from stationary values is either exponential growth or decay. If K is <u>antisymmetric</u>, this time dependence is oscillatory. (Both of these statements follow from the fact that a self adjoint matrix has real eigenvalues.) The symmetric case corresponds to several species competing for the same food supply.[17] The diagonal term is made larger than the off diagonal terms and stability is achieved when each species finds its own "ecological niche". The antisymmetric case corresponds to several species engaged in predator-prey relations and has recently been reviewed in considerable detail.[17,18] In a general ecological problem, of course, one expects a mixture of these two basic activities.

The laser oscillator

The laser oscillator provides an interesting example for both the effects just discussed. The interaction between the inverted atomic population and a single optical mode is predator (optical mode)-prey (inverted atoms).[20,21] The corresponding oscillation is known as "spiking". The interaction of several modes competing for the same inverted population can be described by a symmetric K matrix.[22]

Biological nerve networks

The cube shown in Fig. 9 which can be perceived in either of two orientations suggests a neural mechanism involving several stable dynamic states. Such a theory involving normal modes has been outlined by Greene[23] and also by Ricciardi and Umezawa.[24] Recently Wilson and Cowan[25] have presented an analysis which treats the excitatory and inhibitory neurons as two distinct populations. Equations for firing rates then correspond to a general two species interaction for which both the multistate behavior (Fig. 9) and oscillation (short term memory) are obtained.

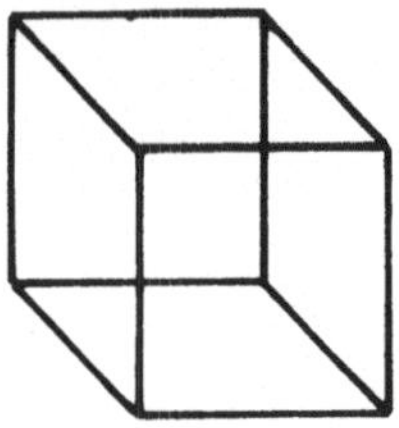

Fig. 9. A cube which can be perceived in two orientations.

REFERENCES

1. Gabor, D., Microscopy by reconstructed wave fronts, Proc. Roy. Soc. London A, 197, 454, 1949.

2. Gabor, D., Microscopy by reconstructed wave fronts, Proc. Phy. Soc. B, 64, 449, 1951.

3. van Heerden, P.J., Theory of optical information storage in solids, Applied Optics, 2, 393, 1963.

4. Longuet-Higgins, H.C., Holographic model of temporal recall, Nature, 217, 204, 1968.

5. Pribram, K.H., The neurophysiology of remembering, Scientific American, 73, 1969.

6. Gabor, D., Associative holographic memories, IBM J. Res. Develop., 156, 1969.

7. van der Pol, B., On relaxation oscillations, Phil. Mag., 43, 700, 1922.

8. Kryloff, N. and Bogoliuboff, N., Introduction to nonlinear mechanics, Princeton U. Press, Princeton, New Jersey, 1947.

9. Lotka, A.J., Elements of mathematical biology, Dover, New York, 1956, Ch. 7.

10. Scott, A.C., Distributed multimode oscillators of one and two spatial dimensions, Trans. IEEE on Circuit Theory, CT-17, 55, 1970.

11. Scott, A.C., Tunnel diode arrays for information processing and storage, Trans. IEEE on Systems, Man and Cybernetics, SMC-1, 267, 1971.

12. Parmentier, R.D., Lumped multimode oscillators in the continuum approximation, Trans. IEEE on Circuit Theory, CT-19, 142, 1972.

13. Hoel, L.S., Keller, W.H., Nordman, J.E. and Scott, A.C., Niobium superconductive tunnel diode integrated circuit arrays, Solid State Electronics, 15, 1167, 1972.

14. Aumann, H.M., Standing waves on a multimode ladder oscillator, Trans. IEEE on Circuit Theory (in press). See also University of Wisconsin Thesis, Active RC Oscillator Arrays, 1973.

15. Parmentier, R.D., Effect of resistive losses on superconductive multimode oscillators (to be published).

16. Daymond, S.D., The principle frequencies of vibrating systems with elliptic boundaries, Quart. J. Mech. Appl. Math., 8, 361, 1955.

17. Hutchinson, G.E., The ecological theater and the evolutionary play, Yale U. Press, New Haven, 1965.

18. Goel, N.S., Maitra, S.C. and Montroll, E.W., On the Volterra and other nonlinear models of interacting populations, Rev. Mod. Phys., 43, 231, 1971.

19. Gasiz, D.C., Montroll, E.W. and Ryniker, J.E., Age-specific, deterministic model of predator-prey populations: Application to Isle Royal, IBM J. Res. Develop., 47, 1973.

20. Sinnett, D.M., An analysis of the maser oscillator equations, J. Appl. Phys. 33, 1578, 1962.

21. Tang, C.L., Statz, H. and de Mars, F., Spectral output and spiking behavior of solid-state lasers, J. Appl. Phys. 34, 2289, 1963.

22. Haken, H. and Sauermann, H., Nonlinear interaction of laser modes, Zeits, f. Physik, 173, 261, 1963.

23. Greene, P.H., On looking for neural networks and "cell assemblies" that underlie behavior, Bull. Math. Biophys. 24, 247 and 395, 1962.

24. Ricciardi, L.M. and Umezawa, H., Brain and physics of many body problems, Kybernetic 4, 44, 1967.

25. Wilson, H.R. and Cowan, J.D., Excitatory and inhibitory interactions in localized populations of model neurons, Biophys. J., 12, 1, 1972.

TRANSMISSION OF INFORMATION BY SOLITARY WAVES

Alwyn C. Scott

Department of Electrical and Computer Engineering
The University of Wisconsin
Madison, Wisconsin

TRAVELING WAVE THEORY

Often we must consider nonlinear partial differential equations for which it is not appropriate to make "quasiharmonic" assumptions. A useful technique is then to seek traveling wave solutions of the form

$$\phi(x,t) = \phi_T(x-ut) \tag{1}$$

Then $\partial\phi/\partial x = \partial\phi_T/\partial x$ and $\partial\phi/\partial t = -u(\partial\phi_T/\partial x)$ and the original p.d.e. for ϕ reduces to an o.d.e. for ϕ_T which can be solved by conventional analytic or phase plane techniques.[1] Note that the wave velocity appears as an adjustable parameter in the o.d.e. There are two types of solitary traveling waves depending upon the nature of the nonlinearity.

Energy exchanging

The simplest example is the ordinary candle. If E is the chemical energy stored per unit length and P is the power required to feed the flame, then the velocity of the flame is given by

$$u = P/E \tag{2}$$

It is fixed at the value for which energy is "eaten" (uE) at the same rate it is "digested" (P). Such solarity waves generally destroy each other upon collision. Other examples for which the pulse velocity has a fixed value include the nerve axon[2] and the neuristor[3] or electronic analog of the nerve axon.

It has recently been suggested that an array superconducting neuristor lines[4-7] may be useful as a detector of relativistic charged particles with high temporal and spatial resolution.[8]

Energy conserving

If a wave system which conserves energy is nonlinear, the velocity (or in some cases the shape) of a solitary wave becomes an adjustable parameter in the family of solutions.[1] In addition many such solitary waves preserve their shape and velocity under collision and are called solitons.[9] Thus a soliton can also be used to carry a "bit" of information.

THE SINE-GORDON EQUATION

A typical example of an energy conserving nonlinear wave equation with soliton solutions is the sine-Gordon equation[10]

$$\phi_{xx} - \phi_{tt} = \sin\phi \tag{3}$$

This equation represents propagation of magnetic flux along a large "Josephson-type" superconducting tunnel junction where we consider velocity measured in units of about 1/20 c, distance in units of the Josephson length (~0.1 mm), and take $\phi = 2\pi$ (magnetic flux)/(flux quantum), where the flux quantum

$$\Phi_o \approx 2\times 10^{-15} \text{ volt-seconds.} \tag{4}$$

Using the assumptions associated with (1), (3) reduces to the pendulum equation which can be integrated twice to yield the traveling wave solution[11]

$$\int_{\phi_i}^{\phi_T} \frac{d\phi}{\sqrt{2(F-\cos\phi}} = \frac{x-ut}{\sqrt{1-u^2}} \tag{5}$$

where $\phi_i = \phi_T(0,0)$ and F is an adjustable constant of the first integration. Thus (5) has two adjustable constants, F and u. For F>1, ϕ_T is a monotone increasing function of x which can be easily demonstrated on a mechanical analog of (3).[12] This solution leads to the observation of "displaced linear slope" in the volt-ampere characteristics of large Josephson junctions.[13,14]

For F=1, (5) becomes

$$\phi_T = 4\tan^{-1}\left[\exp\left(\pm\frac{x-ut}{\sqrt{1-u^2}}\right)\right] \tag{6}$$

which represents the propagation of a single quantum of magnetic flux along a Josephson junction as indicated in Fig. 1. Recently Fulton and Dynes have reported the direct experimental observation of such a single flux quantum.[15]

An early and important study of (3) was published between 1950 and 1953 by Kochendörfer, Seeger and Donth in connection with the propagation of a crystal dislocation.[16] Using Bäcklund transformation techniques they showed that (3) admits the solution

$$\phi(x,t) = 4 \tan^{-1}\left[\frac{u \sinh\,(x/\sqrt{1-u^2})}{\cosh\,(ut/\sqrt{1-u^2})}\right] \tag{7}$$

which represents the nondestructive collision of two of the solitary waves of (6). Thus these solitary waves are solitons.[9] Later (7) was rediscovered by Perring and Skryme[17] during a numerical study of the application of the sine-Gordon equation to model elementary particle collisions.

THE QUANTUM FLUX SHUTTLE

Fulton, Dynes and Anderson have recently described a shift register, called the flux shuttle, in which an individual flux quantum (or soliton) represents a bit of information.[18] The operation of this device is best understood in connection with the mechanical analog[12] shown in Fig. 3. The soliton of (6) is represented by a single loop of the pendula from ϕ=0 to ϕ=2π. An electronic circuit is shown in Fig. 2 where inductors of value L (analogous to the spring) are employed to interconnect discrete Josephson junctions (analogous to the pendula). These Josephson junctions have capacitance C (analogous to inertia), shunt conductance G (analogous to friction), and Josephson current

$$i_J = I_c \sin\phi \tag{8}$$

where

$$\frac{\partial\phi}{\partial t} = \frac{2e}{h} v \tag{9}$$

is analogous to the force of gravity in the mechanical model. Force to move a flux quantum can be applied through the bias currents, I_B, which is analogous to applying torques to the pendula.

The condition $LI_c \ll \Phi_o$ implies a small change in ϕ between adjacent Josephson junctions (or pendula) and, therefore, that a p.d.e. approximation is appropriate as in (3). The condition $LI_c \gg \Phi_o$ implies that many flux quanta can be stored in the

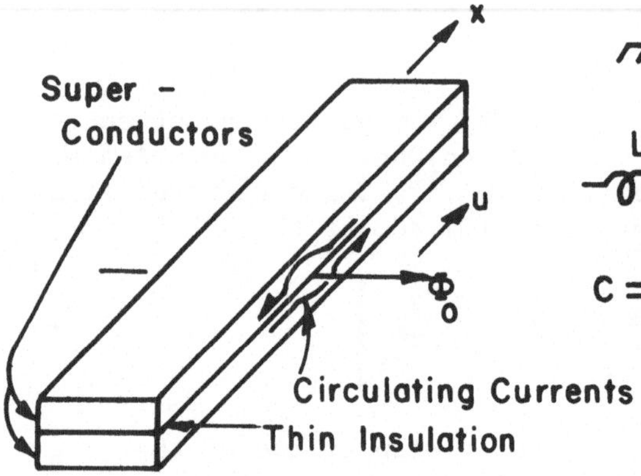

Fig. 1. Propagation of a single flux quantum along a large Josephson function.

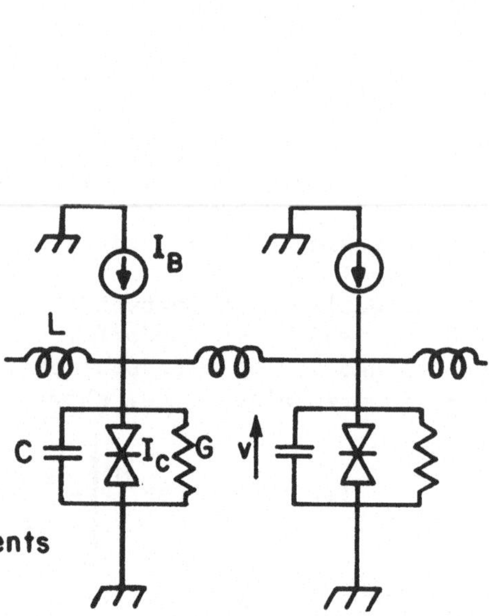

Fig. 2. Electronic circuit for the flux shuttle

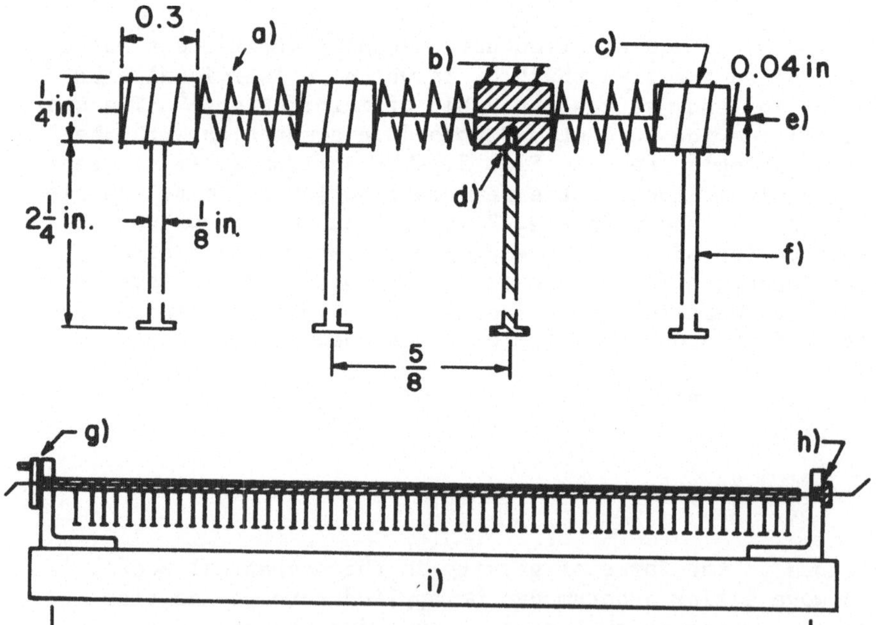

Fig. 3. A mechanical analog of the long Josephson junction. (a) spring, 0.2", (b) solder, (c) brass bushing, (d) tap and thread, (e) piano wire support, (f) nail, (g) and (h) ball bearings, (i) wood or aluminum base.

inductors between adjacent Josephson junctions. This corresponds to a very weak spring (elastic) which can twist many times between adjacent pendula. Fulton, Dynes and Anderson choose

$$LI_c = \Phi_o \tag{10}$$

so a <u>single</u> flux quantum can be stored in the inductor between each pair of junctions. Applying a bias current (torque) of

$$I_B = 0.7\ I_c \tag{11}$$

moves the quantum to the next loop.

For a typical Josephson current density of $2\text{x}10^6$ amp/meter^2, 10 micron x 10 micron junctions would give I_c=200 microamperes and, from (10) and (4), $L=10^{-11}$ henry. Such quanta should transfer in about 40 picoseconds with an energy loss of about $4\text{x}10^{-19}$ joules.[18]

A NONLINEAR SCHRÖDINGER EQUATION

The equation

$$\phi_{xx} + i\phi_t + k|\phi|^2\phi = 0 \tag{12}$$

was introduced by Benney and Newell to describe the propagation of envelope waves in the presence of nonlinearity and dispersion.[19] Equation (12) is somewhat more complicated than it seems for if we write

$$\phi(x,t) = \Phi(x,t)\ e^{i\theta(x,t)} \tag{13}$$

the amplitude, Φ, and phase, ϕ, must satisfy the coupled nonlinear p.d.e.'s

$$\Phi_{xx} - \Phi\theta_x^{\ 2} - \Phi\theta_t + k\Phi^3 = 0$$

$$\Phi\theta_{xx} + 2\Phi_x\theta_x + \Phi_t = 0 \tag{14a,b}$$

Assuming $\theta=\theta(x-u_ct)$ and $\Phi=\Phi(x-u_et)$ where u_c is a carrier velocity and u_e is an envelope velocity and setting

$$\theta_x = u_e/2 \tag{15}$$

satisfies (14b). Equation (14a) can then be integrated to obtain

$$\phi(x,t) = \Phi_M \exp\left[i\frac{u_e}{2}(x-u_ct)\right] \text{sech} \sqrt{\frac{k}{2}}\,\Phi_M(x-u_et) \tag{16}$$

where Φ_M is a constant and $u_e>2u_c$. Recently Zakharov and Shabat have shown that (16) represents a soliton.[20]

More recently Hasegawa and Tappert have suggested that conventional fiber optics may be represented by (12) at power levels of about 1 watt.[21] Initial computer calculations indicate greatly improved transmission characteristics when a bit of information is carried by an envelope soliton. Bundles of such fibers may eventually make it possible to transmit large amounts of information in parallel via pulse trains of optical solitons. Optical information is particularly interesting because it should be suitable for "broadside" input to large scale integrated circuit (LSI) arrays.

REFERENCES

1. Scott, A.C., Active and Nonlinear Wave Propagation in Electronics, Wiley-Interscience, New York, 1970.

2. Nagumo, J., Arimato, S. and Yoshizawa, S., An active pulse transmission line simulating nerve axon, Proc. IRE, 50, 2061, 1962.

3. Crane, H.D., Neuristor - a novel device and system concept, Proc. IRE, 50, 2048, 1962.

4. Scott, A.C., Distributed device applications of the superconducting tunnel junction, Solid State Electronics, 7, 137, 1964.

5. Yuan, H.T. and Scott, A.C., Distributed superconductive oscillator and neuristor, Solid State Electronics, 9, 1149, 1966.

6. Parmentier, R.D., Recoverable neuristor propagation on superconductive tunnel junction strip lines, Solid State Electronics, 12, 287, 1969.

7. Parmentier, R.D., Neuristor analysis techniques for nonlinear distributed systems, Proc. IEEE, 58, 1829, 1970.

8. Drukier, A.K., On the possibility of using superconducting neuristor line as particle detector, Nucl. Inst. and Methods, 104, 593, 1972.

9. Scott, A.C., Chu, F.Y.F. and McLaughlin, D.W., The soliton: A new concept in applied science, Proc. IEEE (to be published).

10. Barone, A., Esposito, F., Magee, C.J. and Scott, A.C., Theory and applications of the sine-Gordon equation, Riv. del Nuovo Cimento, 1, 227, 1971.

11. Scott, A.C., A nonlinear Klein-Gordon equation, Bull. Am. Phys: Soc., 12, 308, 1967.

12. Scott, A.C., A nonlinear Klein-Gordon equation, Am. J. Phys. 37, 316, 1969.

13. Scott, A.C. and Johnson, W.J., Internal flux motion in large Josephson junctions, Appl. Phys. Letters, 14, 316, 1969.

14. Barone, A., Flux-flow effect in Josephson tunnel junction, J. Appl. Phys., 42, 2747, 1971.

15. Fulton, T.A. and Dynes, R.C., Vortex propagation in long Josephson junctions, Bull. Am. Phys. Soc., 17, 47, 1972.

16. Kochendörfer, A., Seeger, A. and Donth, H., Theorie der Versetzungen in eindimensionalen Atomreihen, Zeits. F. Physik, 127, 533, 1950; 130, 321, 1951; and 134, 173, 1953.

17. Perring, J.K. and Skyrme, T.H.R., A model unified field equation, Nucl. Phys., 31, 550, 1962.

18. Fulton, T.A., Dynes, R.C. and Anderson, P.W., The flux shuttle - a Josephson junction shift register employing single flux quanta, Proc. IEEE, 61, 28, 1973.

19. Benney, D.J. and Newell, A.C., The propagation of nonlinear wave envelopes, J. Math. and Phys., 46, 133, 1967.

20. Zakharov, V.E. and Shabat, A.B., Exact theory of two-dimensional self-focusing and one-dimensional self-focusing and one-dimensional self-modulation of waves in nonlinear media, Soviet Physics JETP, 34, 62, 1972.

21. Hasegawa, A. and Tappert, F., Transmission of stationary nonlinear optical pulses in dispersive dielectric fibers, (to be published).

PARALLEL PROCESSING ASPECTS OF SYNTHETIC APERTURE HOLOGRAM TECHNIQUES

WINSTON E. KOCK

VISITING PROFESSOR, THE UNIVERSITY OF CINCINNATI AND CONSULTANT, THE BENDIX CORPORATION

INTRODUCTION

One of the most extensive uses of non-optical holograms has occurred in the microwave radar field in the form of synthetic aperture radar (1,2,3,4). Holography inventor Dennis Gabor has commented (5), "unknown to me, a most interesting branch of holography was developing from 1955 onwards at the Willow Run Laboratory attached to the University of Michigan. It was holography with electromagnetic waves, and reconstruction by light, which was called 'Side-Looking Radar' or 'Synthetic Aerials.' It was classified work; the first publication by Cutrona, Leith, Palermo and Porcello occurred in 1960." More recently, the use of the synthetic aperture concept in acoustic applications has been considered (6,7), with medical imaging being one of these. We first describe the concept, and then examine the parallel nature of its accepting, processing, and display functions.

ABSTRACT

The parallel aspects of the detecting, processing, and displaying functions of synthetic aperture (hologram) systems are discussed and compared with the more usual serial procedures employed in present radars and sonars. In the detecting (accepting) function, the parallel nature is exemplified (1), in the multi-beam forming process (as against the serial beam-scanning technique), (2) in the parallel phase-curvature generating process for near-field points (versus the serial phase-adjusting, focussing procedure), (3) in the automatic parallel adjusting of aperture size so as to cause the metric resolution to be independent of range, and (4) in the providing of an angular resolution which is twice that of an ordinary, radar or sonar of equal aperture. In the processing function, the parallel nature is evident in the simultaneous generation of myriads of zone plates for points at all ranges and in the photographic process which records these zone plates. In the displaying function, the parallel process is manifest in the holographic reconstruction of the reflecting points through the illumination, with laser light, of the photographic record.

THE SYNTHETIC APERTURE CONCEPT

In a synthetic aperture radar system, an aircraft moving along a very straight path continually emits successive microwave pulses (4). The frequency of the microwave signal is very constant (the signal remains coherent with itself for very long periods). During these periods, the aircraft may have traveled several thousand feet, but because the signals are coherent, all the many echoes which return during this period can be processed as though a single antenna as long as the flight path had been used. The effective antenna length is thus quite large, and, because such a long array aperture has the capability of forming an extremely large number of extremely sharp beams, this large "synthetic" aperture provides records having extremely fine details.

As in a hologram, the radar record, which consists of the combination, with a reference wave, of the reflections from the terrain, is recorded photographically. Also, as in a hologram, the record can be looked upon as a super-position of myriads of zone plates, one for each reflecting point in the terrain. When illuminated by laser light, it reconstructs the reflecting points just as a hologram does.

PARALLEL PROCESSING ASPECTS OF SYNTHETIC APERTURE SYSTEMS

(a) Accepting Function

As was noted above, the great equivalent length of the synthetic (linear) aperture permits many extremely sharp beams to be generated. In most radars which have such large apertures, it is an extremely complicated procedure to cause all of these beams to be formed simultaneously. Instead, one sharp beam is formed and this beam is scanned (serially) over the many possible directions in which it can be pointed. In the synthetic system, all beams are automatically formed, (and information is accepted by all) simultaneously.

Any optical or radar aperture whose dimensions extend over many thousands of wavelengths, has a near field which extends outward to very great distances. For such an aperture to have maximum effectiveness in delineating the large number of objects in its near field, the elements must be phased so as to exhibit a curved wave-front property. This curvature is very great for near objects, is less so for mid-range objects, and becomes flat for far field objects. For the standard form of radar, such a phase adjustment procedure, differing for each element of range, would be a very complicated one, and if possible at all, it would surely be done in a serial manner. In the synthetic system, such curvature is automatically introduced through the zone plate action. Information is thus accepted from all range points in a highly efficient manner and in parallel.

In certain radar applications it is desirable to have the metric (not the angular) resolution constant for all ranges. For this to occur, a much larger aperture must be utilized for more distant points (thereby providing a much higher angular resolution) than for near points. In an ordinary radar, the procedure to accomplish this is again a very complicated one. In synthetic aperture systems, this change of aperture length for the various ranges also occurs automatically because of the finite (small) aperture size of the actual antenna employed. Its constant angular resolution versus range (usually tens of degrees) causes its metric resolution to be less at great ranges. This provides longer zone plates for the more distant objects and shorter ones for the nearby objects. The result is a metric resolution which is independent of range. Parallel acceptance of information is thereby automatically provided, thereby achieving the desired result.

Finally, in the accepting function, it is of interest to note that the beam width of a synthetic aperture is twice as sharp as that of a standard radar of equal aperture.

(b) Processing Function

The processing of the received data in a synthetic aperture system can be considered to be, as in a hologram, the photographic recording procedure, which results in myriads of superimposed zone plates, and in the developing in parallel, of all of the zone plates in that photographic record.

(c) Displaying Function

As in the holographic reconstruction of a scene, the display function of a synthetic aperture system is accomplished by the illumination of the photographic record with laser light, causing the parallel displaying of myriads of objects located at many different points in range.

SUMMARY

Through optical procedures, the synthetic aperture system offers a convenient technique by which parallel information processing can be accomplished in microwave and acoustic applications. Its use continues to be extended, in acoustics, in the fields of medicine and underwater sound.

REFERENCES

1. L. J. Cutrona, E. N. Leith, L. J. Porcello, and W. E. Vivian, "On the Application of Coherent Optical Processing Techniques to Synthetic-Aperture Radar," Proc. IEEE, Vol. 54, pp. 1026-1032, Aug., 1966.

2. W. E. Kock, "Side-Looking Radar, Holography, and Doppler-Free Coherent Radar," Proc. IEEE, Vol. 56, pp. 238-239, Feb., 1968.

3. E. N. Leith and A. L. Ingalls, "Synthetic Antenna Data Processing by Wavefront Reconstruction," Appl. Opt., Vol. 7, Mar. 1968, pp. 539-544.

4. W. E. Kock, "Radar and Microwave Applications of Holography," pp. 323-356, in Applications of Holography, New York, Plenum Press, 1971.

5. E. Camatini, Editor, Optical and Acoustical Holography, Plenum Press, 1972, p. 11.

6. J. J. Flaherty, K. R. Erikson, and Van Metre Lund, "Synthetic Aperture Ultrasonic Imagine Systems," U. S. Patent No. 3,548,642, December 22, 1970 (filed March 2, 1967).

7. D. F. Pekau and R. Diehl, "Recording of One Dimensional Holograms as a Function of Object Range," presented at the Int. Symp. Applications of Holography, Besancon, France, July 6-11, 1970.

THE IMAGE FEEDBACK COMPUTER

Stephen P. McGrew

Pacific Research Associates
Post Office Box 9131
Seattle, Washington, U.S.A. 98109

ABSTRACT

An attractive possible future development in the field of optical data processing is the image feedback computer, a multidimensional application of the concepts behind the electronic analog computer. The image feedback computer would use the methods of existing coherent and incoherent image processors, but would include real time feedback provided by the image converters and electrophotographic systems presently under development in various laboratories. It would be capable of generating solutions to linear and nonlinear partial differential equations in three variables as directly as an electronic analog computer is capable of generating solutions to ordinary differential equations in one variable. Working principles, programming techniques, applications, and feasibility are discussed.

INTRODUCTION

Within the field of optical data processing, there presently exists a wide variety of techniques for performing mathematical operations upon a two dimensional data field. Among the real time image operations which have been demonstrated are multiplication, division, addition, subtraction, Fourier transformation, spatial differentiation and integration and amplification.

This set of operations already permits construction of passive image processors capable of outperforming digital computers at certain tasks, notably the tasks of image deblurring, real time spectral analysis, and certain types of pattern recognition[1]. Optical data processors gain their advantages over digital computers by virtue of their inherent parallelism and simplicity. A typical optical computer will receive input in the form of a photographic transparency with perhaps a billion revolvable data points, perform a two-dimensional Fourier transformation on the set of points, multiply the billion data points in the transform by the billion data points in another photographic transparency, and form the inverse Fourier transform of the product set. The entire process will take only the length of time necessary to expose a photographic plate at the output (microseconds or milliseconds), whereas a digital computer may require several hours to accomplish the same set of operations[1].

Of course, the chemical processing time required for development of an exposed photographic plate should be included in the total time for data processing in optical computers, and that effectively reduces their speed by a factor of 10^4 to 10^6. However, even with chemical processing time considered, optical computers have a significant advantage over digital computers in many image processing tasks.

In order to take maximum advantage of the speed of optical data processing, it will be necessary to create situations in which output images are used on-line and in real time without the delay of photographic development. Even the use of electronic scanning systems as interfaces between parts of an optical computing system will drastically reduce overall system speed below that of an ideal system, where the computation speed would be limited only by the speed with which light propagates through the computer.

Because suitable techniques have not previously been available, there are very few existing applications where optical computers can perform at anywhere near their full potential. The purpose of this paper is to present the concept of the image feedback computer which should come very close to illustrating the full capabilities of optical data processing. The image feedback computer is conceived as a tool for solving partial differential equations, much as the electronic analog computer is a tool for solving ordinary differential equations. Just as the electronic analog computer is used to provide real time solutions of one dimensional equations, the image feedback computer is intended to provide real time solutions of two and three dimensional equations.

BACKGROUND

Before introducing the concept of the image feedback computer, it will be beneficial to examine the basic mathematical operations that may be performed on images in an optical processor, with particular attention to their potential use in truly real time systems.

multiplication and division

Given a fixed spatial function g(x,y) and a time dependent function f(x,y,t), it is easy to form the product f·g by simply transmitting the time dependent image which represents f through a photographic transparency representing the function g. The pattern of light emergent from the transparency will, in fact, be the product f·g, performed in real time. In order to divide f by g, a photographic transparency is used which represents the function g^{-1}; this is the reversed contrast transparency or "negative" of g.

In order to form the product of two time dependent functions f(x,y,t) and h(x,y,t), it will in effect be necessary to generate a "real time photographic transparency" or real time spatial filter (RTSF) to represent h(x,y,t), then transmit f(x,y,t) through it. Scanned image transducers such as the Lumitron[2] would suffice, but are not ideal for this application. What is needed is an image convertor that produces an optically addressable image from an optical image directly, without scanning. Among the devices under development which have the required capability of direct image conversion are the strain-biased ferroelectric-photoconductor imaging devices[3], the photoconductive PROM[4], (Pockets Read Out Modulator) Eden's image intensifier[5], the image amplifier or Satake et al[6], and Reizman's photosensitive membrane light modulator[7].

Essentially, these devices use a layer of photoconductor or photosensitive semiconductor to produce an electric field whose spatial dependence is directly related to the spatial dependence of the intensity of an incident light pattern. The electric field in turn controls the transmittance or reflectance characteristic of a second layer of material. In the strain-biased devices, the birefringence of a ceramic plate is controlled. The PROM combines both the photosensitive layer and the second layer into a single layer of an electro-optic photoconductive crystal such as ZnS. Eden's device is a photoconductor/electro-optic crystal sandwich structure. Satake's device includes an area photocathode and electron multiplier to provide electronic amplification, followed by an electron collector and an electro-optic crystal. Reizman's device uses an array of photosensitive p-n junctions backed by thin reflective membranes which deform by an amount dependent upon the voltage drops across the junctions. In the operation of these devices, a light pattern incident upon the photosensitive surface induces a pattern of transmittance or reflectance variation in the second layer. A beam of light incident upon the second layer is thus spatially modulated in a pattern corresponding to the light pattern incident upon the photosensitive surface.

It is possible to form the product of two time dependent functions f and h by imaging the function h upon the photosensitive surface of one of these real time spatial filters, and imaging the function f onto the second layer. This is directly analogous to passing the function f through a photographic transparency of h, and it forms the desired product in the form of a transmitted or reflected image. Division is accomplished by contrast reversal: it is possible to construct the real time spatial filters to have the effect of either a photographic negative or a photographic positive, and as previously indicated, a positive transparency of a function f will induce a modulation equivalent to multiplication by f, while the negative transparency will induce a modulation equivalent to division by f.

negative numbers in real time multiplication

For optimum usefulness, an optical computer should be capable of handling negative values of functions. Because the real time spatial filters are sensitive to the intensity pattern of the control image, it would seem that the sign of the local value of a function would inevitably be lost. However, in a real time spatial filter with sufficient resolution, it should be possible to preserve sign information holographically. This would be accomplished as shown in figure 1a, by mixing the control image h(x,y,t) with a reference beam upon the photosensitive surface. A hologram f h will be formed on the second layer. If a duplicate of the reference beam is used to illuminate the second layer, the emergent beam will carry a duplicate of the control image. Negative values of h may be represented by a phase shift of 180° in the control image, and the phase shift will be duplicated in the reconstruction of h.

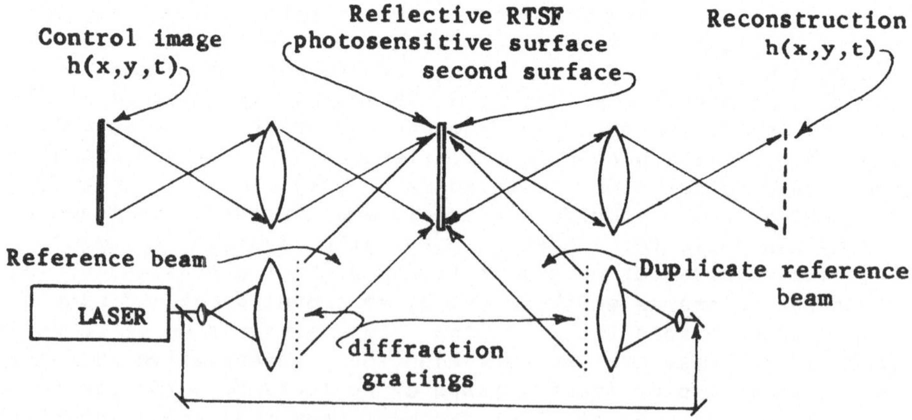

Figure 1a: Real Time Holographic Image Conversion

If, on the other hand, a second image f instead of a duplicate reference beam is used as in figure 1b to illuminate the second layer, the product of h and f will be formed instead of a reconstruction of h. In areas where h and f have opposite sign, the combined phase shift will be equivalent to a 180° shift or a negative sign. Where h and f have the same sign, the combined shift will be equivalent to zero, or a positive sign.

A high resolution real time spatial filter used with a reference beam should provide a good means of handling negative and positive values in real time multiplication of time dependent images.

addition and subtraction

During the last few years an effective technique for the operations of addition and subtraction has been developed[8,9] using diffraction gratings to coherently superimpose two images either in phase for addition or 180° out of phase for subtraction. The technique is real-time, preserves sign information without difficulty, and is well suited for use in an image feedback computer.

Fourier transformation and Fourier plane operations

The Fourier transform is easy to perform optically and is a basis for most optical computers that have been built. An important consequence of the simple optical Fourier transform is the capability of performing operations in the spatial frequency domain, rather than in the spatial domain. In the frequency domain, the operations of differentiation, integration, and convolution become algebraic operations of multiplication by the proper functio

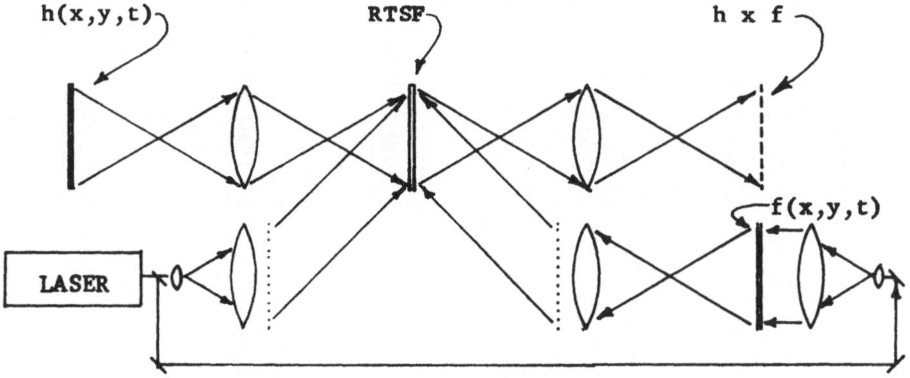

Figure 1b: Real Time Holographic Image Multiplication

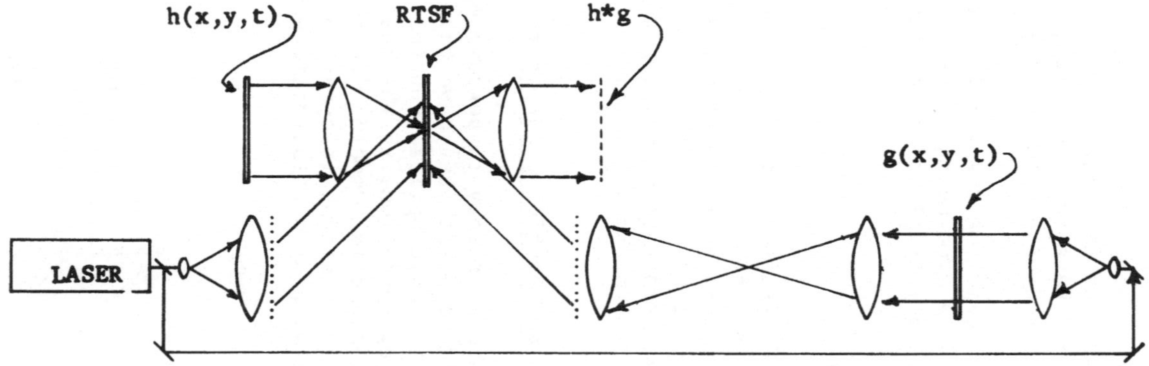

Figure 2: Real Time Holographic Image Convolution

Convolution requires multiplication of the Fourier transforms of two functions, followed by the inverse Fourier transform. The partial differential is formed by multiplying the Fourier transform of a function by the function jx (or jy), and taking the inverse Fourier transform of the product. Partial integration is performed by obtaining the inverse Fourier transform of the product of 1/jx with the Fourier transform of the function.

A major problem encountered when performing differentiation in an optical computer is the error introduced by thickness variations in the filters used. Thickness variations lead to effective phase modulation, which can in turn produce a large noise component in the output. Presumably this difficulty can be overcome in the future, possibly by means of a holographic aberration correction technique. The Fourier plane operations of differentiation[10] and integration are inherently real time and preserve phase or sign information. Real time convolution of two time dependent images will require a real time spatial filter as shown in figure 2.

temporal integration

Any real time photographic system will act as a temporal integrator inasmuch as total exposure in a photographic process is the time integral of the incident light intensity. A number of real time image storage devices have been developed, including a photosensitive electrolytic cell, and several image storing PROM-type devices. The electrolytic cell stores an image by depositing onto a surface a material which attenuates or reflects light. The rate at which the material is plated is controlled by the electrical current through the cell; and the current is locally controlled by the resistivity of a layer of photoconductor. A pattern of light incident upon the photoconducting layer thus controls the deposition rate. Light transmitted through or reflected from the area where material is deposited is thereby spatially modulated according to the total exposure at each point. Because the total exposure is the time integral of the incident light intensity, the transmitted or reflected light intensity is related to the time integral of the image to which the photoconductor is exposed.

Because the photoconductor responds to light intensity rather than amplitude, phase or sign information will be lost unless a reference beam is used to provide holographic temporal integration of the stored image, as shown in figure 3. The holographic temporal integrator in effect forms the real time sum of a great number of time increment holograms, and the transmitted or reflected beam will receive contributions from each of the incremental holograms so that the time integral is formed in real time.

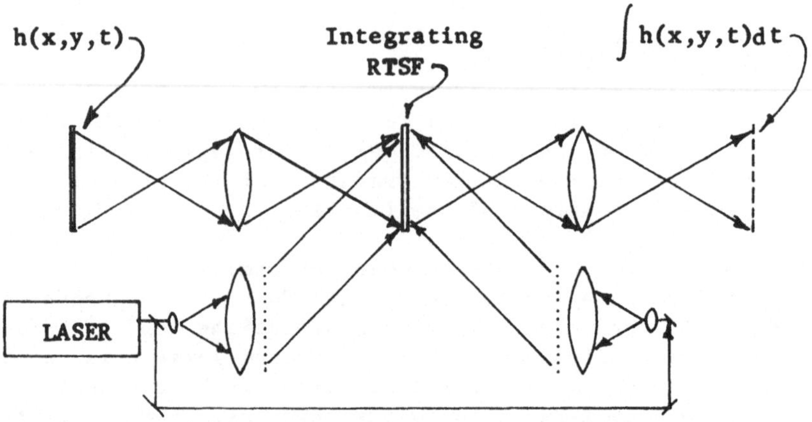

Figure 3: Real Time Holographic Temporal Integration

The image storing PROM-type devices are constructed similarly to the PROM-type real time spatial filters except that the conductivity of the electro-optic layer is made sufficiently low that charge passing through the photoconductive layer is stored next to the electro-optic crystal layer rather than being conducted through it. Lateral mobility of the charge carriers is kept minimal to prevent lateral charge migration. A charge pattern is built up when light is incident upon the photoconductive layer, resulting in a spatial pattern of voltage across the crystal layer and a consequent spatial modulation of light passing through the layer. Holographic integration should be possible with both the PROM-type devices and the electrolytic image cell.

amplification

Any of the real time spatial filters may be used as an image amplifier, merely by illuminating the second layer with a spatially uniform light beam sufficiently brighter than the control image. Another kind of image amplifier has been developed, however, which will preserve phase information without holography. This is the laser image amplifier[11], in which the optical wavefront which is to be amplified is transmitted through a pumped lasing medium and is amplified directly by the medium.

The laser image amplifier has the advantage of being almost free of alignment errors, but has the disadvantage of having relatively low gain. A real time spatial filter used as an image amplifier has the advantage of providing controllable gain over an extremely wide range, but the disadvantage of requiring precise and stable alignment of reference beams in order to preserve phase information. It appears that the laser image amplifier is capable of diffraction limited resolution, whereas the resolution of photoconductive image amplifiers seems limited by the point-spread function in the photoconductor to about 300 line pairs per millimeter[12].

At the moment it appears that the photoconductive real time spatial filter used as an amplifier will be more practical than the laser image amplifier for real time optical computer applications, because a constant amplification factor is needed, and the amplification factor of the laser amplifier is difficult to control.

THE IMAGE FEEDBACK COMPUTER

The image feedback computer would use the set of real time mathematical operations available in optical data processing to construct active optical analogues to partial differential equations in the variables of x,y, and t. The techniques would be quite similar to the techniques of electronic analog computation, so a review of electronic analog computer techniques is in order.

electronic analog computer techniques

In order to solve an ordinary differential equation with an electronic analog computer, an electronic circuit must be constructed such that one of the voltages or currents in the circuit obeys an equation identical in form to the equation which is to be solved. Appropriate relationships are determined to relate the constants and variables in the circuit equation to the constants and variables in the equation to be solved, and the solution is found by direct measurement of the evolution of the voltage or current which represents the equation being solved.

Electronic analog computers are constructed from components which perform the operations of multiplication, inversion, addition, subtraction, temporal integration, and a few other operations, all in real time. These components will, for example, form the sum of a pair of voltages as an output voltage, or produce a voltage which is proportional to the time integral of a single voltage. The operations may be considered to be zero-dimensional operations in the sense that at any instant the result of an operation may be represented by a single number. In contrast, the result of an operation in an optical computer at any instant must be represented by a two-variable function; an optical computer therefore may be said to use two-dimensional operations. An electronic analog computer produces an output which is a function of time, so its output may be said to be one-dimensional.

A circuit configuration which forms an electronic analog computer to solve a given equation is usually quite simple to find, though the most obvious such circuit is not always the best because components are never ideal. However, assuming "sufficiently ideal" components, designing an electronic analog computer to solve a complicated ordinary differential equation is almost as easy as merely writing the equation. For example, the equation

$$\frac{d^2}{dt^2} F + \frac{d}{dt} F + F^2 = 0 \qquad \text{Eq. 1.0}$$

is solved by the analog computer diagrammed in figure 4. A way to arrive at this design is to begin by "assuming" that F(t) is available. The equation may be integrated twice to obtain

$$F(t) + \int_{t_0}^{t} F(t')dt' + \int_{t_0}^{t} \int_{t''_0}^{t''} F^2(t')dt'dt'' = 0 \qquad \text{Eq. 1.1}$$

Assuming that F(t) is available, the second term may be formed by simply feeding F(t) into an integrator. The third term may be formed by feeding F(t) into both inputs of a multiplier, and feeding the product through two integrators in series. From equation 1.1, we have the fact that F(t) is the negative of the sum of the second and third terms, so we can form the sum with an adder, invert it, and we have obtained F(t) again. But now we can feed the output F(t) back into the input, and thereby form a circuit whose evolution is precisely described by the equation. The constants of integration are determined by the initial values of the outputs of the integrators, and the integrators are designed so that the initial values may be preset.

Nonlinear and linear differential equations may be solved by electronic analog computers. The speed with which an equation may be solved is determined by the frequency range over which the computer components have "sufficiently ideal" behavior. The maximum accuracy of a solution is limited by both the dynamic range and the frequency response of the components.

Perhaps the most serious limitation of electronic analog computers, aside from their limited accuracy, is their inability to solve equations for functions of more than one variable (i.e., partial differential equations). On the other hand, they have several distinct advantages over digital computers in many applications, particularly those applications where real time solutions are desired. A significant point in favor of electronic analog computers is the conceptual simplicity of their programming: to program an analog computer is simply to translate an equation into a circuit diagram having the same "form."

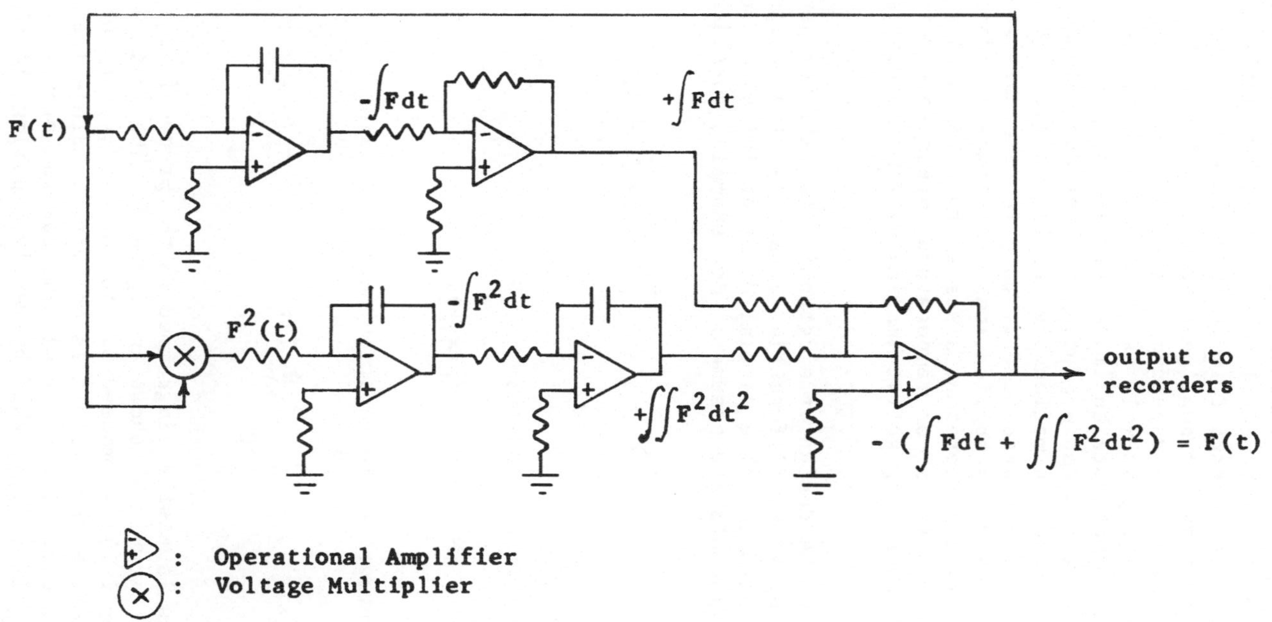

Figure 4: Electronic Analog Computer to Solve Eq. 1.0

Given the real time, two dimensional operations of multiplication, division, addition, subtraction, partial differentiation, partial integration, temporal integration, and amplification, it will be possible to construct image feedback circuits which are direct analogs of linear or nonlinear partial differential equations. In these circuits, the dependent variable is a function of two spatial variables x and y as well as of time. Such an image feedback circuit would constitute an image feedback computer, using two-dimensional operations and producing a three-dimensional output: a time-dependent image. A motion picture is, in this view, a three-dimensional function represented by many incremental two-dimensional samples. In computing applications, the time-dependent image amplitude or intensity can represent a scalar function of any three variables such as x, y, and z; or temperature, pressure and ion density. Earlier work regarding active imaging systems is described in references 13-17.

As in the design of electronic circuits for generating solutions to ordinary differential equations, the design of image feedback circuits to solve partial differential equations is simple if ideal components are assumed. For example, consider the equation

$$\frac{\partial}{\partial t} F(x,y,t) = \left[\frac{\partial}{\partial x} F(x,y,t)\right]^2 + \left[\frac{\partial}{\partial y} F(x,y,t)\right]^2, \qquad \text{Eq. 2.0}$$

with the initial condition

$$F(x,y,0) = F_0(x,y). \qquad \text{Eq. 2.1}$$

We can integrate the equation once to obtain

$$F(x,y,t) = \int_{t_0}^{t} \left(\left[\frac{\partial}{\partial x} F(x,y,t')\right]^2 + \left[\frac{\partial}{\partial y} F(x,y,t')\right]^2\right) dt' \qquad \text{Eq. 2.2}$$

This is easily diagrammed in the proper form for an image feedback circuit as shown in figure 5. The initial condition (equation 2.1) is imposed by storing $F_0(x,y)$ on the temporal integrator before the circuit is activated. Since the temporal integrator is effectively a real time camera, storage of the initial condition is accomplished by merely exposing the photosensitive surface to an image identical to the function $F_0(x,y)$.

A further example is given, where an equation with boundary conditions is solved. The equation describes heat flow in an irregularly shaped, two-dimensional slab of material which is thermally insulated from the rest of the universe. The slab is given an initial temperature distribution

$$T(x,y,0) = T_0(x,y), \qquad \text{Eq. 2.3}$$

and the boundary condition is that no heat can flow across the border of the slab,

$$\hat{n}\cdot\vec{\nabla}T(x,y,t) = 0. \qquad \text{Eq. 2.4}$$

Here $\hat{n}$ is the unit vector perpendicular to the slab boundary.

For a slab with uniform thermal conductivity, the heat equation has the form

$$\frac{\partial}{\partial t} kT = c\nabla^2 T, \qquad \text{Eq. 2.5}$$

where k is the heat capacity of the slab and c is the thermal conductivity. The equation may be put into a more convenient form by treating c as a function of position, in which case we have

$$k\frac{\partial}{\partial t}T = \frac{\partial}{\partial x}c\frac{\partial}{\partial x}T + \frac{\partial}{\partial y}c\frac{\partial}{\partial y}T + c\frac{\partial^2}{\partial x^2}T + c\frac{\partial^2}{\partial y^2}T, \qquad \text{Eq. 2.6}$$

which may be integrated with respect to time to obtain

$$kT = \int_{t_0}^{t}\left(\frac{\partial}{\partial x}c\frac{\partial}{\partial x}T + \frac{\partial}{\partial y}c\frac{\partial}{\partial y}T + c\frac{\partial^2}{\partial x^2}T + c\frac{\partial^2}{\partial y^2}T\right)dt'. \qquad \text{Eq. 2.7}$$

This equation is diagrammed as shown in figure 6. Again, the initial condition is imposed by storing $T_0(x,y)$ on the temporal integrator before the circuit is activated.

A final example is that of coupled partial differential equations. Consider the set of equations

$$\frac{\partial}{\partial t} F(x,y,t) = g(x,y,t)F(x,y,t) \qquad \text{Eq. 2.8}$$

$$\frac{\partial}{\partial t} g(x,y,t) = F(x,y,t)\nabla^2 g(x,y,t) \qquad \text{Eq. 2.9}$$

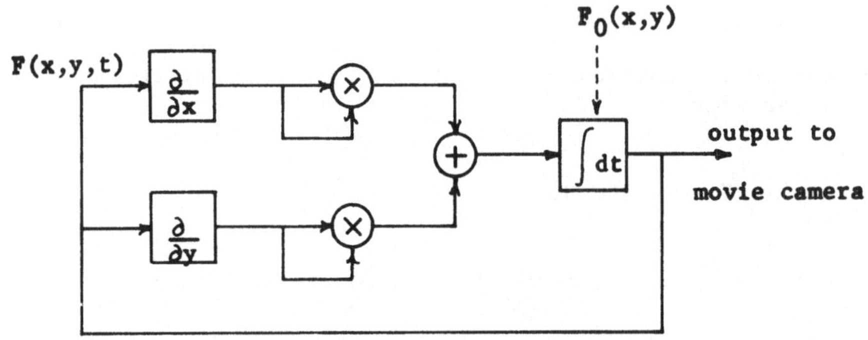

Figure 5: Image Feedback Computer Diagram to Solve Eq. 2.0.

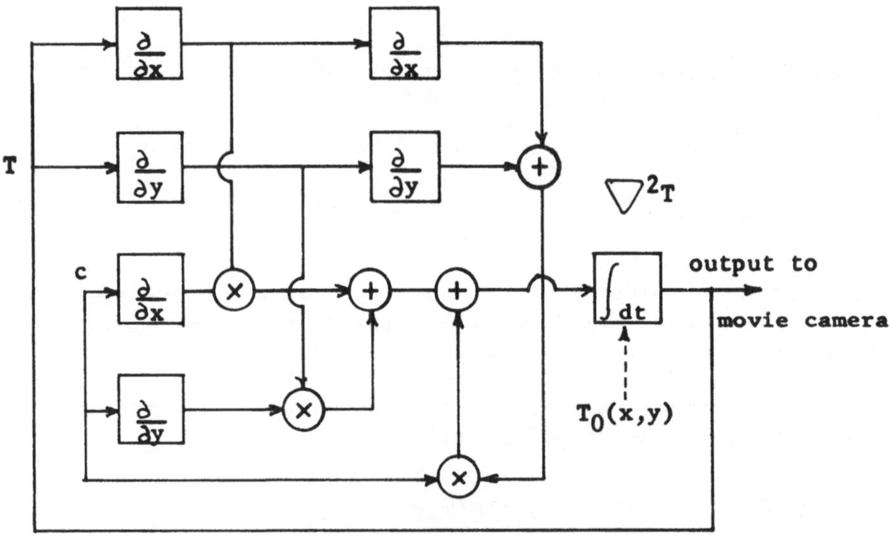

Figure 6: Diagram of Image Feedback Computer to Solve the Equation $\frac{\partial}{\partial t} T = c \nabla^2 T$ (Eq. 2.5) with boundary conditions.

This set of equations is diagrammed as shown in figure 7 and is really no more difficult to realize in an image feedback circuit than the previous examples.

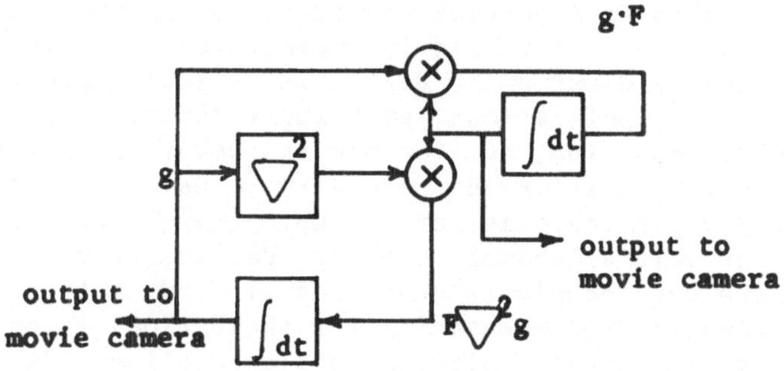

Figure 7: Simplified Diagram of Image Feedback Computer to Solve a Coupled Set of Partial Differential Equations (Eq. 2.8 & 2.9)

feasibility of image feedback computers

Among the factors affecting the feasibility of image feedback computers are:

1) the spatial and temporal resolution attainable in optically active components (multipliers and temporal integrators),
2) the attainable linearity and dynamic range of components,
3) the problem of noise due to surface and volume imperfections in the materials and components used,
4) the value of the image feedback computer in the applications where it will be used compared with its cost.

As stated earlier, the resolution attainable in a photoconductive image amplifier/multiplier seems limited to about 300 line pairs per millimeter. A photoconductive temporal integrator is similarly limited. This resolution is sufficient for many applications, including handling positive and negative values holographically. It would be highly desirable, however, to have an available resolution on the order of 1000 line pairs per millimeter.

Linearity and dynamic range of passive components of an image feedback computer are not likely to place important constraints on its usefulness. Existing techniques using diffraction limited optics and multiple photographic emulsions can provide filters useful over a range of better than six decades. Active components such as the multiplier and temporal integrator are likely, on the other hand, to strongly limit the performance of image feedback computers until their range of linear response is increased considerably.

Noise is likely to be an important consideration in the design of image feedback computers. It is nearly impossible to fabricate perfectly flat surfaces, and extremely difficult to avoid material inhomogeneities. In some applications, existing techniques will provide adequate noise reduction, but for precise solution of partial differential equations, it may be necessary to develop techniques for compensating for the phase errors which result from surface imperfections and volume inhomogeneities. Particularly promising in this regard are the techniques described by Ward, Auth and Carlson[18] for holographically compensating for aberrations in optical systems. By treating phase noise as lens aberrations, it should be possible to generate holographic masks to correct the phase noise in each component of the computer.

The problems of alignment in image feedback circuits have not been explored.

The value of real time solution of partial differential equations is difficult to estimate, but it is potentially very great. In the image feedback computer diagrammed above for solving a heat equation, it is possible to vary the shape of the slab and the initial temperature distribution by simply inserting different photographic transparencies representing $c(x,y)$ and $T_0(x,y)$. It can easily take hundreds or thousands of man-hours to write a program to solve a nonlinear partial differential equation on a digital system, while an image feedback circuit analog of a partial differential equation takes only a few minutes to diagram.

A sequential digital system is very poorly equipped to provide I/0 in the case of multidimensional problems. Reading images into digital form and plotting digital information in the form of images can easily take hours of digital computer time, while the image feedback computer could accept and output images practically instantaneously in the form of photographs. To merely display the three-dimensional function which is the solution of a three-dimensional partial differential equation, a digital system could easily be tied up for a year, while a movie camera could record and display the same information from an image feedback computer in an hour.

For practically any application requiring simulation, analysis or control of multidimensional physical processes in real time, a sequential digital system is hopelessly inadequate, if only because of the I/0 bottleneck. It is, in principle, possible to construct parallel digital systems which would approach the speed of an image feedback computer, but the costs of such a system would be astronomical. So, for such applications, the image feedback computer may prove to be the only workable system.

It should perhaps be pointed out here that there are other systems beside the image feedback computer which could be called multidimensional active analog computers. Wind tunnels with scale models of aircraft provide solutions to the equations of air flow with boundary conditions imposed by aircraft shape. Models of river systems provide solutions to equations governing erosion and fluid flow. There are certain solid-state devices which can simulate the behavior of three-dimensional gaseous plasmas. The image feedback computer is conceived as a general-purpose system which may eventually be capable of serving some of the same purposes as these physical-analog computers, as well as handling more abstract problems.

CONCLUSIONS

The potential value of an image feedback computer is clear enough. Problems which are impossible to solve with digital systems due to time and cost constraints may prove to be solvable by an image feedback computer. Questions which must be explored in determining the feasibility of the concept are:

1) To what extent can noise sources be eliminated or compensated for?

2) To what extent can the resolution, speed and dynamic range of active image transducers be extended?

3) How will alignment difficulties affect operation of the computer in specific applications?

Regardless of the answers which develop to these questions, there will be some applications for image feedback computers which can be constructed with presently existing techniques and hardware. If, in fact, the envisioned general purpose system becomes feasible, we will have at our disposal new and valuable tools useful in solution of partial differential equations, modeling of physical systems and processes, and real time analysis and control of multidimensional processes.

BIBLIOGRAPHY

1. Stroke. "Optical Computing" IEEE Spectrum Dec. 1972.

2. R. J. Doyle and W. E. Glenn, "Lumatron: A High-Resolution Storage and Projection Display Device." IEEE Trans. Electron Devices, Vol. ED-18, Sept. 1971.

3. J. R. Aldonado and A. H. Meitzler, "Strain-Biased Ferroelectric-Photoconductor Image Storage and Display Devices" Proc. IEEE March 1971.

4. J. Feinleib and D. S. Oliver. "Reusable Optical Image Storage and Processing Device" Applied Optics Dec. 1972.

5. D. D. Eden, "Photoconductive Electro-Optic Image Intensifier Utilizing Polarized Light" U.S. Patent #3,449,583.

6. Kimihiko Satake et al., "Light Intensity Amplifying Device Utilizing a Semiconductor Electron Sensitive Variable Resistance Layer" U.S. Patent #3,499,157.

7. F. Reitzman, "Optical Spatial Phase Modulator Array" Proc. Electro Opt. Syst. Design Conf., New York Sept. 1969.

8. S. K. Yao and S. H. Lee, "Synthesis of a Spatial Filter for the Combined Operations of Subtraction and Correlation" Applied Optics May 1971.

9. S. H. Lee, S. K. Yao, and A. G. Milnes, "Optical Image Synthesis (Complex Amplitude Addition and Subtraction) in Real Time by Diffraction Grating Interferometric Method" J. Opt. Soc. Amer. 60, 1037 (1970).

10. R. G. Eguchi and F. P. Carlson, "Coherent Optical Gradient System" Univ. of Wash. Tech. Rept. T. R. 127 Nov. 1968.

11. T. W. Hansh, F. Varsanyi, and A. L. Schawlow, "Image Amplification by Dye Lasers" App. Phys. Lett. 15 Feb. 1971.

12. P. Nisenson and Sato Iwasa, "Real Time Optical Processing with $Bi_{12}SiO_{20}$ PROM" Applied Optics Dec. 1972.

13. M. O. Hagler, "Active Synthesis of Inverse Spatial Filters: Applied Optics Dec. 1971.

14. R. V. Pole and R. A. Myers, "Wide Field Active Imaging" IEEE J. of Quantum Electronics QE-2, 270 (1966)

15. H. Wieder and R. V. Pole, "Reactive Optical Information Processing II: Factors Affecting the Applicability and Efficiency of the Method" Applied Optics 1761 (1967).

16. H. Wieder and R. V. Pole and P. Heidrich, "Electron Beam Writing of Spatial Filters" IBM J. Res. Dev. 13, 1691 (1969)

17. E. S. Barrekette in Applications of Holography, Barrekette, Koch, Ose, Tsujichi, Stroke, eds. Pleuum, N.Y. 1971, p. 309

18. J. E. Ward, D. C. Auth, and F. P. Carlson, "Lens Aberration Correction by Holography" Applied Optics April 1971.

CLIP 3: A CELLULAR LOGIC IMAGE PROCESSOR

M.J.B. Duff and D.M. Watson

Image Processing Group, Department of Physics and Astronomy, University College London

INTRODUCTION

CLIP 3 is the third in a series of cellular logic image processors being developed by the Image Processing Group in the Department of Physics and Astronomy at University College London. Two earlier processors operating on different principles have been described elsewhere in the literature.[1,2] The present series implement and extend arrays proposed by Unger,[3,4] McCormick,[5] and Levialdi[6] and develop further ideas which emerged during the study of the first two arrays constructed at University College.

CLIP 3 consists of a 16 by 12 array of processors, usually referred to as 'cells'. Each cell acts on the corresponding element of a binary image which is stored in a 192 bit shift register. The shift register provides parallel connections to every bit so that once an image has been loaded into the register, the array processors can obtain access simultaneously to all elements of the image. Every cell is connected to each of its immediate neighbours in the array.

The operation of the array will be described with reference to Figure 1. The central core of each cell is a Boolean processor which produces two independent binary outputs D and N from its two binary inputs A and P (equivalent to B + T). Input A is the value of the image element corresponding to the location of cell in the array; output D will be the value of the processed image at the same location. N is a binary output which fans out to all adjacent or neighbouring cells in the array, entering each cell via a summing and thresholding gate, as shown in the figure. Control lines are used to operate further gates to select which of the eight interconnection directions will take part in the summations. After summing and thresholding the selected inputs to form the binary quantity T, a second shift register, which can contain another image B, combines with T through an OR-gate to provide the input P, defined as B + T.

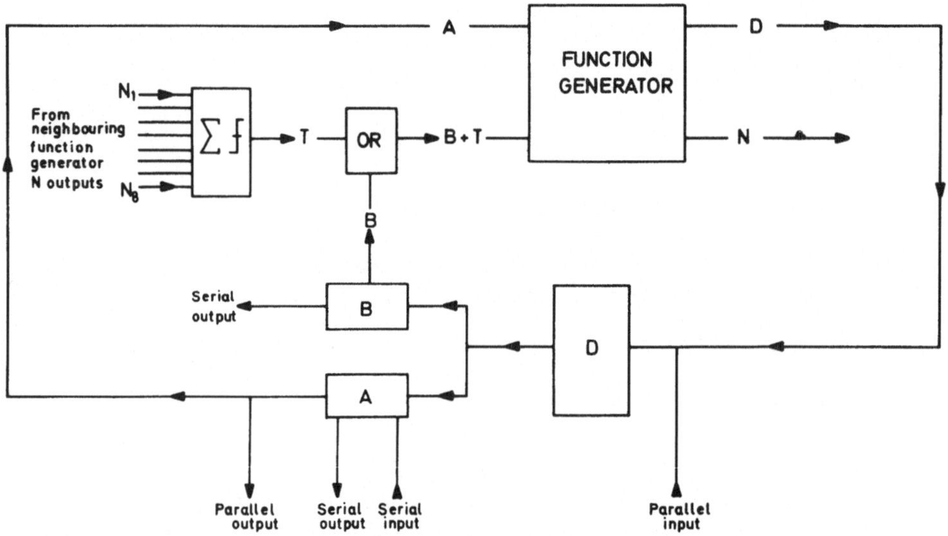

Figure 1 Schematic for CLIP 3 system

The processed image D can be stored in any one of 16 locations referred to as the D memories and labelled D_0 to D_{17} in octal notation. Shift registers A and B can be loaded in parallel from any D location. In addition, by cycling the A register, it can be loaded with an image obtained from a flying spot scanner, and its contents displayed as a dot array on a cathode ray oscilloscope. A light pen is used in conjunction with this display to erase from or write into the register. The contents of the B register are similarly displayed but cannot be modified with the light pen. Conventionally, black parts of an image are given a binary value 1 and appear as bright points in the oscilloscope display.

Every processing cell in the array is required to carry out simultaneously the same operation on corresponding elements of the images being processed, consequently the 16 control lines which determine the Boolean functions for N and D, and which select the interconnection directions, are common to all the cells in the array. Three more control lines select the threshold value for all the threshold gates. Since the array has been wired so as to permit both square and hexagonal architecture, another control line selects which interconnection structure is operative. Finally, all interconnection inputs (N_1 to N_8 in the figure) which come from outside the array to cells in the array margins, are linked together and set at 1 or 0 depending on the state of one more control line.

Figure 2 shows the complete logic diagram for one cell.

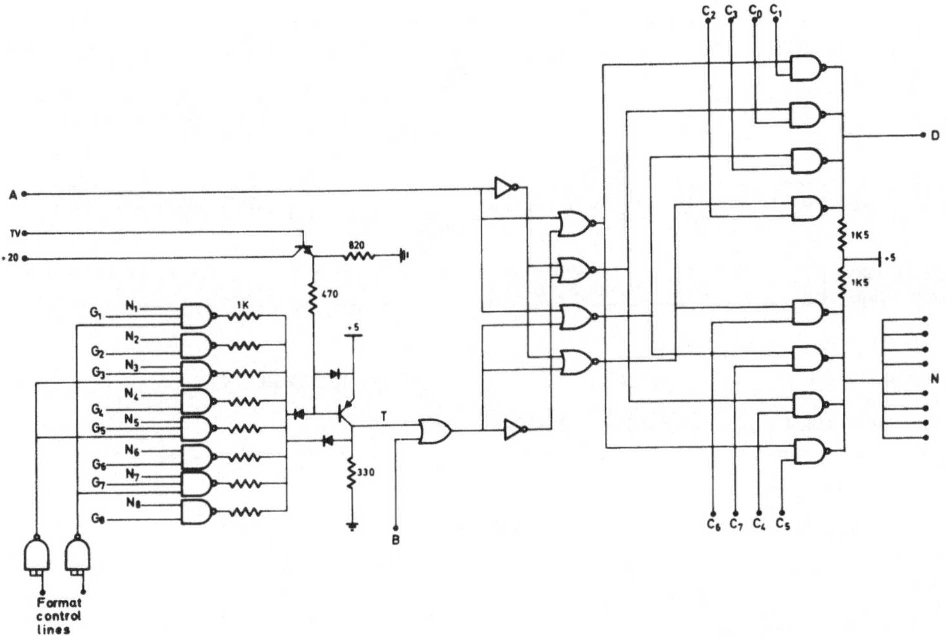

Figure 2 CLIP 3 cell logic

OPERATING THE ARRAY

The processor is controlled by sequences of 24 bit instruction words which are stored in a 256 word random access memory. During processing, a counter steps from the selected start address and extracts each instruction from the memory sequentially. Five types of instruction are possible (see Figure 3). A LOAD instruction loads registers A and B from chosen D memory locations and specifies the D memory destination for the D output resulting from a subsequent PROCESS instruction. In the same instruction the mode in which A and B are to be loaded is specified (normally, replacing their contents with a new image from D, otherwise displaying their original contents and replacing the contents as before, clearing the registers by replacing the contents with 0's, or leaving the contents unchanged). The array architecture is set at square or hexagonal in this instruction.

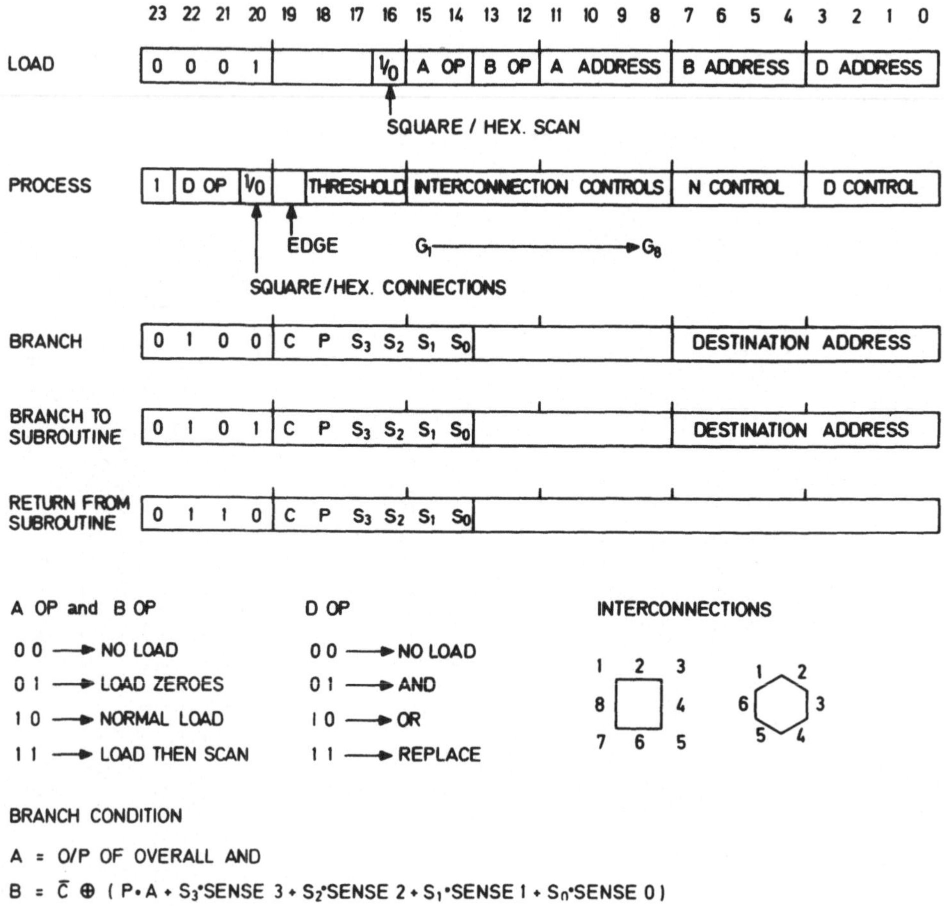

Figure 3 CLIP 3 instruction format

The PROCESS instruction sets up the required threshold and interconnection directions, selects the array margin interconnections value, respecifies the architecture and determines the Boolean functions for N and D. At the same time, the method of loading the selected D location with a new image from the D output is also specified. The new image can AND or OR with the original contents of the chosen D location, can replace the original contents or can be formed but not loaded into any memory location. It is after this operation that a 192 input AND gate sets a one bit register P to 1 if all points in D take the value 1. The 'no-load' mode for D is executed in this context. The value of P can be used in subsequent BRANCH instructions (note that P used in this context should not be confused with $P \equiv B + T$).

Three types of BRANCH instruction are provided. Branches are conditional on a 1 appearing in P (the array AND gate register) or in the states of any of S_0 to S_3 (four sense switches on the control panel). The instruction specifies which out of the five points will be inspected; a 1 appearing on any point selected will cause the program counter to jump to the destination address specified in the BRANCH instruction. An additional bit C in the instruction word reverses this decision, i.e., if C is set at 0 then the BRANCH will be executed when a 1 does *not* appear in any of the inspected locations. BRANCHES TO or FROM a SUBROUTINE are similar in operation except that a return address is stored to allow the normal program sequence to be followed after executing the subroutine. Subroutines can be nested to 16 levels.

THE MNEMONIC CODE

At the time of writing, instructions to CLIP 3 must be entered in machine code, as sequences of 24 bit instruction words. However, programs are written in mnemonic code and it has been found convenient to use this code as a low level language during the design of image processing algorithms, since the code provides a link between the operation of the array and the logical operations involved in image processing. Details of the translation from mnemonic code to machine code have been published in an internal report[7] and will not be repeated here.

'Load' Instructions

These are of the form:

```
I   LD   a,   b,   d,   s
```

where I = the instruction address

a = the D location from which A is to be loaded

b = the D location from which B is to be loaded

d = the D location into which the processed pattern is to be loaded

s = S or is omitted for square or hexagonal operation respectively

The letter D placed after a location (at a or b) causes the relevant register to cycle. If C replaces a location, then the relevant register is cleared.

Example: 2   LD   3D,   C,   6

implies 'Instruction 2 will display the contents of register A and reload it with the contents of location D_3, clear register B and after the subsequent PROCESS instruction, load location D_6 with the processed image appearing at D. The array will operate with hexagonal architecture.' If the next instruction is not a PROCESS instruction, or if it is a 'no-load' process instruction, the d field is omitted.

'Process' Instructions

These are of the form:

$$I \quad D \quad = \quad t\ (n_j, n_k, \ldots n_z) \quad f_n, \quad f_d, \quad es$$

where I = the instruction address

t = the binary threshold which when exceeded gives T = 1

n_j to n_z = the interconnection directions selected from directions 1 to 8

f_n = the Boolean function of P and A determining N

f_d = the Boolean function of P and A determining D

e = E or is omitted for margin interconnections 1 or 0 respectively

s = S or is omitted for square or hexagonal operation respectively

Example: 0 D = 1 (2, 3, 4) A, P + A, ES

implies 'Instruction 0 will set the margin interconnections at 1 and use square architecture, use threshold 1 and interconnection directions 2, 3 and 4, use the Boolean functions N = A, D = P + A, having loaded A and B as specified in the foregoing LOAD instruction, and load the resulting D output into the location specified in the LOAD instruction.'

To allow the various loading procedures for D to be indicated, the following notation has been devised. If the central PROCESS required is

$$D = f\ (t, n_j, f_n, f_d) \equiv f(v)$$

we can write one of the following forms:

$D = f(v)$

$D = D. f(v)$

$D = D + f(v)$

$D \neq f(v)$

The last form is used only when the operation is carried out in order to set the one bit register P. Contents of the D memories remain unchanged.

'Branch' Instructions

These are of the form:

I BR C,P, S_j S_k ..., I_D

where I = the instruction address

C = 1 or 0 depending on whether the branch is or is not to be taken if the conditions apply

P = P or is omitted if the state of the array AND gate is or is not one of the branch conditions

S_j, S_k = a selection from 3, 2, 1 or 0, being the sense switches which will be inspected during the BRANCH operation

I_D = the destination instruction address for the BRANCH

Example: 16 BR 0, P, 3, 27

implies 'Instruction 16 branches to instruction 27 when neither the AND gate register P nor sense switch S_3 are 1, otherwise instruction 17 is the next to be executed.'

BRANCH to SUBROUTINE instructions are similar and take the form:

I BTS C, P, S_j S_k ..., I_D

BRANCH from SUBROUTINE instructions omit the destination address thus:

I BFS C, P, S_j S_k ...

CLIP 3 ALGORITHMS

When a PROCESS instruction is executed in the array, it is helpful to consider the operation of the cells in two stages. In the first stage, an interconnection signal flows laterally in the array in a manner which is determined by the nature of the Boolean function for N. The first requirement is to identify the source of the interconnection signal.

Referring to cells for which A = 1 as 'black' cells and those for which A = 0 as 'white' cells, then the following possibilities can be considered:

White cells generate an interconnection signal ($N = \bar{A}$)

Black cells generate an interconnection signal ($N = A$)

White cells transmit an interconnection signal ($N = P.\bar{A}$)

Black cells transmit an interconnection signal ($N = P.A$)

There is no generation or transmission of interconnection signals ($N = 0$)

Also, since the P input is composed of B + T, then black cells in the pattern in B will always act as signal generators.

The only remaining possible source for the interconnection signal is the margin condition E. Thus by considering the Boolean function for N, the contents of the B register, and the state of the margin interconnections, it is possible to predict the complete pattern of signal flow in the N - P channels of the array.

The next stage is to regard each cell as having two inputs which are now fully determined. These are A, the 'colour' of the cell and P, the interconnection input which has just been established. By choosing an appropriate form for the Boolean function for D, the output image D can be made to contain any combination of the following:

White cells receiving an interconnection signal ($D = P.\bar{A}$)

Black cells receiving an interconnection signal ($D = P.A$)

White cells *not* receiving an interconnection signal ($D = \bar{P}.\bar{A}$)

Black cells *not* receiving an interconnection signal ($D = \bar{P}.A$)

PROGRAMMING CLIP 3

An image processor with an array size as small as 16 by 12 cells is unlikely to be useful for 'production' image processing. Nevertheless, it is quite large enough to aid the design of algorithms which should become effective in a larger array. In order to indicate the nature of the sort of process available in CLIP 3, and so as to illustrate the use of the mnemonic code, one program will be described in detail. This program checks whether all parts of the black figure in the input image are connected and outputs the image only when it is completely connected. The flow diagram for the program is shown in Figure 4.

The principles employed in the program are as follows: the highest black point in the image is found (if there are two points both equally near the top of the array, then the left-most is selected). Propagation from this point removes from the image all parts of the figure connected to the point. The array is then inspected to see if any figure remains. If not, then all the figure is connected and it is displayed in the output. Otherwise, the program returns to the first instruction without displaying in the output array.

```
 0   LD   OD,   1D   0

 1   D = A

 2   BR   0,  3,   0

 3   LD   0,   C,   1

 4   D = 0 (3, 6) P + A,   P

 5   LD   1,   C,   1

 6   D = 1(1, 2) P.Ā,   P.A,   E

 7   LD   1,  C,   1

10   D = 0  (6)  P.A,   E

11   LD   0,   C,   2

12   D = A

13   LD   2,   1,   1

14   D = 0(6)   P.Ā,   P.A

15   LD   2,   1,   1
```

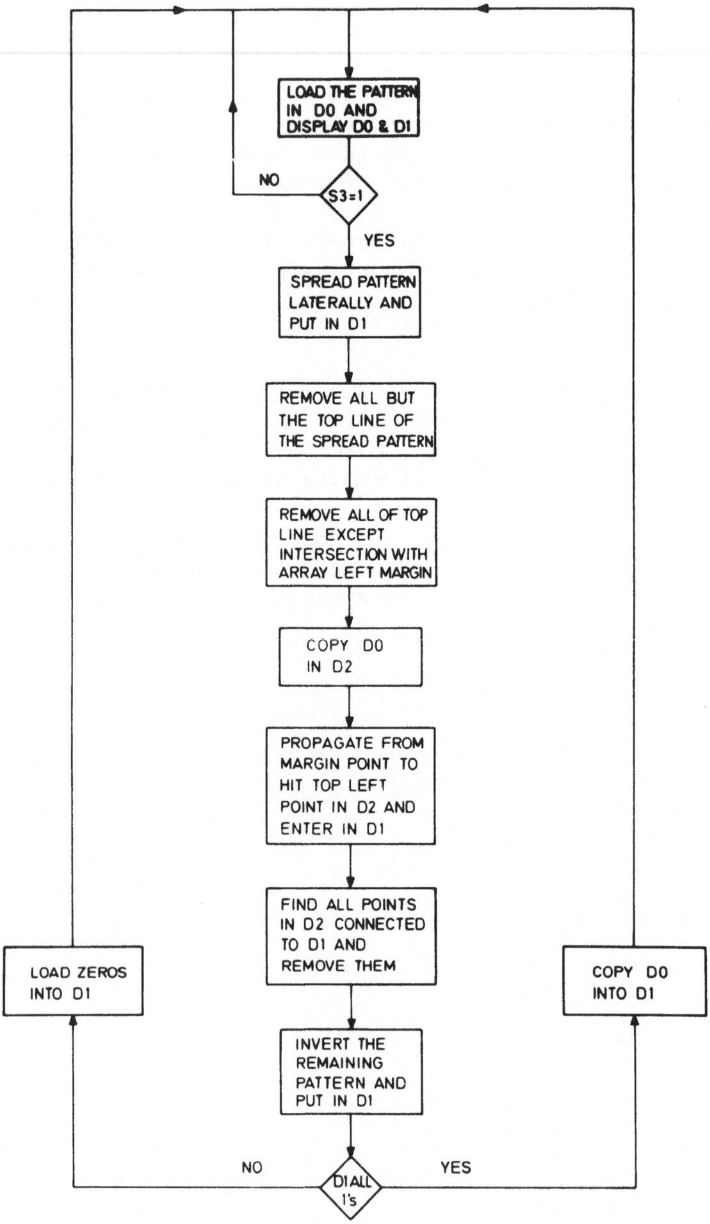

Figure 4 CONNECTED figure program

16 D = 0(1 → 6) P.A, $\overline{\bar{P}.A}$

17 BR 0, P, 23

20 LD 0, C, 1

21 D = A

22 BR 0

23 LD C, C, 1

24 D = 0

25 BR 0

Notes: (a) Instruction 6: the use of threshold 1 and directions (1, 2) is obligatory because interconnections of either direction alone can be found in the side margins of the array; (b) Instruction 16: the direction list (1 → 6) implies all directions 1 to 6 inclusive; (c) Instructions 22 and 25: since the remaining 234 registers in the instruction store may contain arbitrary instructions, the unconditional BR 0 is used to return the program to its first instruction; (d) Instructions 0, 1 and 2: this LOAD instruction displays D_0 and permits the new contents of A to be altered with the light pen. The PROCESS instruction sets D_0 equal to the new contents of A. The BRANCH instruction returns to the LOAD instruction to display (and, if necessary, further modify) the contents of D_0 until such time as the operator puts sense switch S_3 at 1 when the rest of the program is executed.

CLIP 3 PROGRAMS

A large number of CLIP 3 programs have now been written, largely with a view to acquiring skill in the manipulation of parallel processing algorithms.

A non-propagating instruction takes 1 μsec and, in the 16 by 12 array, a propagating instruction is allowed 3 μsec on the assumption that propagation paths more than 30 cells long are unlikely to occur. As a consequence, in all the programs which have been written so far, the processing appears to be instantaneous. In order to produce delays and protracted display times, the LOAD instruction in a display mode is repeated many times in succession. Each display takes 10 msec so that delays of up to $2\frac{1}{2}$ secs can be obtained without difficulty. Longer delays can be produced by writing iterative loops containing display instructions.

It is not sensible in a short paper to try to review the many types of program which can be executed on CLIP 3; this review will be published elsewhere. However, it can be stated that it has been found possible to write CLIP 3 programs implementing a very varied collection of algorithms including, recently, grey level manipulation algorithms. Also, the system will soon be augmented by a PDP 11/10 computer which will permit the control and programming of CLIP 3 in high level languages, and also by a system which is just being constructed which will interface CLIP 3 with a TV camera. CIMU (the Camera Interface Memory Unit) will scan the 16 by 12 array into 48 positions, thereby simulating a 96 by 96 array, and will extract 8 grey levels from the TV camera image. As stated above, some grey level algorithms have been simulated on CLIP 3 and propagational processes in a scanned array mode such as will occur in CIMU have also been simulated. The next development which is anticipated will be the implementation of the array in large scale integrated circuitry; discussions with possible manufacturers are soon to take place.

REFERENCES

1. Duff, M.J.B., Jones, B.M. and Townsend, L.J., Parallel processing pattern recognition system UCPR1, *Nucl. Instrum. Meth.*, 52, 284, 1967.

2. Duff, M.J.B., Cellular logic and its significance in pattern recognition, *AGARD Conf. Proc. No. 94 on Artificial Intelligence*, 25-1, 1971.

3. Unger, S.H., A computer orientated toward spatial problems, *Proc. IRE*, 46, 1744, 1958.

4. Unger, S.H., Pattern detection and recognition, *Proc. IRE*, 47, 1737, 1959.

5. McCormick, B.H., The Illinois pattern recognition computer - ILLIAC III, *Trans. IEEE*, EC-12, 791, 1963.

6. LEVIALDI, S., Parallel counting of binary patterns, *Electr. Letters*, 6, 798, 1970.

7. Duff, M.J.B. and Watson, D.M., CLIP 3 operating manual, *UCL Image Processing Group Report*, 73/4, 1973.

SOME ASPECTS OF THE LOGIC FUNCTIONS OF CLIP 3

C. D. Stamopoulos
Inst. of Computer Science, Univ. of London

M. J. B. Duff
Physics Dept. Univ. College London

D. M. Watson
Physics Dept. Univ. College London

The Cellular Logic Image Processor, CLIP 3 , is a parallel processor with a 12 × 16 logic cell array permitting hexagonal (H) or square (S) tesselations.[1,2]

A general idea of CLIP 3 is given in Fig. 1.

The function generator has two inputs, P and A, and two outputs, N and D.

The D output may be stored in the D memory, the contents of which may be displayed in the A or B displays. The A display is connected to the A input so that the image displayed in A is also the input image.

The N output propagates a signal to the 8 neighbours of each cell. Thus, each cell may send information to its 8 neighbours and may receive information from the same (N_1 - N_8 inputs).

The information propagated by the neighbours is thresholded by Σf and the result, T, may be ORed with the contents of the B display which in general is used to display the output.

Fig. 2 gives an idea of the interconnections between one cell and its neighbours, for hexagonal tesselation, and the directions used in the hexagonal and square modes.

Fig. 3 gives details of one logic cell. We can distinguish the Threshold Section, the OR - gate and the Function Generator, which generates the Minterms of the inputs and gates them, by means of the control lines C_0 to C_7 into the outputs N and D.

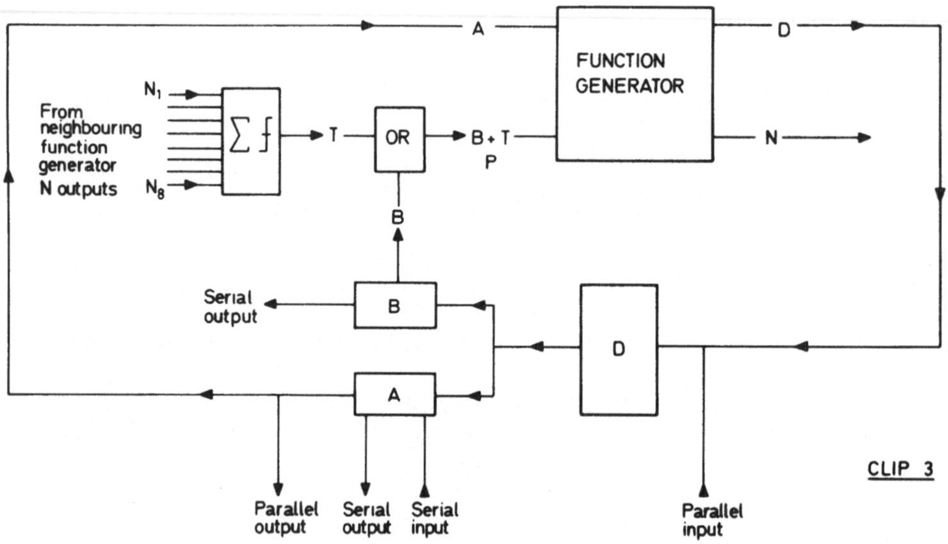
FUNCTION GENERATOR
A
D
N
From neighbouring function generator N outputs
N_1
N_8
T
OR
B + T
P
B
Serial output
B
D
A
Parallel output
Serial output
Serial input
Parallel input
CLIP 3

Fig. 1

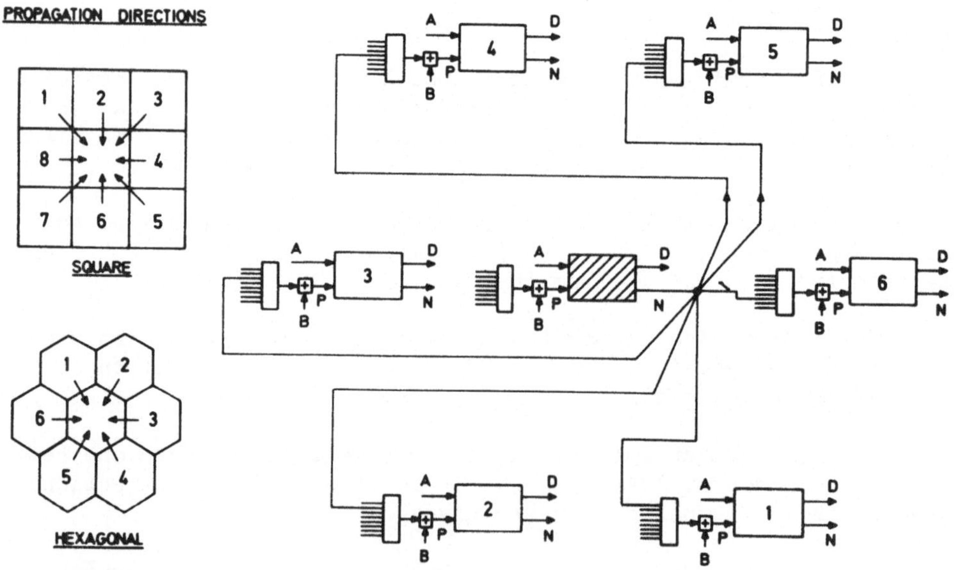
PROPAGATION DIRECTIONS
SQUARE
HEXAGONAL
A
D
N
P
B

Fig. 2

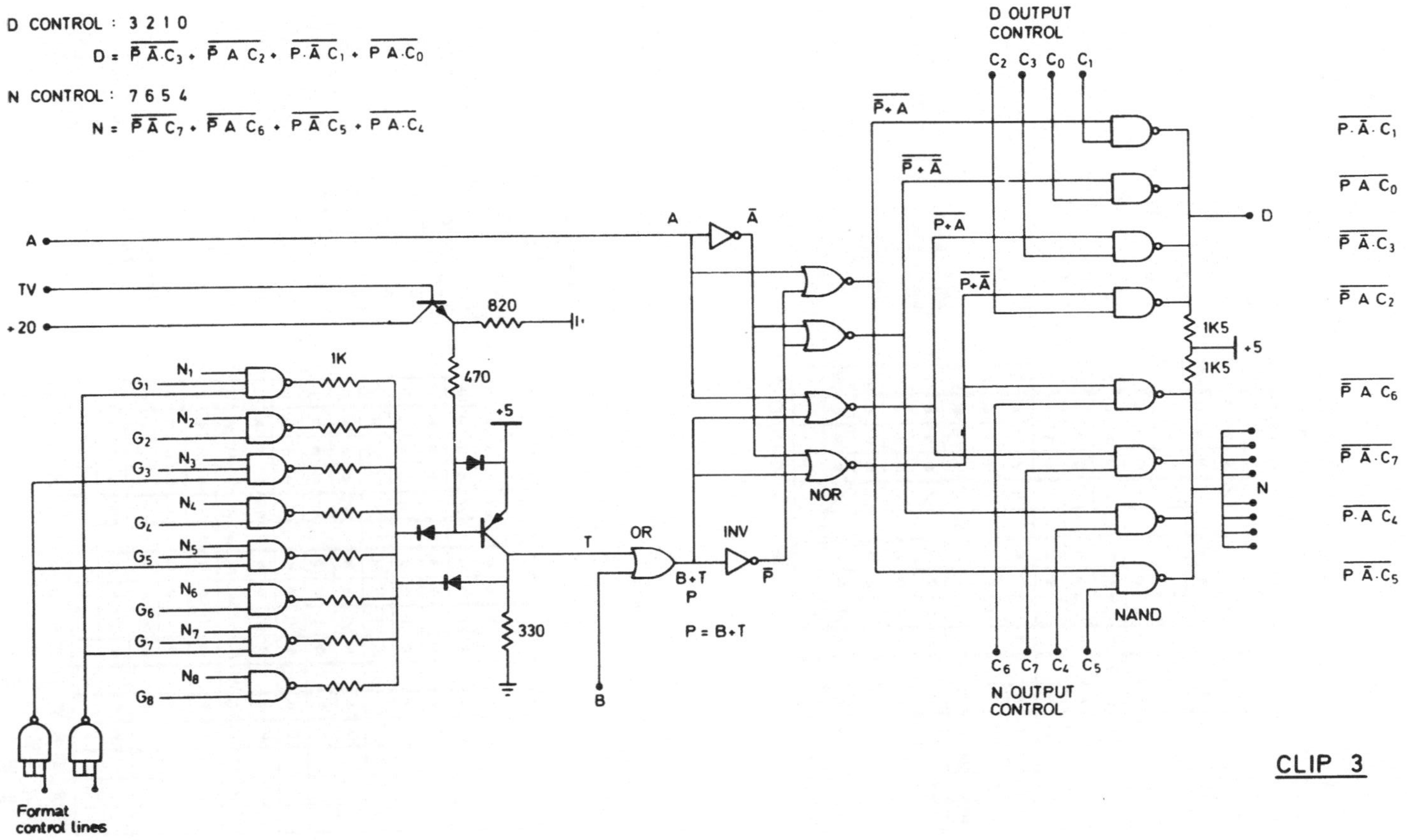
D CONTROL : 3 2 1 0
$D = \overline{\bar{P}\,\bar{A}\cdot C_3} + \overline{\bar{P}\,A\,C_2} + \overline{P\cdot\bar{A}\,C_1} + \overline{P\,A\cdot C_0}$
N CONTROL : 7 6 5 4
$N = \overline{\bar{P}\,\bar{A}\,C_7} + \overline{\bar{P}\,A\,C_6} + \overline{P\,\bar{A}\,C_5} + \overline{P\,A\cdot C_4}$
D OUTPUT CONTROL
C_2 C_3 C_0 C_1
A
TV
+20
820
470
1K
+5
330
T
OR
B+T
P
INV
$\bar{P}$
P = B+T
B
A
$\bar{A}$
NOR
$\overline{\bar{P}+A}$
$\overline{\bar{P}+\bar{A}}$
$\overline{P+A}$
$\overline{P+\bar{A}}$
NAND
1K5
1K5
+5
D
N
C_6 C_7 C_4 C_5
N OUTPUT CONTROL
$\overline{P\cdot\bar{A}\cdot C_1}$
$\overline{P\,A\,C_0}$
$\overline{\bar{P}\,\bar{A}\cdot C_3}$
$\overline{\bar{P}\,A\,C_2}$
$\overline{\bar{P}\,A\,C_6}$
$\overline{\bar{P}\,\bar{A}\cdot C_7}$
$\overline{P\cdot A\,C_4}$
$\overline{P\,\bar{A}\cdot C_5}$
N_1 G_1 N_2 G_2 N_3 G_3 N_4 G_4 N_5 G_5 N_6 G_6 N_7 G_7 N_8 G_8
Format control lines
CLIP 3

Fig. 3

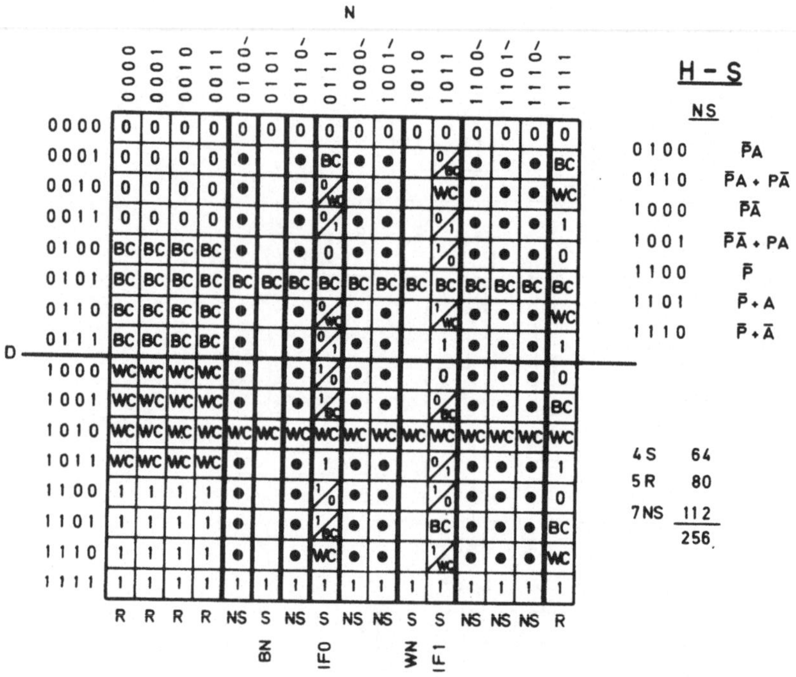

N
H - S
NS
0100 P̄A
0110 P̄A + PĀ
1000 P̄Ā
1001 P̄Ā + PA
1100 P̄
1101 P̄ + A
1110 P̄ + Ā
D
4S 64
5R 80
7NS 112
256
R R R R NS S NS S NS NS S S NS NS NS R
BN IF0 WN IF1

Fig. 4

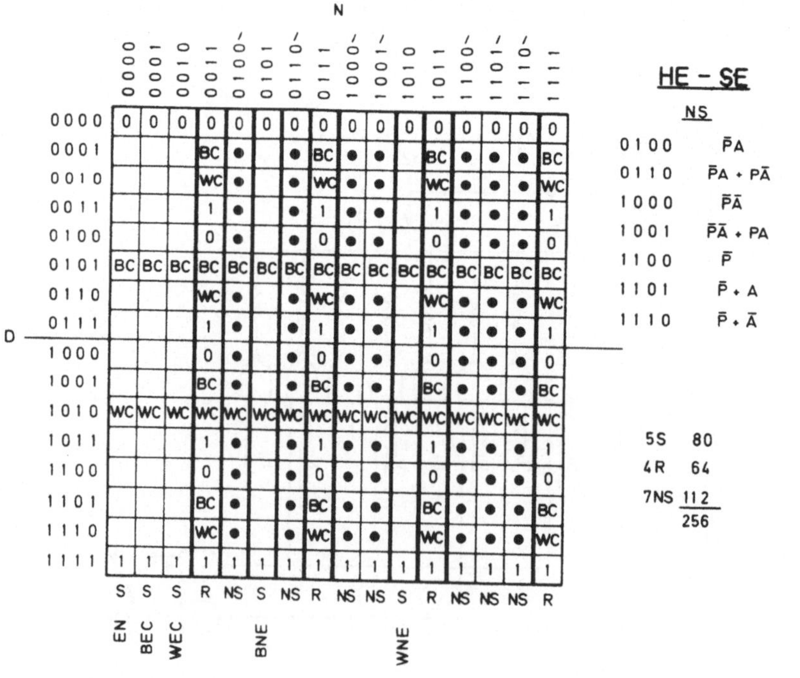

N
HE - SE
NS
0100 P̄A
0110 P̄A + PĀ
1000 P̄Ā
1001 P̄Ā + PA
1100 P̄
1101 P̄ + A
1110 P̄ + Ā
D
5S 80
4R 64
7NS 112
256
S S S R NS S NS R NS NS S R NS NS NS R
EN BEC WEC BNE WNE

Fig. 5

CLIP 3 provides the use of EDGE = 1 or 0. The edge may be thought of as a frame of imaginary cells outside the 192 cells array. These cells may propagate either 0 (E = 0) or 1 (E = 1).

The use of assymetrical or directional functions is also provided.

The functions may be classified as follows:

a) ZERO ORDER -The D output is entirely (0) or (1) and is INDEPENDENT of the A input.

b) FIRST ORDER -The D output is either the A INPUT or its BINARY COMPLEMENT $\bar{A}$.

c) SECOND ORDER -The D output is a function of the A input to each cell and of its IMMEDIATE NEIGHBOURS.

d) THIRD ORDER -The D output is a function of the A input to each cell and of the EDGE CONNECTIVITY of its NEIGHBOURS.

e) FOURTH ORDER -The D output is UNDETERMINED due to logical inconsistency in the cell function.

f) CONDITIONAL ORDER -The D output is (0, 1, BC, WC) CONDITIONAL on the A input.

As there are two values for E (0 or 1) there is a total of $2 \times 256 = 512$ functions, all falling into one of the above orders. However, the distinct operations are much fewer than the number of possible functions.

Fig. 4 is a 16×16 table of the H-S functions, i.e. hexagonal or square functions with E = 0.

We can distinguish 4 groups with stable functions (S), 5 groups of stable but redundant (R) and 7 groups containing the unstable functions (NS).

The redundant groups are not of interest because they result in operations which give 0, BC, WC, 1, operations also available in the stable groups.

Similarly, Fig. 5 is a 16×16 table of the HE-SE functions, i.e. functions with E = 1.

Here we can distinguish 5 stable groups (S), 4 redundant groups (R) and 7 unstable groups (NS), the latter being exactly the same as in the H-S case.

What is immediately evident is that the necessary condition for instability is the presence of $\overline{P}$ in the N output. Some of the functions in the unstable groups are actually stable but all of them are redundant (0, BC, WC, 1).

Another feature is the duality shown between the upper and lower halves of the tables of Fig. 4 and 5. Thus, the lower half may be described in terms of its dual in the upper half and vice-versa. This permits description by using the most convenient way.

The stable groups include functions with a certain common characteristic appearing repeatedly in each group. This results in another way of classifying the various functions.

Code	Description	No. of groups	
BN	BLACK'S NEIGHBOUR	1	
IFO	IF ZERO	1	
WN	WHITE'S NEIGHBOUR	1	
IF1	IF ONE	1	
EN	EDGE'S NEIGHBOUR	1	
BEC	BLACK EDGE CONNECTED	1	
WEC	WHITE EDGE CONNECTED	1	
BNE	BLACK'S NEIGHBOUR (EDGE)	1	
WNE	WHITE'S NEIGHBOUR (EDGE)	1	
R	REDUNDANT	9	
NS	UNSTABLE	14	32

To avoid long descriptions with words we may use the code shown above. Some additional useful items are shown below.

C	CELL
E	EDGE
N	NEAR, NEIGHBOUR
F	FIGURE
G	GROUND
BC	BLACK CELL
WC	WHITE CELL
CON	CONNECTIVITY
$\wedge$	AND
$\vee$	OR
$\sim$	NOT
+	PLUS
-	MINUS
CONT	CONTOUR
INS	INSIDE
OUTS	OUTSIDE
1NO	BLACK NOISE
ONO	WHITE NOISE

An idea of the number of possible operations with symmetrical functions is obtained by the following numbers:

Functions - Total	2 × 256 =	512
Functions in unstable groups	14 × 16 =	224
Functions in stable groups		288
Other apparently stable	14 × 4 =	56
Stable in general		344

Functions - Total	512
Stable in general	344
Unstable	168

HE-SE	Distinct	5 × 12 = 60	
	Basic (0, BC, WC, 1)	1 × 4 = 4	64
H - S	Distinct	2 × 12 = 24	
	Conditional	2 × 4 = 8	32
	Total - Distinct		96

Stable in general	344
Distinct	96
Redundant	248

Order	Total	Distinct	Redundant
ZERO	122	2	120
FIRST	122	2	120
SECOND	48	48	0
THIRD	36	36	0
FOURTH	168	-	-
CONDITIONAL	16	8	8
	512	96	248

As we see, of the 512 possible functions only a small number is distinct and this is further reduced if we consider their usefulness.

Some experimental results with symmetrical functions are shown in Figs. 6 - 11 (lower part: INPUT, upper part: OUTPUT).

These results are by no means exhaustive but are typical of what can be expected and may be of help in understanding the underlying logic.

An implication of the unstable groups would appear to be the possibility of simplification by complete elimination of $\overline{P}$ from the N output. However, we have found that under certain conditions, especially when using directional functions, unstable functions may become stabilized and result in some interesting outputs, like checkerboards and other figures, which may be used with advantage under certain circumstances. A better study of these functions appears necessary.

Interesting outputs are also obtained when using assymetrical functions of the redundant groups.

Some experimental results obtained with directional and stabilized functions are shown in Figs. 12 - 23.

Although the number of possible combinations in CLIP 3 logic is enormous (10^{308}), as in the case of chess, only a small part of them is meaningful, this making it easier to program and to experiment with this kind of processor.

REFERENCES

1. CLIP 3 Technical Specification (1), Report No. 73/1, Image Processing Group, Department of Physics and Astronomy, University College London.

2. CLIP 3 Operating Manual, Report No. 73/4, Image Processing Group, Department of Physics and Astronomy, University College London.

BIBLIOGRAPHY

1. Duff, M. J. B., Cellular logic and its significance to pattern recognition, AGARD AVIONICS PANEL, XXIst Technical Symposium "Artificial Intelligence", Rome, Italy, 24-28 May 1971.

2. Duff, M. J. B., Watson, D. M., Fountain, T. J. and Shaw, G. K., A cellular logic array for image processing, to be published in Pattern Recognition, Pergamon Press, 1973.

3. Image processing in cellular arrays, Report No. 72/1, Image Processing Group, Department of Physics and Astronomy, University College London.

4. Rosenfeld, A., Connectivity in digital picture, J. ACM, 17, No. 1, 146, 1970.

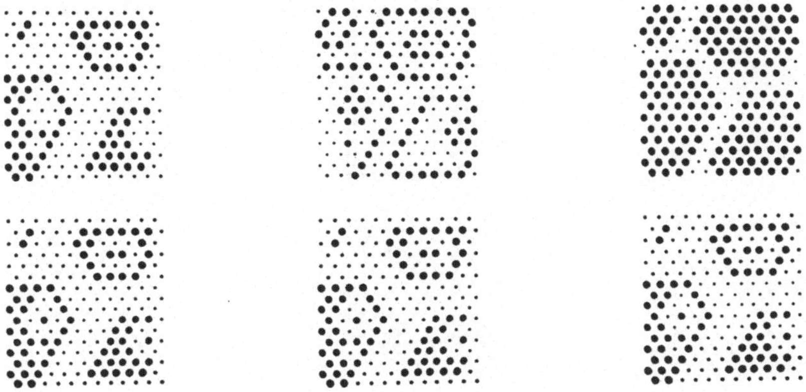

$D = O(1\text{-}6)A,\ PA$ $D = O(1\text{-}6)A,\ \overline{P}A+P\overline{A}$ $D = O(1\text{-}6)A,\ P + A$

Fig. 6

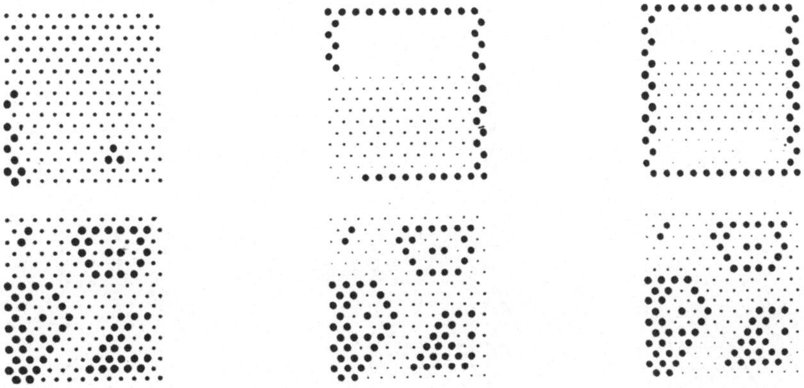

$D = O(1\text{-}6)\overline{A},\ \overline{P}A$ $D = O(1\text{-}6)O,\ P\overline{A},\ E$ $D = O(1\text{-}6)O,\ P,\ E$

Fig. 7

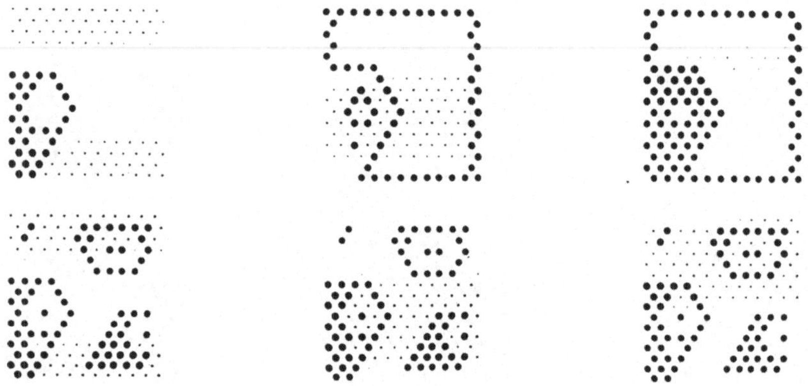

D = O(1-6)PA, PA, E   D = O(1-6)PA, $P\overline{A}$, E   D = O(1-6)PA, P, E

Fig. 8

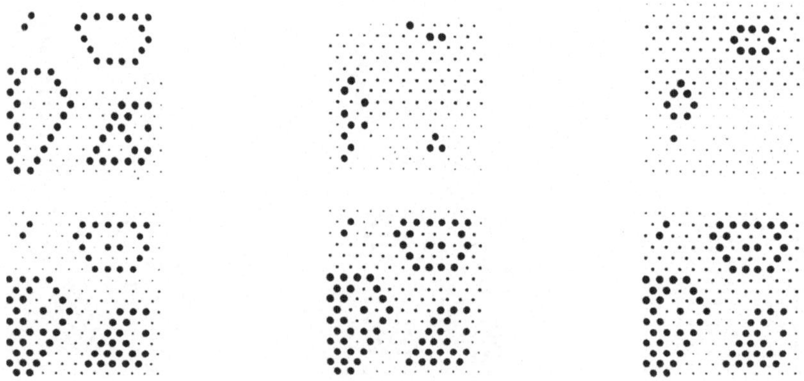

D = O(1-6)$P\overline{A}$, PA, E   D = O(1-6)$P\overline{A}$, $\overline{P}A$, E   D = O(1-6)$P\overline{A}$, $\overline{P}\ \overline{A}$, E

Fig. 9

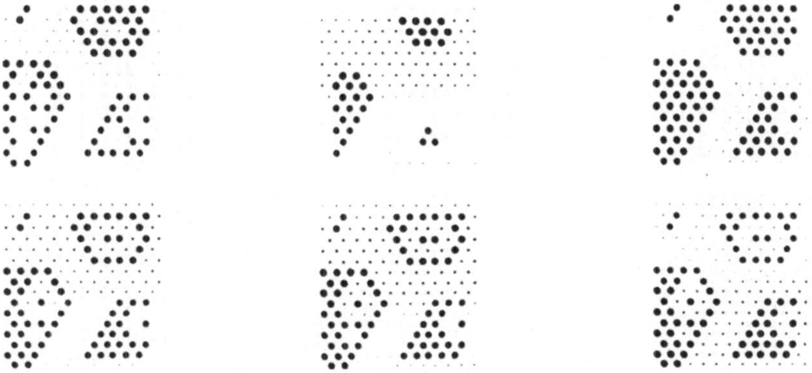

$D=O(1-6)P\overline{A},\ \overline{P}\ \overline{A}+PA, E$ $D = O(1-6)P\overline{A},\ \overline{P},\ E$ $D = O(1-6)P\overline{A},\ \overline{P} + A,\ E$

Fig. 10

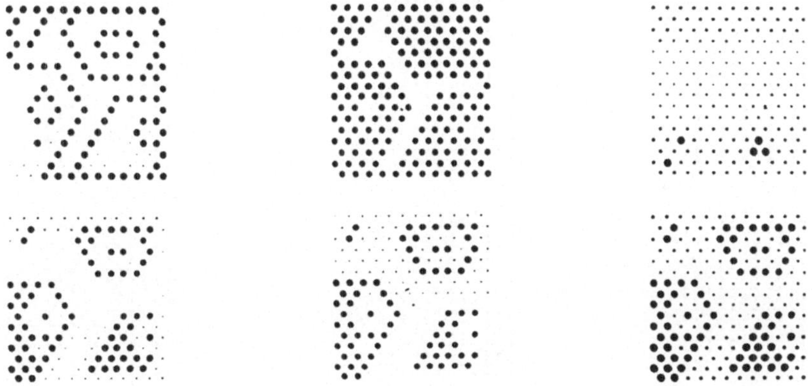

$D = O(1-6)A,\ P\overline{A},\ E$ $D = O(1-6)A,\ P,\ E$ $D = O(1-6)\overline{A},\ \overline{P},\ E$

Fig. 11

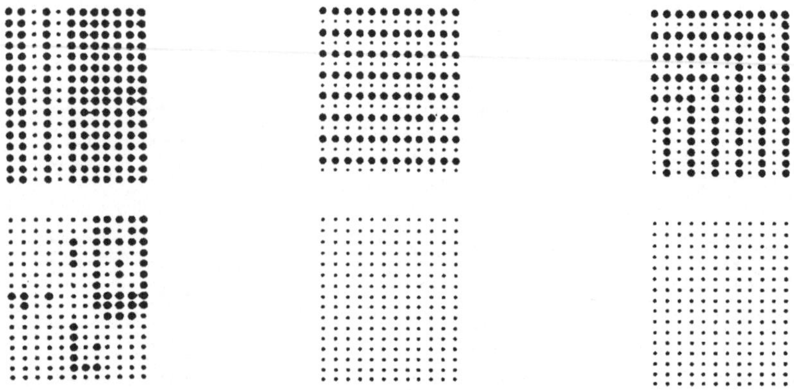

D = O(2,6)P+A, P, S   D = O(2)$\overline{P}$, $\overline{P}$, S   D = O(3)$\overline{P}$, $\overline{P}$, S

Fig. 12

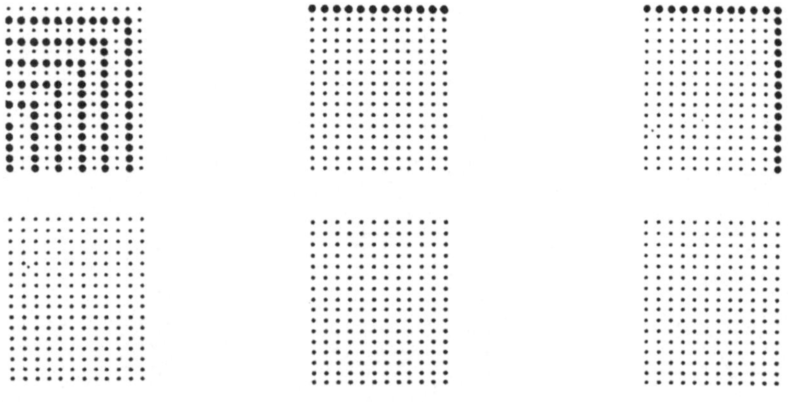

D = O(3)$\overline{P}$, P, S   D = O(2)1, $\overline{P}$, S   D = O(3)1, $\overline{P}$, S

Fig. 13

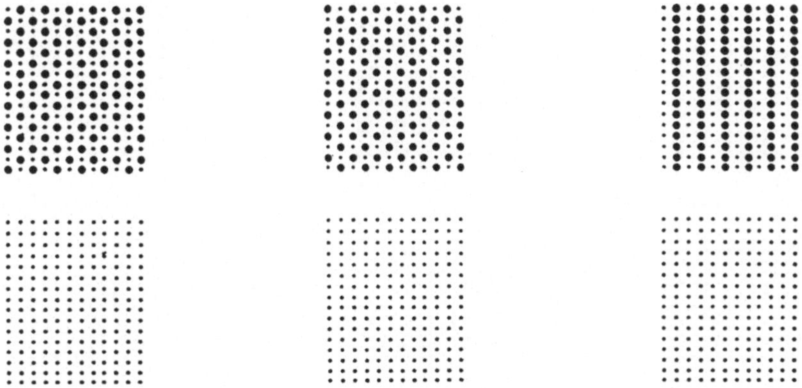

$D = O(2,4)\overline{P}, \overline{P}, S$ $D = O(2,4)\overline{P}, P, S$ $D = O(3,5)\overline{P}, \overline{P}, S$

Fig. 14

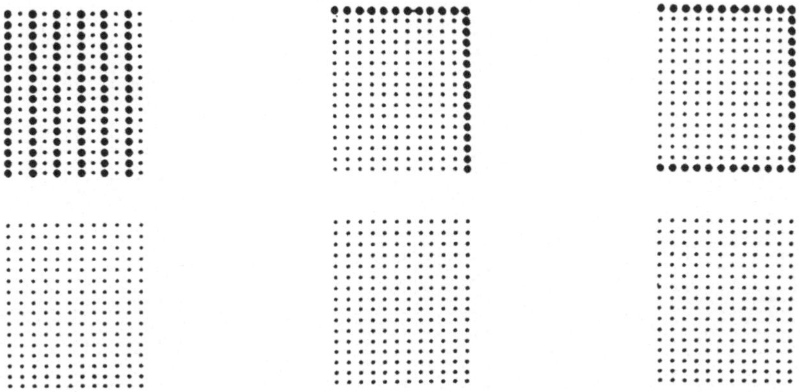

$D = O(3,5)\overline{P}, P, S$ $D = 1(2,4)1, \overline{P}, S$ $D = 1(3,5)1, \overline{P}, S$

Fig. 15

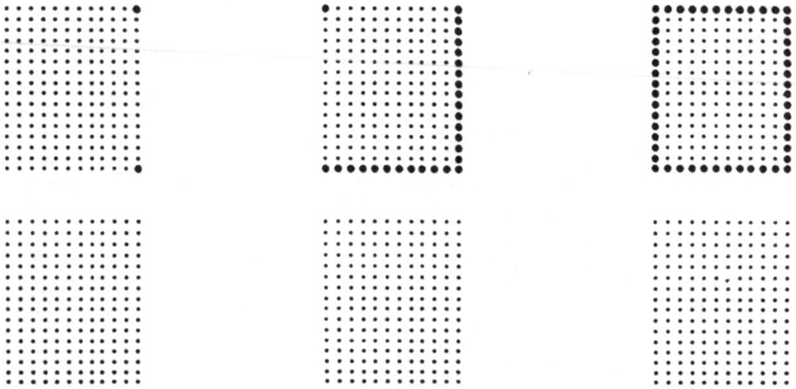

D = 1(2,4,6)1, $\overline{P}$, S   D = 1(3,5,7)1, $\overline{P}$, S   D = 2(1,3,5,7)1, $\overline{P}$, S

Fig. 16

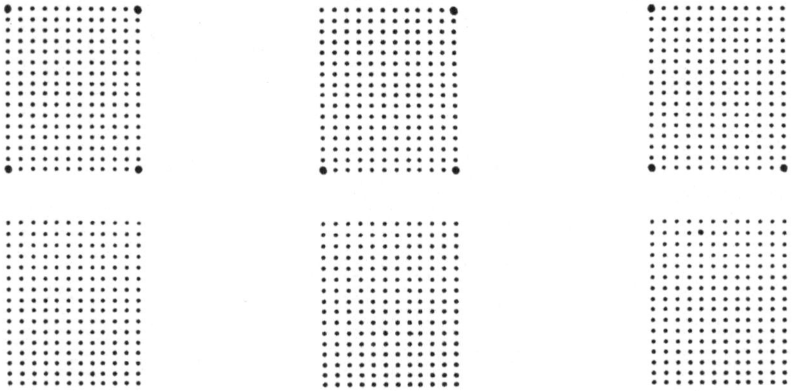

D = 2(2,4,6,8)1,$\overline{P}$,S   D = 2(2,4-6,8)1,$\overline{P}$,S   D = 2(2,4,6-8)1,$\overline{P}$,S

Fig. 17

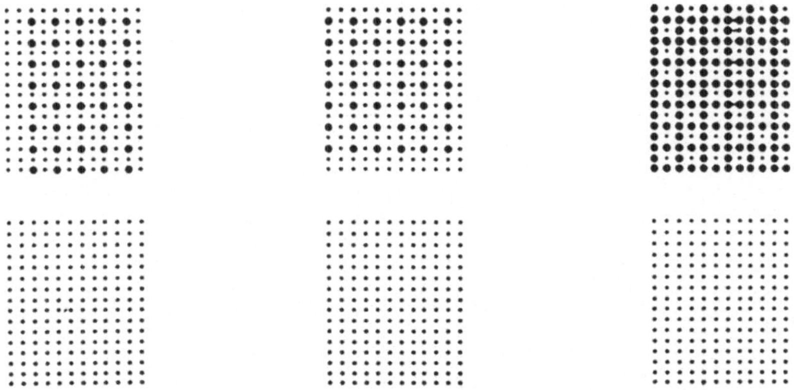

D = 3(1-4)$\overline{P}$, P, S   D = 3(2-5)$\overline{P}$, P, S   D = 0(1-4)$\overline{P}$, P, S

Fig. 18

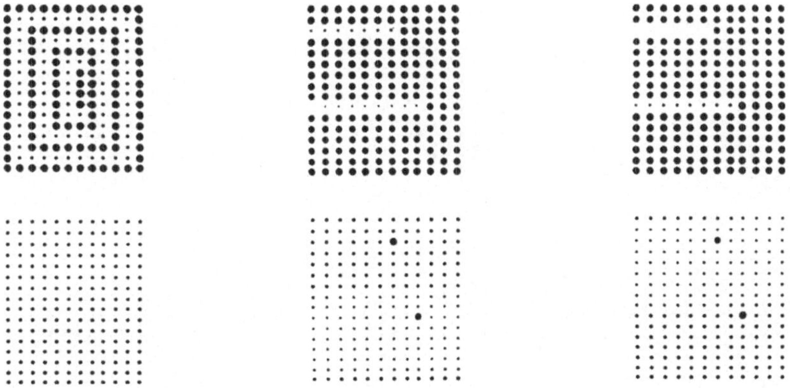

D = 2(1,3,5,7)$\overline{P}$,$\overline{P}$,S   D = 0(4)P$\overline{A}$,P$\overline{A}$,SE   D = 0(4)P$\overline{A}$, P, SE

Fig. 19

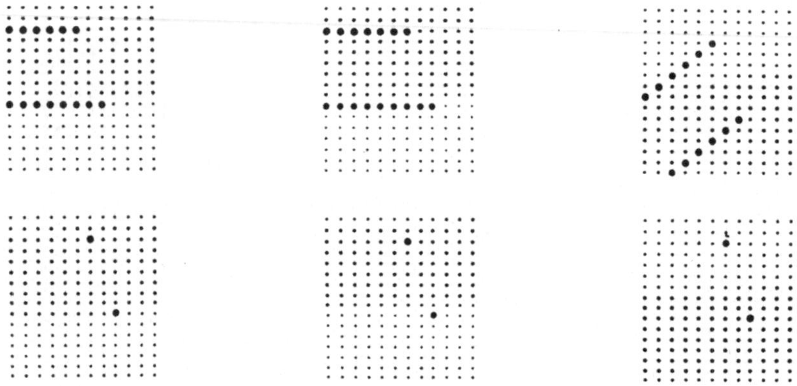

$D = 0(4)P\overline{A}$, $\overline{P}$, SE $D = 0(4)P\overline{A}, \overline{P}+A$, SE $D = 0(3)P\overline{A}$, $\overline{P}$, SE

Fig. 20

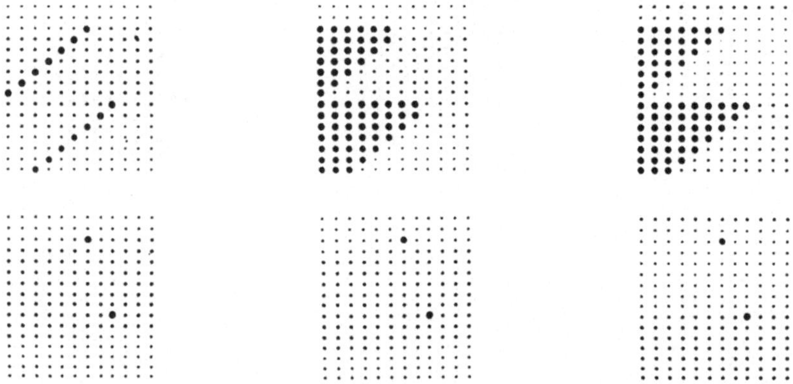

$D = 0(3)P\overline{A}, \overline{P}+A$, SE $D = 1(3,4)P\overline{A}$, $\overline{P}$, SE $D = 1(3,4)P\overline{A}, \overline{P}+A$, SE

Fig. 21

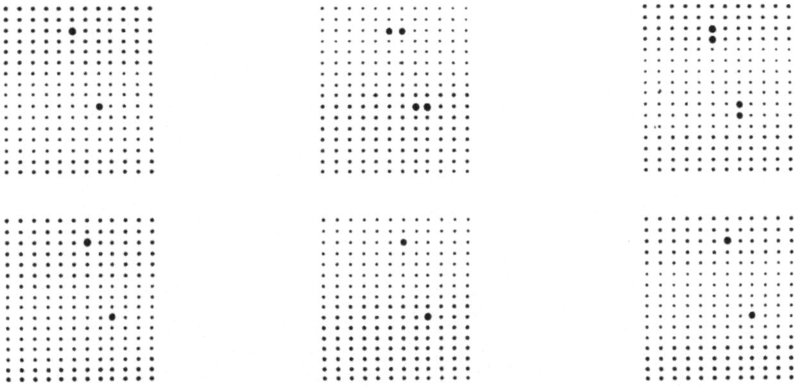

D = O(4)$\overline{A}$, $\overline{P}$, SE   D = O(4)$\overline{A}$, $\overline{P}$+A, SE   D = 1(3,4)$\overline{A}$, $\overline{P}$, SE

Fig. 22

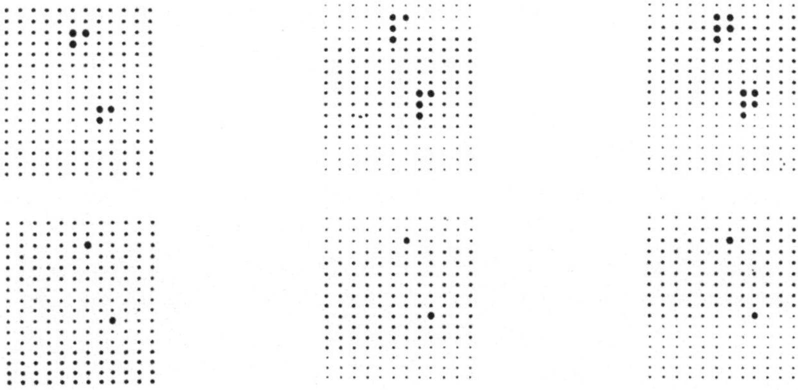

D = 1(3,4)$\overline{A}$,$\overline{P}$+A,SE   D = 3(3-6)$\overline{A}$, $\overline{P}$, SE   D = 3(3-6)$\overline{A}$,$\overline{P}$+A,SE

Fig. 23

PARALLEL PROCESSING FOR IMAGE ANALYSIS

C. Arcelli, L. Cordella and S. Levialdi

Laboratorio di Cibernetica del CNR
Arco Felice, Naples, Italy.

Introduction.

In the field of picture processing the majority of algorithms that have been proposed to extract characteristic subsets of the input (i.e. the features) are sequential in nature so as to be profitably implemented on standard digital computers. Nevertheless, several attempts have been made in the last fifteen years to perform the same tasks by means of special purpose machines that operate in parallel. In fact, since the Euclidean plane may be regularly tessellated using triangles, squares or hexagons, it seemed natural to associate to each element of the discrete plane a logical element (cell) which would have been able to compute the transformed state of that element as a function of its own state and that of its neighbourhood. A finite set of such identical cells, is usually referred to in the literature as an iterative array. Ever since S.H. Unger[1] proposed in 1959 the use of a parallel computer to process pictures so as to obtain from them a significant description for their recognition, a number of authors (see Murtha[2] for any early review) have proposed and/or built parallel devices for the same purposes using in their machines the latest available technology. Although significant improvements were achieved in this way, some standing problems remained, such as the excessive over-all volume for a machine to be of practical use, the interconnection problem, and last but not least, the cost. These problems will be overcome when the use of LSI circuitry will be commercially available although such time is not as near as it was thought of some years ago.

The problems that were faced during the development of these processors may be roughly divided into two kinds: the first kind includes the design of cells with increasing computational power, the flexibility of inter-cell communication and the possibility of easily controlling the input-output and processing operations of the array. By easily we mean both that the instruction repertoire of the program is operator-oriented and that each instruction may be executed with the minimum amount of hardware. The second kind of problems is concerned with the introduction of parallel algorithms for picture processing and with the proofs that these really worked. These proofs are relatively more complex than the corresponding ones for the sequential case since they must often ensure the preservation of global properties when only local transformations are considered during the process. (See for instance Beyer[3], Rosenfeld,[4] Levialdi[5]).

Finally we would like to remark that one of the reasons that has hindered the development of parallel picture processing was also the fact that even if these devices were supposed to be designed in the most flexible way it often happened that the practical implementation of the algorithms proposed in the literature was cumbersome. Furthermore the weak interaction with the people that were faced with real picture processing problems (for instance in the areas of biomedicine, defense, space research,etc.) prevented researchers from hardware and software interests to join forces for the solution of specific practical tasks.

It now seems that the problems which have just been mentioned may be overcome in the near future: in fact the number of people interested in picture processing research has greatly increased and large scale integration promises cheaper and better machines. Furthermore a new scientific trend exists which considers the solution of practical problems a necessary requirement for the development of the field.

In the following we will present two algorithms which have been developed keeping in mind the characteristics of a given parallel machine (CLIP, Cellular Logic Image Processor). An earlier version of this machine (CLIP 2) has already been fully described elsewhere[6] whilst a more powerful version (CLIP 3) will be presented at this meeting. Such a machine can be programmed to perform a parallel operation on a given input binary pattern by a single instruction. The transformed state of any cell within the array depends on its state and on that of a certain number of neighbours.

One of the problems that appears in the processing of pictures is the automatic (interactive) extraction of every (chosen) connected component which lies within the picture.

Once a component has been extracted, a set of measurements is performed on it (e.g. area, perimeter) and suitable transformations are made so as to obtain an adequate description useful for classification. After this has been accomplished the component will be removed from the original picture and a new component will be considered.

In sequential processing, images to be analyzed come, either from a microscope through a television camera, or from film through a flying spot scanner. The digitized version of such images is stored in the memory of a computer so as to be available for processing. On the other hand when a fully parallel system is considered a different acquisition means should be employed. This means is a highly integrated photomatrix in which every photoelement produces an electric signal depending on the light intensity coming from its corresponding covered area. For several tasks it is sufficient to consider a binary output from the photomatrix and this is readily done by choosing a suitable threshold. Such an input appears as the most desirable to enhance the possibilities of a complete parallel processing system.

Figure 1 shows a schematized version of a system on which our algorithms will be implemented. Block A represents the photomatrix input, Block B the cellular array (e.g. a CLIP machine) and block C a set of registers each of which will be used either as input or as output register. Block D is a control unit which is responsible for the handling of the input-output and processing operations of the array. Finally block E contains a numerical unit designed to give either the values of specific measurements performed by the array, or to provide to the control unit logical evidence of a certain property existing in the pattern.

We will now describe the set of symbols occurring in the instructions used to program such a system. In the choice of the instructions we have tried to make the program easy to read and interpret for a human operator. We have also assumed that a specific translator could be easily built since it would be only necessary to establish an equivalence function between each instruction and a certain neighbourhood for each element of the array.

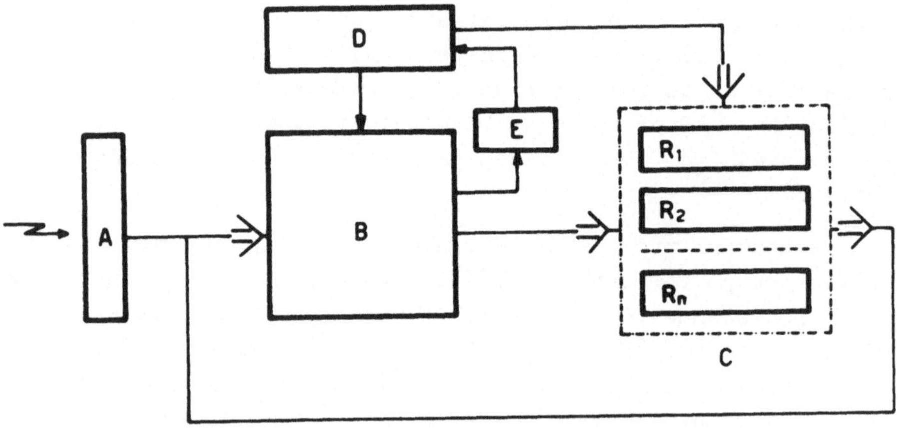

A = Input photomatrix
B = Cellular Array
C = Register
D = Control unit
E = Numerical unit

Fig. 1 - Block diagram of a parallel machine

These instructions define operations which will be used to process a binary picture of 1-elements and 0-elements. The set of 0-elements which is connected to the border of the picture will be called background and will be assumed to be a connected one whilst all other sets of 0-elements will be called holes. A set of elements S will be 4-(8-) connected respectively, if for any two elements in S a path exists joining them through successive elements ..,(i,j), (m,n),...all in S, such that

$$|i-m| + |j-n| \leqslant 1 \quad \text{(4-connectivity)} \qquad (1)$$

$$\max\Big(|i-m| , |j-n|\Big) \leqslant 1 \quad \text{(8-connectivity)} \qquad (2)$$

The symbols used within the instructions are the following

$\rightarrow$, $>$, $<$, $=$, $+$, $\cdot$, $-$

and imply respectively: a store instruction which replaces the contents of the right hand register by those of the left hand one without clearing it; greater, smaller, equal, logical or, logical and; finally a subtraction sign which stands for the following logical operation: $A-B = A\cdot\bar{B}$.

The main processing instruction in the programming system is EXTRACT. Such an instruction transfers into a chosen register all the elements which satisfy a given neighbourhood condition. The complete instruction will be written as follows:

$$\text{EXTRACT P, } (c_1, c_2 \ldots c_k) \text{ N} \gtrless t, \ R_i \rightarrow R_j$$

where P may assume the value 1 or 0 according to the present state of the elements which must be considered for extraction, $c_1 \ldots c_k$ are numbers which identify the neighbours which lie in one of the coded directions shown on Fig. 2; N stands for the number of neighbours just indicated which are in state 1 and t is the threshold value for N. When P is missing both 0 and 1-elements are considered. Finally, the pattern to be processed is in register R_i and the extracted elements will be stored in R_j overriding its contents. As an example we give the instruction which singles out the 4-connected contour of a pattern given in R_1:

$$\text{EXTRACT 1, (2,4,6,8) N} < 4, \quad R_1 \rightarrow R_2$$

Since every processing task may be thought of as a sequence of instructions organized in procedures and subprocedures, it is necessary to automatically establish whether a specific procedure (or subprocedure) should be iterated or stopped. The natural way to do this is by using a logical termination test.

Fig. 2 - Coded directions

The particular test we have considered here uses a given cell of the array as a detector. Such a cell will be in state 1 if at least one element belonging to the pattern stored in a given register is in state 1, elsewise such a cell will be in state 0. The logical value of such a test is obtained in the numerical unit and sent to the control unit so as to either iterate the procedure or stop the iteration. In practice this test uses a particular EXTRACT instruction which generates a propagation process. Such a propagation, after having started from a cell (or a set of cells) that is in state 1, will terminate only when all the remaining cells have reached the same state 1. If we now consider a given cell within the array, for example the one lying in the center of the array, such a cell will, after a given time, be reached by the propagation wave and will assume the same state as the cells which have started the propagation. This will happen in a number of steps equal at most to max $(\frac{m}{2},\frac{n}{2})$ where m, n are the dimensions of the array. At the same time the numerical unit will be enabled to read the state of the central cell and will send this logic value to the control unit. The operations so far described will be synthetized in the instruction TEST R_i , n_1, n_2 where R_i stands for the register to be tested and n_1 and n_2 are the addresses of the instructions that should follow if R_i is empty or not respectively.

The Scanner.

The first algorithm that we propose is intended for the automatic extraction of all the connected components of 1-elements contained in a picture. This task is accomplished by defining a scanning cell C^*, which will explore the matrix row by row from left to right, up to bottom. Whenever such a cell overlaps a 1-element, a propagation process starting from C^* and involving all the 1-elements connected to C^* takes place. When this is achieved the component is transferred to a specified register and erased from the array. The scanning cell will now proceed in its raster fashion until it overlaps a new 1-element, in this way a new component will be extracted and stored in a second register. When the right bottom element will be reached by C^* we will have stored in a set of different registers all the components of the input picture so as to be able to perform on them suitable measurements.

The implementation of a scanning program using a parallel machine might seem an inadequate task, since one instruction is necessary for every displacement of C^*, i.e. $n \cdot m$ steps are required so that all the elements of the array are considered. Nevertheless such a process is necessary whenever all the connected components must be separately extracted.

To start the process two registers R_1 and R_2 are required. On each register the top left element is set in state 1 and corresponds to the initial C^*. We will consider R_1 as the input of the array and so as to move the scanning cell from left to right by one step (horizontal shift), an EXTRACT instruction will be employed. This instruction will change every 0-element that has a 1-element in position 4 into a 1-element. To establish whether this instruction should be iterated (this happens if C^* has not overflown from the right of the array) a TEST instruction must be provided on the contents of R_1. If R_1 is empty an instruction is required to enable to scan the following row, otherwise a new horizontal movement is performed. To scan the following row, a vertical movement (vertical shift) of the scanning cell C^* present in R_2, must be provided. This is accomplished with an EXTRACT instruction that will change every 0-element which has a 1-element in position 2 into a 1-element. In order to know whether the scanning cell has overflown the bottom of the matrix, a TEST will be inserted at this point in the program, on the contents of register R_2. The scanning process will be finished only when R_2 is empty. Otherwise the scanning of the following row is started after storing the contents of R_2 in R_1 (see fig. 3).

The object extraction previously mentioned is a propagation process extended to all the 1-elements which are connected to the first 1-element overlapped by the scanner cell. This process will be indicated in the program by the instruction CONN. Three registers are involved in the execution of the instruction CONN. The first one (R_o) contains the input pattern, the second one (R_i) has only one 1-element (the marker) obtained by an AND operation between the register R_1, containing the scanning cell, and the register R_o containing the input pattern. The instruction will be written as follows:

R_o CONN $R_i \rightarrow R_j$

The extracted component after being stored will be subtracted from R_o.
We will now give the list of instructions that implements the overall program we have just described.

1) $R_2 \rightarrow R_1$

2) $R_1 \cdot R_o \rightarrow R_3$

3) TEST R_3, 4,8

4) EXTRACT 0, (4) N = 1, $R_1 \rightarrow R_1$

5) TEST R_1, 6,2

6) EXTRACT 0, (2) N = 1, $R_2 \rightarrow R_1$

7) TEST R_2, 10, 1

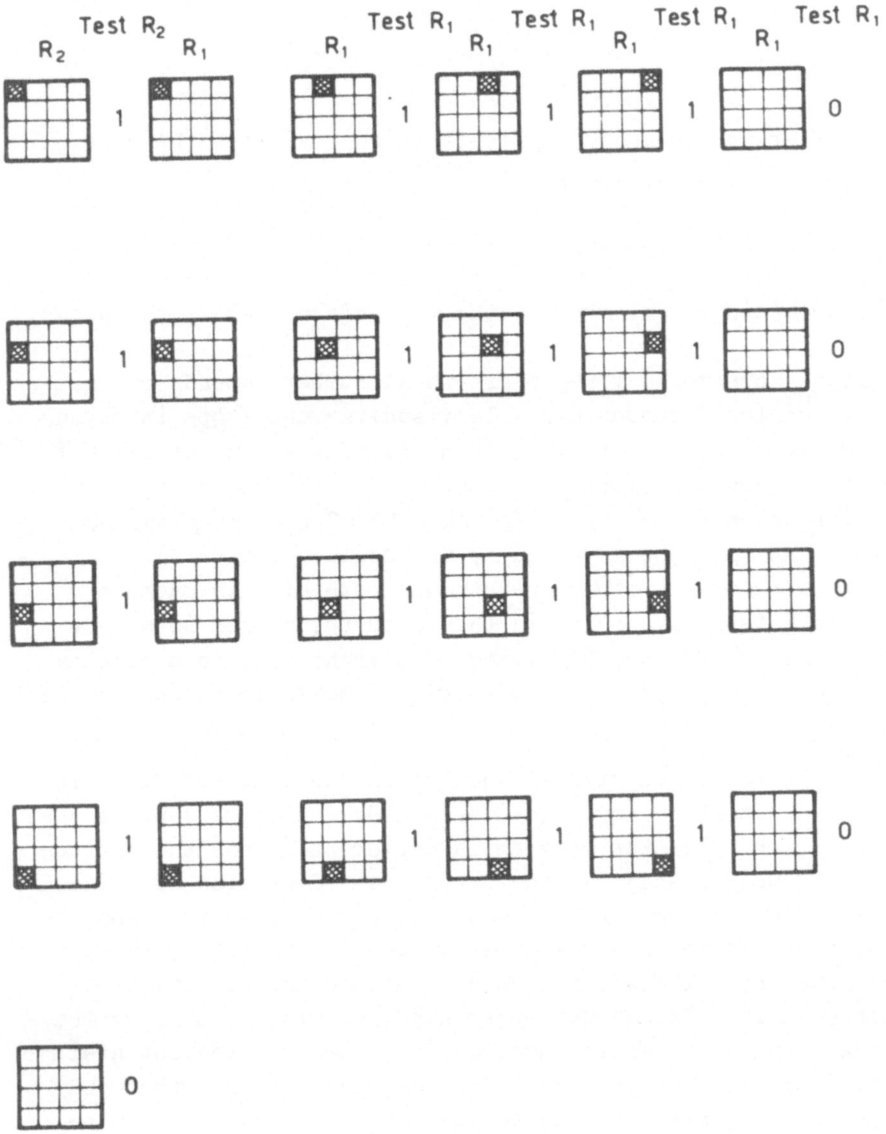

Fig. 3 - Scanning process.

8) R_o CONN $R_3 \rightarrow R_n$

9) $R_o - R_n \rightarrow R_o$, 4

10) STOP

A number after the comma in any instruction indicates the next instruction to be executed by the machine.

R_o contains the input pattern.

R_1 and R_2 contain a 1-element in the top left corner of the array.

$R_3, R_4, \ldots, R_n$ are working registers whose number depends on the number of connected components. To visualize the loops involved in this program refer to fig. 4. In practice since the size of the array is known a priori, the instructions 5) and 7) which prevent the scanning cell from "going out" of the array may be eliminated and this job executed by the control unit.
As suggested before, another operational mode of this machine could be an interactive one; in this case the marker is set by a human operator directly (by means of a light pen) on a display showing the input pattern so that a given chosen component may be extracted[8].

We will try now to give an insight in the times involved in performing the task we have considered above. Let us suppose that the machine, on which the program is implemented, has the characteristics of CLIP 3 and that the array has a dimension of 100x100 elements. The time required to scan the whole array is $\sim$ 2ms. while at most a time of $\sim$0.4 ms is needed to extract every connected component. These times do not include the operation of the control unit. If, on the other hand, we wish to perform the same task using a sequential machine (e.g. today's fastest available computer) with cycle time of 100 ns and we assume that only LOAD instructions are required to scan the elements of the picture, we will obtain a time of $10^{-7} \times 10^4 = 1$ msec. In addition we should consider the time insumed for the extraction of a connected object from such a picture and this is not a short task since it implies filling in, labelling and sometimes even relabelling so that more than one other millisecond for every connected component will surely be employed.

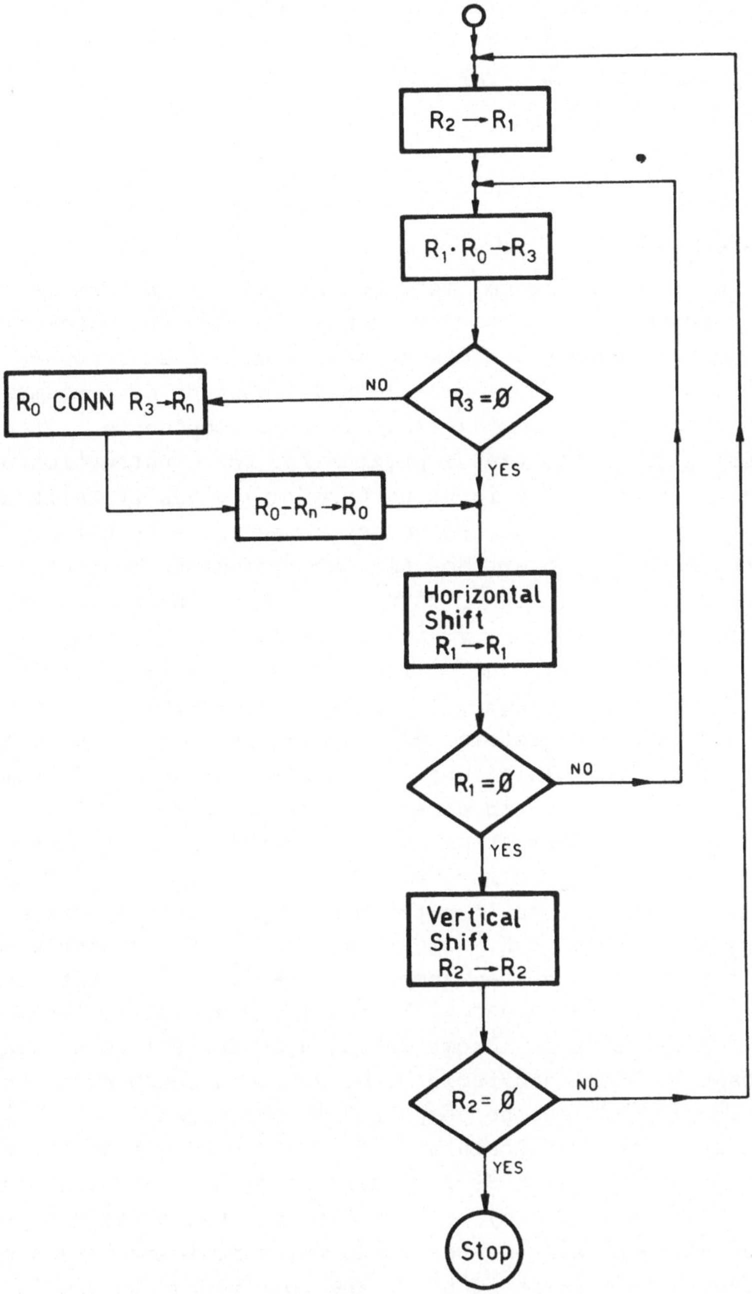

Fig. 4 - Flow chart of the scanning program for extracting connected components.

Summarizing we might note that, for the extraction of a single component, both machines insume an amount of time of the same order of magnitude but, the use of parallel machines becomes specially profitable when a large number of components is present in the picture.

The adjacency tree.

Whenever the topological information about a pattern is required, a convenient means to represent it is by its adjacency tree. In such a graph the nodes stand for either an 8-connected component of 0-elements or a 4-connected component of 1-elements, and the arcs represent the adjacency of such components. It might be interesting to notice that a program for the construction of the adjacency tree from an input pattern may be easily built up on the basis of the previous connected component extraction. In fact we are also able to extract all the 0-elements connected components of the pattern, either using an EXTRACT instruction referred to the 0-elements or by giving to the previous program an input which is the complement of the picture. For clarity's sake we will use in the sequel the second alternative. We will so have, in separate registers, all the 1- and 0-connected components of the picture. This subject has been investigated from a theoretical point of view in a recent paper by A.Rosenfeld [9]. In this paper, amongst other things, it is proved that this graph is a tree. Furthermore two algorithms for extracting such a tree are included. They reflect the two more typical approaches employed to solve such a task. The first ones uses a raster scan to detect for every connected component the first top left element which will be marked and will constitute a node in the searched tree. After this is accomplished, a border following algorithm is used to label and isolate the so found component. In addition a scanning inside the component is performed to establish the presence of other components. After these have been found (if they were present) a new node for every connected component is generated. In this way every node of the tree has exactly a predecessor node corresponding to the component containing it and one or more successors corresponding to the components contained by it. The second approach considers a left to right scanning sequence in which a comparison is made between every $(k+1)^{th}$ row and the kth row previously scanned. The connected components present in the

$(k+1)^{th}$ row are checked with the components present in the previous row. During the process two different situations might appear: 1) the $(k+1)^{th}$ row is identical with the kth row or 2) one of the following cases may happen: i) appearance of new connected components; ii) disappearance of connected components; iii) splitting of a component; iv) merging of components. The tree is built, row by row, updating the components at each step as well as their adjacency relations. The adjacency tree is represented in this case by a well formed parenthesis string.

Iterative arrays have also been proposed to accomplish the connected component's extraction (see for instance T. Beyer[3] and S. Selkow[10]), but the strategies employed were not significantly different from the ones just mentioned. Finally we would like to remark that the most delicate step in the process is to establish the arcs of the tree i.e. the adjacency relations among the connected components. In the method we suggest here, this step is achieved by two operations once every connected component has been extracted and stored in separate registers. The two operations are the following: 1) 4-expansion of the first extracted component and 2) the intersection between such an expansion and all the remaining components. By 4-expansion of a component we mean the set of the elements of the array which are 4-connected to the component. After iteration of this procedure on all the components, a string of 1's and 0's is obtained from the numerical unit of the array. This string can be arranged as an incidence matrix which is just another form of writing the adjacency tree.

To visualize the process involved in constructing the adjacency tree of a picture an example is included in figure 5. We may see in figure 5a) the input pattern contained in the register R_o and the complement of such a pattern in $\overline{R}_o$. From each one of these patterns all the connected components of 1-elements are successively extracted starting from the complemented input; in this way the first extracted component corresponds to the background i.e. the root of the tree. We should remark here that owing to the chosen connectedness (the set of 1-elements is 4-connected in R_o and 8-connected in $\overline{R}_o$) we must extract every connected component with a propagation process that considers for every cell 4 neighbours in R_o and 8 neighbours in $\overline{R}_o$. In figure 5b) we show the 4-expansion of the component existing in register R_2, its AND with the component in register R_4 and the TEST performed to establish the logic value of the intersection, which is

1 for this particular case. Finally in figure 5c) we give the incidence matrix after rearranging the string of 0's and 1's obtained from the numerical unit once the number of connected components has been counted.

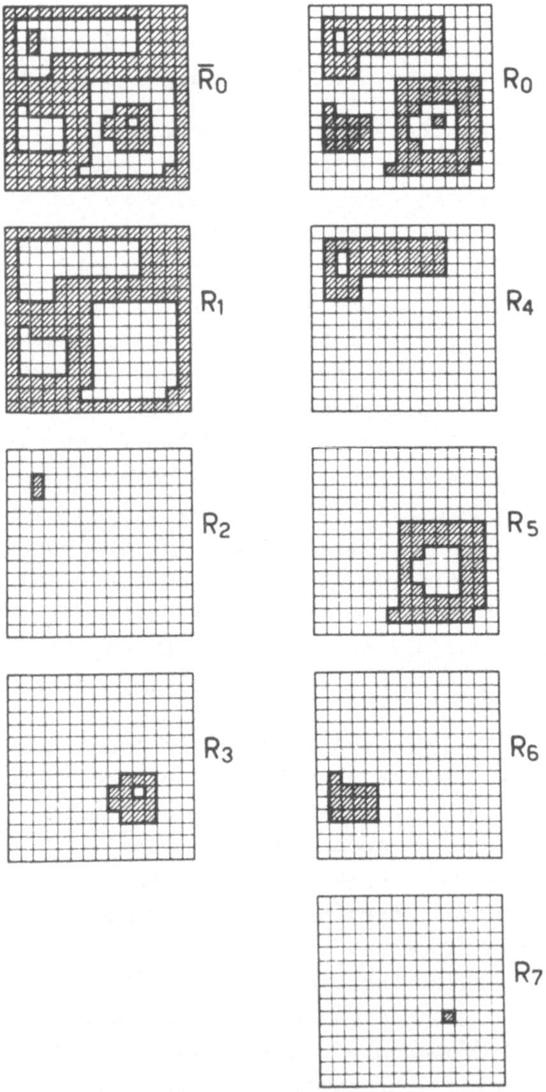

Fig. 5a) Connected component's extraction

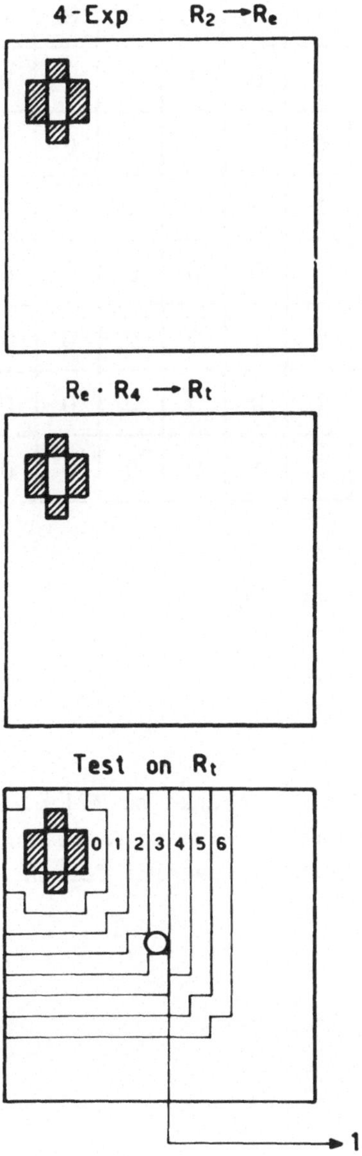

Fig. 5b) Extraction of adjacent component

	R_1	R_2	R_3	R_4	R_5	R_6	R_7
R_1	0	0	0	1	1	1	0
R_2	0	0	0	1	0	0	0
R_3	0	0	0	0	1	0	1
R_4	1	1	0	0	0	0	0
R_5	1	0	1	0	0	0	0
R_6	1	0	0	0	0	0	0
R_7	0	0	1	0	0	0	0

Fig. 5c) - Incidence matrix

Acknowledgments.

We wish to express our thanks to M.J.B. Duff for many interesting discussions about CLIP 3, to A. Rosenfeld for useful comments on the paper and to U. Cascini for the preparation and layout of the figures.

REFERENCES

1. Unger, S.H., Pattern detection and recognition, Proc.IRE, 47, 1737, 1959.

2. Murtha, J.C., Highly Parallel Information Processing Systems, Advances In Computers, Academic Press, New York, 1966.

3. Beyer, W.T., Recognition of topological invariants by iterative arrays, Ph.D. Thesis, M.I.T. Cambridge, 1969.

4. Rosenfeld, A., Connectivity in digital pictures, Journal of ACM, 17, 146, 1970

5. Levialdi, S., On shrinking binary picture patterns, Communications of ACM, 15, 7, 1972.

6. Duff, M.J.B., Image processing in cellular arrays, to appear in Pattern Recognition.

7. Konigsmark, B.W., Methods for the counting of neurons, from Contemporary Research Methods in Neuroanatomy, Springer-Verlag, Berlin, 1970.

8. Arcelli, C., Duff, M.J.B., Levialdi, S., Watson, D., Connectivity extraction by hexagonal iterative arrays, Proceedings II Congress of Cybernetics, Casciana Terme (Pisa), 1972.

9. Rosenfeld, A., Adjacency in digital pictures, preprint Computer Science Centre, University of Maryland, October 1972

10. Selkow, S.M., One-pass complexity of digital picture properties, Journal of ACM, 19, 283, 1972.

SEQUENTIAL AND PARALLEL AUTOMATA

Azriel Rosenfeld

Computer Science Center
University of Maryland
College Park, Maryland 20742, U.S.A.

This paper reviews some elementary properties of sequential and parallel automata, with emphasis on their relative capabilities and efficiencies as regards simple recognition tasks. Automata that accept strings -- i.e., that have one-dimensional tapes -- are first discussed, and automata that accept rectangular arrays are then briefly considered.

1. TAPE-BOUNDED AUTOMATA

Informally, a (sequential) tape-bounded automaton is a sort of "bug" that, at any given time, is in one of a finite (nonempty) set of states. The bug is intially placed on some symbol, say the leftmost, of a given (finite) input string, and it is initially in a special state, called the starting state. It can move around on the string, and change from one state to another, in discrete steps of the following sort: At any given step, the bug

a) Reads the symbol at its current position, and (possibly) erases that symbol and replaces it by a new one
b) Moves to the neighboring symbol on the left or right (or possibly does not move)
c) Changes to a (possibly) new state.

The bug cannot move off its input string; when it is at one end of the string, it cannot move further in that direction.

For any given bug, only certain combinations of rewritings, moves, and state changes are allowed; we define specific bugs by specifying the permissible combinations. The bug is said to accept its input string if it ever enters one of a distinguished set of states called the final states. The set of strings accepted by the bug T is called the language of T. More specifically, if V_o is any (finite nonempty) set of symbols, then the set of input strings of symbols in V_o that T accepts will be called the V_o-language of T.

In more formal terms, let V be the (finite, nonempty) set of symbols that the bug can read and write, where $V_o \subseteq V$; let Q be the set of the bug's states; and let Δ be the set of movement directions {L,R,N} ("left", "right", "no move"). Then the behavior of the bug is defined by specifying a function

$$\delta_T: V \times Q \to 2^{V \times Q \times \Delta}$$

which takes (symbol, state) pairs into sets of (symbol, state, direction) triples. If the triple (v,q,d) is in the set $\delta_T(u,p)$, where u,v are symbols, p,q are states, and d is a direction, this means that one possible behavior for the bug, when it is in state p and reads symbol u, is to replace u by v, move in direction d, and change to state q. The bug is called deterministic if it has only one possible behavior in any given situation -- i.e., if no set $\delta_T(u,p)$ has more than one element; otherwise, the bug is called nondeterministic.

To keep a formal bug from moving off its string, we shall stipulate that every input string must begin with the special symbol ℓ and end with the special symbol r (these symbols occur only at the beginning and end of a string, respectively); and we require that, for all states p,

$$(v,q,d) \in \delta_T(\ell,p) \text{ implies } v = \ell,\ d = R$$

$$(v,q,d) \in \delta_T(r,p) \text{ implies } v = r,\ d = L$$

In other words, when T reads ℓ, it cannot replace the ℓ by any other symbol, and it can only move to the right; and similarly for r. We shall assume from now on that the input string always has at least one other symbol besides its ℓ and its r.

In order to formally define acceptance of an input string by a bug, we introduce the notion of an "instantaneous description" (ID). This is a triple (σ,i,q), where σ is a string of symbols in V, say of length n; i is a positive integer, $1 \le i \le n$; and q is a state. The initial ID is $(\sigma_o,1,q_s)$, where σ_o is the initial string and q_s is the starting state. Given the current ID (σ,i,p),

let u be the ith symbol in σ;

let (v,q,d) be any triple in $\delta_T(u,p)$;

let σ' be σ with its ith symbol changed from u to v;

and let i' = i-1, i, or i+1, according to whether d = L, N or R.

Then (σ',i',q) is a possible next ID. We say that T accepts σ_o if the initial ID can ever give rise, by repeated use of δ_T in this way, to an ID whose third term is a final state.

2. BOUNDED CELLULAR AUTOMATA

We have just defined a simple class of sequential automata. Our next goal is to define a class of parallel automata, and then to show that the two classes are equally powerful -- i.e., that any language accepted by an automaton in one class is also accepted by some automaton in the other class, and vice versa.

Informally, a (parallel) bounded cellular automaton is a string of "bugs" each of which, at any given time, is in some state. Initially, the bugs are placed in a pattern of states which constitute the automaton's input string; these initial states would normally be required to belong to a subset Q_o of the set Q of all the automaton's possible states. All of the bugs change their states simultaneously, in discrete steps of the following sort: At any given step, each bug "reads" the states of its left and right neighbors, and changes to a (possibly) new state.

By convention, the first bug in the string is permanently in a special state ℓ, and the last bug in a special state r; our discussion of state changes applies only to the intermediate bugs, at least one of which will be assumed to exist. The leftmost bug other than the one in state ℓ will be called the chief bug. The automaton is said to accept its input string if its chief bug ever enters a final state.

Formally, such an automaton is defined by specifying a function

$$\delta_C: Q \times Q \times Q \to 2^Q$$

which takes triples of states (namely, the states of a bug and its two neighbors) into sets of states. If state q is in the set $\delta_C(p_L,p,p_R)$, then one possible behavior of a bug which is in state p, and whose left and right neighbors are in states p_L,p_R, respectively, is to go into state q. The automaton is called deterministic if no $\delta_C(p_L,p,p_R)$ consists of more than a single state q; otherwise, nondeterministic. The formal definition of acceptance requires no special explanation.

The definitions of Q and δ_C above do not depend on the number of bugs in the automaton. For any positive integer n, the function δ_C defines an n-bug automaton C_n (we do not count the two bugs that are permanently in states ℓ and r). Given $Q_o \subseteq Q$, the set of input strings of states in Q_o that C_n accepts will be called the Q_o-

language of C_n. The union of all of these languages, for $n = 1,2,...,$ will be called the Q_o-language "of C".

3. TAPE AUTOMATA CAN IMITATE CELLULAR AUTOMATA

Let $\mathfrak{L}$ be the language of any bounded cellular automaton C, for some given input vocabulary Q_o. In this section, we informally describe a tape-bounded automaton T whose Q_o-language is exactly $\mathfrak{L}$. The states of T are triples from the set $Q' \times Q' \times \{R,L,R',L'\}$, where Q' is Q (the state set of C) together with a special symbol 0.

T begins in state (0,0,R) on the leftmost symbol ℓ of its input string. It leaves the ℓ unchanged, moves to the right, and goes into state $(\ell,0,R)$. Whatever symbol it now reads, say x, is left unchanged; T again moves to the right and goes into state $(\ell,0,L)$. T now reads symbol y (say), leaves it unchanged, and moves to the left, going into state (ℓ,y,R). T is now back on x, and is in a state that contains information about x's left and right neighbors, ℓ and y. Thus T can now imitate C's first (non-ℓ) bug, and can rewrite x as any symbol in the set $\delta_C(\ell,x,y)$; T then moves to the right and goes into state (x,0,R).

The remaining moves of T are similar. It is now on symbol y; it leaves y unchanged, moves right, and goes into state (x,0,L). It now reads the symbol, say z, to the right of y, does not rewrite it, moves left, and goes into state (x,z,R). T is now back on y, and knows the states of y's (original) left and right neighbors, x and z. Hence T can imitate C's second bug, and can rewrite y as any symbol in the set $\delta_C(x,y,z)$. T then moves right and goes into state (y,0,R).

This three-step process is repeated for each successive symbol in the input string. Note that T moves right whenever its (old) state ends in R, left when it ends in L. Note also that T always stays in an R state for two consecutive moves; it changes the R to L when the second term of its state is 0 (unless the first term is also 0, which can only happen initially).

When T, operating in this way, reaches the right end of the string, it has imitated what C could have done in its first time step; the string now looks just like C could have looked after a single step. T now switches to a set of states ending in R' or L', rather than R or L, and performs a mirror-image set of operations while scanning the string from right to left, thus imitating the second time step of C. (The switch from R and L states to R' and L' states is triggered by reading the symbol r at the right end of the string.) When T gets back to the left end, it reverts to the (R,L) states (this is triggered by reading ℓ), and does another left-to-right scan to imitate C's third time step; and so on. If T is in a state of the form $(\ell,p_R,)$, and reads a symbol p such that $\delta_C(\ell,p,p_R)$ contains a final state, then instead of continuing to imitate C, T goes into its own final state, say (ℓ,r,R); this can only happen if the chief bug of C could have gone into a final state at that time step.

It will be noted that, to imitate a single time step of C, T requires about 3n time steps, where n is the length of the input string. Note also that if C is deterministic, so is T.

4. CELLULAR AUTOMATA CAN IMITATE TAPE AUTOMATA

Let $\mathcal{L}$ be the language of any tape-bounded automaton T, for some given input vocabulary V_o. We now describe a bounded cellular automaton C whose V_o-language is exactly $\mathcal{L}$. Let T have symbol set V and state set Q, and let $V' = V \cup \{0\}$, $Q' = Q \cup \{0\}$. Then the state set of C consists of V' together with

a) All pairs of the form $V \times Q'$
b) All triples of the form $V \times Q \times \Delta$

Initially, C is in a string of states in $V_o \subseteq V$. At the first time step, C's chief bug goes into state (v,q_s), where v was its previous state; no other bugs change state. From now on, as we shall see, exactly one bug will have a state which is a pair in $V \times Q$ (or, at alternating time steps, a triple); we shall call this bug the <u>head bug</u>.

At each odd-numbered time step, the head bug's state will be a pair, say (u,p). This bug changes state to any triple in $\delta_T(u,p)$, say to (v,q,d); no other bugs change. At the next time step,

If $d = N$, the head bug goes into state (v,q).
If $d = L$, the head bug goes into state v. The bug whose right-hand neighbor is the head bug goes into state (w,q), where it was previously in state w or $(w,0)$; this latter bug has thus become the head bug.
If $d = R$, the head bug goes into state v (or $(v,0)$, if it is the chief bug). The bug whose left-hand neighbor is the head bug goes into state (w,q), where it was previously in state w; it has thus become the head bug.

No other bugs change. Note that the chief bug is always in a state of the form $(w,0)$, unless it is the head bug; all other non-head bugs have states in V.

If the head bug enters a state (v,q,d) whose q term is a final state of T, then it stops the imitation of T, and goes into the special state 0, which is C's final state. Any bug whose right neighbor is in state 0 goes into state 0; thus at the succeeding time steps, 0's "propagate" leftward, so that eventually the chief bug enters state 0.

Note that this imitation process requires only about two time steps of C for each step of T, plus at most n steps (the string length) at the end, when the 0's propagate. Note also that if T is deterministic, so is C.

5. SPEED LIMITS

Except in trivial cases, the time required for an automaton to accept a string must increase with the length of that string. Suppose, in fact, that a tape-bounded automaton T accepts a string σ of length n in $m < n$ time steps. T cannot have made more than m rightward moves in m time steps; hence its state at time m cannot depend on the last n-m symbols of σ, nor on how much bigger n is than m. Thus T's language includes every possible string which begins with the first m symbols of σ.

Similar remarks apply to a bounded cellular automaton C, since acceptance is defined by the state of C's chief bug, and this bug's state at time m cannot depend on the initial states of bugs more than distance m away from it. Thus if C accepts a string σ of length n in $m < n$ time steps, it also accepts every possible string which begins with the first m symbols of σ.

These observations suggest that, for "interesting" languages $\mathcal{L}$, the time required for the automaton to accept strings of length n in $\mathcal{L}$ must be at least n. There exist many simple languages for which this lower bound can, in fact, be achieved. For example, let $\mathcal{L}_x$ be the set of strings in which some symbol x never appears. A deterministic tape-bounded automaton T that accepts just $\mathcal{L}_x$ is defined by

$$
\begin{aligned}
&v, q_s \rightarrow v, q_s, R \quad \text{for all } v \neq x \text{ or } r\\
&x, q_s \rightarrow x, q_o, N\\
&x, q_o \rightarrow x, q_o, N\\
&r, q_s \rightarrow r, q_f, L
\end{aligned}
$$

where, e.g., the first line is short for "$\delta_T(v,q_s) = (v,q_s,R)$". Informally, T moves to the right, remaining in its starting state q_s, until it finds an x. If this happens, T goes into a "dead" (non-final) state q_o, from which it never emerges. If T reaches r without finding an x, it goes into a final state q_f. Evidently, T accepts a string in $\mathcal{L}_x$ of length n (counting the ℓ and r) in just n steps.

A deterministic bounded cellular automaton C that accepts $\mathcal{L}_x$ can be defined as follows (where, e.g., "uvr $\rightarrow q_f$" is short for "$\delta_C(u,v,r) = q_f$"):

$$\left.\begin{array}{l} uvr \rightarrow q_f \\ uvq_f \rightarrow q_f \end{array}\right\} \text{ for all u, and all } v \neq x \text{ or } q_o$$

$$\left.\begin{array}{l} u\,x\,r \rightarrow q_o \\ u\,x\,q_f \rightarrow q_o \end{array}\right\} \text{ for all u}$$

(all other triples go into their center terms). Informally, C propagates q_f's leftward, starting at r. If an x is reached, it turns into a "dead" q_o, which never changes. Otherwise, the propagation reaches the chief bug, which means that the string is accepted, since q_f is a final state. Note that this takes just n steps (not counting the ℓ and r) for a string of length n.

6. SPEED DIFFERENCES

In Section 5 we considered a language, $\mathcal{L}_x$, for which nothing significant is gained by using a parallel automaton to accept it, since even a sequential automaton can accept $\mathcal{L}_x$ in the shortest possible time. In this section we give a simple example of a language for which parallel acceptance is much faster than sequential. This language, $\mathcal{L}_{sym}$, is the set of <u>symmetrical</u> strings of odd length made up of a's and b's, with the center symbol of the string always a unique symbol c. For example, ℓabaabacabaabar is in $\mathcal{L}_{sym}$, but ℓabcabr is not.

It can be shown that for a sequential automaton T having language $\mathcal{L}_{sym}$, the time required by T to accept a string in $\mathcal{L}_{sym}$ of length $2n+1$ is at least of order n^2. For example, we can define such a T as follows:

a) $\ell,q_s \rightarrow \ell,q_s,R$
$a,q_s \rightarrow z,q_a,R$
$b,q_s \rightarrow z,q_b,R$
$$\left.\begin{array}{l} x,q_a \rightarrow x,q_a,R \\ x,q_b \rightarrow x,q_b,R \end{array}\right\} \text{ for } x = a,b,c$$

T memorizes the leftmost symbol ($\neq\ell$), erases it (i.e., rewrites it as z), and moves to the right.

b) $$\left.\begin{array}{l} x,q_a \rightarrow x,q_a',L \\ x,q_b \rightarrow x,q_b',L \end{array}\right\} \text{ for } x = r \text{ or } z$$
$a,q_a' \rightarrow z,p_s,L$
$b,q_b' \rightarrow z,p_s,L$
$x,q_a' \rightarrow x,q_o,N$ for $x \neq a$
$x,q_b' \rightarrow x,q_o,N$ for $x \neq b$

When T reaches the right end of the string of a's and b's, it checks the last of these. If it is the same as the memorized symbol, T erases it. Otherwise, T goes into a "dead" state q_o.

c) $a,p_s \rightarrow z,p_a,L$
$b,p_s \rightarrow z,p_b,L$

$x,p_a \to x,p_a,L$
$x,p_b \to x,p_b,L$ } for x = a,b,c

T then memorizes the rightmost unerased symbol and moves to the left.

d) $z,p_a \to z,p_a',R$
$z,p_b \to z,p_b',R$
$a,p_a' \to z,q_s,R$
$b,p_b' \to z,q_s,R$
$x,p_a' \to x,q_o,N$ for $x \neq a$
$x,p_b' \to x,q_o,N$ for $x \neq b$

When T reaches the left end of the a's and b's, it checks the last one. If it is the same as the memorized symbol, T erases it and the entire process begins again; otherwise, T "dies".

e) $c,q_s \to c,q_1,R$
$z,q_1 \to z,q_f,N$
$r,q_1 \to r,q_f,L$

$c,p_s \to c,q_2,L$
$z,q_2 \to z,q_f,N$

If, when T is ready to memorize a new symbol, that symbol is c rather than a or b, then T checks that only z's remain on the other side of the c, and if so, T goes into its final state and accepts the string. Readily, acceptance takes about $(2n+3) + \ldots + 1 = (n+2)^2$ time steps (plus an additional step for the final check).

We now exhibit a parallel automaton C that accepts $\mathcal{L}_{sym}$ in time proportional to string length. For all x,

a)

xac → z	cax → z	xa'c → z	ca"x → z
xbc → z	cbx → z	xb'c → z	cb"x → z

a's and b's adjacent to the c become z's

b)

azx → a'	a'zx → a'	xza → a"	xza" → a"
xaz → z	xa'z → z	zax → z	za"x → z
bzx → b'	b'zx → b'	xzb → b"	xzb" → b"
xbz → z	xb'z → z	zbx → z	zb"x → z

a's and b's can shift to the right or left through z's. Once an a or b has shifted to the right, it becomes primed, and can thereafter shift only to the right; similarly, once it shifts left, it becomes double-primed and can only shift left.

c) acb → d
bca → d

If the shifting ever brings an a and a b adjacent to the c, it becomes a d.

d) $xzr \to q$ $\quad$ $xcq \to q_f$
$xzq \to q$ $\quad$ $xzq_f \to q_f$

When z's reach the r, a q is created that can "propagate" to the left through z's. If it finds a c, a q_f is created which propagates leftward through z's until it reaches the chief bug, which thereby accepts. This can only happen if all the pairs of a's and b's that became z's, by shifting until they reached the c, were like pairs, i.e., the string was symmetric. Readily, the entire process takes only about 3n time steps (n for the first z's to reach the ends of the string; n for the last a's and b's to change to z's and the q to reach the c; n for the q_f to reach the chief bug).

7. TWO-DIMENSIONAL AUTOMATA

In this section we generalize the concepts of tape-bounded and bounded cellular automata so that they can be used to accept rectangular arrays, rather than strings.

A two-dimensional tape-bounded automaton T is a bug that can move left, right, up or down on a rectangular array of symbols. T is defined by specifying a function

$$\delta_T\colon\ V \times Q \to 2^{V \times Q \times \Delta}$$

exactly as in Section 1, except that now Δ is a set of five directions {L,R,U,D,N} (U = "up", D = "down"). When T is in state p and reads symbol u, then for any triple (v,q,d) in $\delta_T(u,p)$, T can write symbol v, change to state q, and move in direction d.

To keep T from moving off the array, we assume that all arrays are bordered by a layer of special symbols -- t's on top, ℓ's on the left, r's on the right, b's on the bottom. (The four corners can be arbitrarily assigned to either of the sides that meet at these corners; in any case, T will never see a corner except initially, when we start it at the upper left corner.) We then simply require that, for all (v,q,d) in any

$\delta_T(\ell,p)$, we have $v = \ell$, $d = R$
$\delta_T(r,p)$, we have $v = r$, $d = L$
$\delta_T(t,p)$, we have $v = t$, $d = D$
$\delta_T(b,p)$, we have $v = b$, $d = U$

The formal definition of acceptance, using instantaneous descriptions, is analogous to that in Section 1. Here the ID is a triple $(\alpha,(i,j),q)$, where α is an array, say m by n, and (i,j) is a pair of positive integers, $1 \le i \le m$, $1 \le j \le n$. When T moves right or left, i increases or decreases by 1, and j is unchanged; when T moves up or down, j increases or decreases by 1, and i is unchanged.

A two-dimensional bounded cellular automaton C is a rectangular array of bugs, with the top and bottom rows, and the left and right columns, permanently in special states t,b,ℓ and r, respectively. C is defined by specifying a function

$$\delta_C\colon\ Q \times Q \times Q \times Q \times Q \to 2^Q$$

exactly as in Section 2, except that now the new state of a bug depends on its old state and on the old states of its neighboring bugs above, below, to the left, and to the right. If a bug is in state p, and these neighbors are in states p_T, p_B, p_L, p_R respectively, then the given bug can go into any state q in the set $\delta_C(p_T, p_L, p, p_R, p_B)$. Acceptance, and the language, are defined exactly as in the one-dimensional case. Here the chief bug is the unique bug whose upper neighbor is in state t and whose left neighbor is in state ℓ.

The proofs that T's and C's can simulate one another are similar to those in Sections 3-4. C simulates T by moving a "head bug" around. If this bug ever enters a final state, that state is propagated, e.g., upward until it hits just below the t's, then leftward until it reaches the chief bug. The simulation takes two time steps of C for every one of T; the propagation takes at most m + n time steps, where m,n are the dimensions of the array.

T simulates a single time step of C by doing a systematic scan of the array. It can scan, for example, by starting at the upper left non-border symbol, moving right until it bounces off an r, then down one step; then (changing to a "left-moving" state) moving left until it bounces off an ℓ, then down one step; then (changing back to the "right-moving" state) moving right until it bounces off an r, then down one step; and so on, until it bounces off a b at one of the downward moves.

At each step of this scan, T also (using a small additional set of states to control its motion) visits the four neighbors of the symbol currently being scanned, memorizes their (old) values (p_T, p_L, p_R, p_B, say), and rewrites the current symbol as a pair (p,q), where p was the original symbol, and q is any symbol in $\delta_C(p_T, p_L, p, p_R, p_B)$. [T cannot simply rewrite p as q, since when it processes the symbol below p later in the scan, it will need to know p. T cannot memorize the values of an entire row of symbols in order to process the row below, since it has only finitely many states, and the row can be arbitrarily long.] When T gets to the symbol below p, it can move up, read the pair (p,q) which is now stored there, rewrite it as q, memorize p, and use the memorized value to rewrite the symbol below p.

When the scan is complete, every p has been rewritten as a possible next state q of C at that position, except for the p's on the bottom row, which are still pairs. Finally, T can rescan the bottom row, replacing each pair (p,q) by its second term q. This completes the simulation of an entire time step of C; the number of time steps required is approximately proportional to mn, the area of the array. T can now do another scan (from bottom to top, if desired) to simulate the next time step of C; and so on, until the simulation of C is complete.

8. SPEED COMPARISONS

It is clear, as in Section 5, that if a two-dimensional tape-bounded automaton T accepts an m-by-n array in time less than mn, it cannot have seen the entire array. Thus acceptance of "interesting" array languages by a sequential automaton requires at least time proportional to the *area* of the array being accepted.

Similarly, if a two-dimensional bounded cellular automaton C accepts an m-by-n array in time τ less than m + n, this acceptance cannot depend on the initial states of those bugs whose "city block" distances from the chief bug are greater than τ -- in particular, it cannot depend on the initial state of the bug diagonally opposite the chief bug. Thus parallel acceptance of "interesting" array languages requires at least time proportional to the *diameter* of the accepted array. This lower bound is smaller for C than for T because C is two-dimensionally connected, rather than just being a string of length mn. The T bug can move horizontally *or* vertically, but the C bugs can pass information along horizontally *and* vertically.

As in the one-dimensional case, an example in which these bounds are achieved is the language $\mathcal{L}_x$ of arrays that do not contain the symbol x. A sequential automaton T can accept $\mathcal{L}_x$ by scanning its input array. If T finds an x, it goes into a "dead" state. If it finishes the scan without having found an x, it goes into a final state. This process takes about mn time steps.

A parallel automaton C that accepts $\mathcal{L}_x$ can be defined as follows:

a) $\begin{matrix} & u & \\ w & v & z \\ & b & \end{matrix} \to q$ and $\begin{matrix} & u & \\ w & v & z \\ & q & \end{matrix} \to q$ for all u,w,z and all $v \neq x$

C propagates q's upward, starting from every point of the bottom row; but the propagation is blocked by x's.

b) $\begin{matrix} & t & \\ w & q & r \\ & q & \end{matrix} \to q_f$ and $\begin{matrix} & t & \\ w & q & q_f \\ & q & \end{matrix} \to q_f$ for all w

If the q's all reach the top row, then a final state q_f, generated by the upper right non-border bug, can propagate leftward all the way to the chief bug, which thereby accepts the array. Note that the entire procedure takes just m + n time steps.

A less trivial example which deserves mention here is the language $\mathcal{L}_{conn}$ of arrays containing two symbols, x and y, and in which the set of x's is connected. [The analogous language in one dimension is very easy to handle, since the x's are connected if and only if x and y (or x and r) occur consecutively in the input string exactly once; this property can be checked, even by a sequential automaton, in a number of time steps equal to the string length.] A number of nontrivial methods have been developed for acceptance of $\mathcal{L}_{conn}$ in two dimensions by a parallel automaton C in only m + n time steps, where m,n are the array dimensions. For example, C can use a "shrinking" process developed by Levialdi and by Beyer. This process, which will not be described here in detail, shrinks each connected component of x's to a single isolated point in a number of time steps equal to the sum of the dimensions of the component's circumscribed rectangle. When isolated points appear, they are shifted, say, upward and then leftward, until they reach the chief cell. If two of them reach the chief cell, either simultaneously or successively, they create a "dead" state. While all this is happening, the lower right corner of the array also sends a special message symbol toward the chief cell. When this message has arrived, which takes time m + n, the chief cell knows that enough time has elapsed for all possible isolated points to reach it. If, at that time, exactly one isolated point has arrived, the chief cell goes into the accepting state q_f.

9. CONCLUDING REMARKS

We have only touched here on a few elementary aspects of the comparison between sequential and parallel automata. Many important ideas have not been mentioned, particularly in connection with the parallel machines -- for example:

a) Possible alternative definitions of acceptance; e.g., all bugs enter -- or <u>any</u> bug enters -- a final state.
b) "Speedup" of the automaton by "packing" its input into a fraction of the given space before attempting acceptance (this, of course, requires a much larger set of states).
c) "Speedup" of the automaton by allowing each bug to sense the states of more than just its immediate neighbors.

We have also not discussed systems that generate, rather than accepting, patterns -- e.g., formal grammars. Finally, we have not considered automata that can handle inputs more complex than strings or rectangular arrays -- e.g., arbitrary "blob"-shaped arrays of bugs, or arbitrary networks ("graphs") of bugs.

One particular type of machine that accepts rectangular arrays will be mentioned here. This is a one-dimensional cellular automaton whose length is the width of the input rectangle. The automaton can explore and process the array by moving up and down on it, as a rigid unit; the moves are under the direction, say, of the chief bug. It is easy to see that such a "mixed" automaton M is equivalent in acceptance power to the purely sequential and purely parallel types. A sequential automaton T can imitate M in about m time steps of T for each time step of M, where m is the array width; and M can imitate a parallel automaton C in about n time steps of M for each time step of C, where n is the array height. These mixed automata seem to represent a useful compromise between the long computation times of purely sequential machines, on the one hand, and the large hardware requirement of purely parallel machines, on the other.

REFERENCES

1. Levialdi, S., On shrinking binary patterns, Comm. ACM, 15, 7, 1972.
2. Milgram, D. L., and Rosenfeld, A., Array automata and array grammars, in IFIP 71, North-Holland, Amsterdam, 1972, 69.
3. Rosenfeld, A., and Milgram, D. L., Parallel-sequential array automata, Information Processing Letters, in press.
4. Smith, A. R., Cellular automata and formal languages, in Proc. 11th Symp. on Switching and Automata Theory, 1970, 216.

TO WHAT EXTENT CAN OR MUST COMPUTATIONS BE PARALLELIZED (1)

Corrado Böhm and Mariangiola Dezani-Ciancaglini
Istituto di Scienza dell'Informazione della Università di Torino

ABSTRACT. Every partial recursive function can be expressed by means of a generalized Markov algorithm. Such an algorithm consists of an ordered list of transformation rules; its computation for a given input string is a (possibly infinite) sequence of strings obtained by applications of "determined" transformation rules to "determined" substrings. Neglecting partially the prescriptions on the choice of the rule and the substring, three schemes of parallel computation can be defined:
1) parallel-serial
2) non-deterministic
3) strong-parallel.

The Church-Rosser property gives a principle to decide whether a generalized Markov algorithm can be executed according to one or more of the above parallel computation schemes for a given input string without losing the desired output string.

An interesting decision process of Knuth permits the generalization of the above result to the set of all

---

(1) This research was supported, in part, by the Italian C.N.R. under contract n.72.000246.42 of the Special Program for Informatica.

the possible input strings for which the algorithm terminates. In this way the authors obtain a method for transforming, when it is possible, a given terminating generalized Markov algorithm into a parallel algorithm weakly equivalent to it.

Another case obtains when a given input string yields both terminating and non-terminating computations, depending on the order in which the rules are applied. This is the case observed by Manna et al. for the computation of the partial functions defined recursively. The authors present within the framework of the present exposition a non-trivial example of a function for which only a "parallel" computation scheme can achieve the correct result.

1. INTRODUCTION.

In this paper we consider the possibility of executing a sequential algorithm in a parallel way, leaving the result unchanged.

First (section 2) we choose as a framework the family of generalized Markov algorithms which seems a suitable model of computation to define some possible schemes of parallelism. In fact, a transformation rule of a generalized Markov algorithm specifies uniquely neither the substring to be transformed (as does, for example, an address inside a machine instruction) nor the next transformation rule to be applied (as do , for example, the operators of an equation system obtained by translating a flow-chart.[1]). The generalized Markov algorithm becomes deterministic by the existence of an external computation rule specifying at every step both the substring to be transformed and the rule to be applied. The parallelism can thus be obtained by neglecting one or both of these prescriptions (section 3).

Then (section 4) we show that the problem of when and how to parallelize a generalized Markov algorithm is related to the existence of a Church-Rosser property.[2] Therefore (section 5) we use a semidecision method described by Knuth[3] for testing the Church-Rosser property of a set of transformation rules.

Further (section 6) the same method is applied to test different computation schemes[4] for recursive definitions of partial functions. In this case the Church-Rosser property assures that every computation,if it terminates,gives as result the value of the least fixed-point function. The problem here is to reach that value, if it exists.

The choice of a correct computation scheme depends essentially on the operators occurring in the definition of the functional; different authors[4,5,6] give computation schemes correct only for certain class of functionals. Exhibition of a simple example enables us to prove that only Kleene's computation scheme, which can be thought as the most parallel scheme, assures the overall reaching of the fixed-point function.

2. GENERALIZED MARKOV ALGORITHMS.

As model of sequential computation we choose the generalized Markov algorithms,which are complete with respect to the partial recursive functions.[7] Our definition of generalized Markov algorithm follows that of Cohen, Wegstein[8] and Caracciolo.[7] In order to render this paper self-contained we give extensively all the definitions.

Def.1. A *string structure* relative to an alphabet Σ is a string on the alphabet $V\cup\Sigma$,where $V\cap\Sigma=\Phi$ (1). V is called the alphabet of variables.

Examples. If $V=\{\alpha,\beta\}$ and $\Sigma=\{a,b,c,e,f,g,(,)\}$,the following are string structures:

$$(\alpha_1\alpha_2)$$

$$((\alpha_1(\beta a))\alpha_2)$$

where the subscripts are used to distinguish between occurrences of like variables.

Def.2. An *assertion set* for an alphabet V relative to an

(1) Φ denotes the empty set.

alphabet Σ is a set of context-free production rules according to which every element of V generates a (in general, infinite) language over Σ.

Example: if V and Σ are the same as in the above example, the following is an assertion set:

$$\alpha ::=(\alpha_1\alpha_2)/\beta$$

$$\beta ::=a/b/c/d/e/f/g$$

Def.3. A *string matches with a string structure* according to a given assertion set iff that assertion set contains replacements of variables which generate the string from the structure.

Example: according to the assertion set of the preceding example :

(((dc)(ba))((eg)f)) matches $(\alpha_1\alpha_2)$ with the replacements: α_1=((dc)(ba)) and α_2=((eg)f); it matches also $((\alpha_1(\beta a))\alpha_2)$ with the replacements: α_1=(dc), β=b and α_2=((eg)f).

Def.4. A *transformation rule* is a pair of string structures where all the variables of the right hand side (r.h.s.) occur also in the left hand side (l.h.s.).

Examples: with the same assumptions as in the preceding examples the following are transformation rules:

$$(\alpha_1\alpha_2)\to(\alpha_2\alpha_1)$$

$$((\alpha_1\alpha_2)\alpha_3)\to(\alpha_1(\alpha_2\alpha_3))$$

Def.5. Applying *a given tranformation rule to a given string* according to a given assertion set means:

1) to find a substring matching the l.h.s. string structure.
2) to replace this substring with the string obtained making the replacements just discovered in the l.h.s.

Example: $(\alpha_1\alpha_2)\to(\alpha_2\alpha_1)$ applied to (((dc)(ba))((eg)f)), choosing the substring ((dc)(ba)),and,consequently,the replacements: α_1=(dc) and α_2=(ba) gives :(((ba)(dc))((eg)f)).

We say that a set of transformation rules cannot be applied to a given string iff it is impossible to find a substring matching with the l.h.s. string structures of some rule of the set. Such a string is called "irreducible" with respect to that set.

A *generalized Markov algorithm* is an (ordered) list

of transformation rules together with an assertion set.
Example: the following is a generalized Markov algorithm:

1. $(\alpha_1\alpha_2) \rightarrow (\alpha_2\alpha_1)$
2. $((\alpha_1\alpha_2)\alpha_3) \rightarrow (\alpha_1(\alpha_2\alpha_3))$

$\alpha := (\alpha_1\alpha_2)/\beta$

$\beta := a/b/c/d/e/f/g$

The *computation of a generalized Markov algorithm* for a given initial string is the (possibly infinite) sequence of strings obtained from the initial one applying at every step the first applicable rule of the list to the leftmost-innermost reducible substring. We can think here that the preceding computation scheme can be rendered effective by an ambidexterous man pointing at every step with the left hand to the rule to be applied, if possible, and with the right hand to the substring to be transformed.[9]

Examples: The computation of the generalized Markov algorithm of the preceding example for the string $(((dc)(ab))((eg)f))$ is the following:

$$((\underbrace{(dc)}_{1}(ba))((eg)f)) \rightarrow ((\underbrace{(cd)}_{1}(ba))((eg)f)) \rightarrow ((\underbrace{(dc)}_{1}(ba))((eg)f) \rightarrow$$

....

where we underlined, as in the remaining examples, the substring to be transformed with the number of the rule to be applied. The preceding algorithm can be modified in order to render the computation finite by conditioning the application of the first rule:

1. $(\alpha_1\alpha_2) \rightarrow (\alpha_2\alpha_1)$ if $\alpha_1 > \alpha_2$ (1).

For the same initial string we have the following computation:

$$((\underbrace{(dc)}_{1}(ba))((eg)f)) \rightarrow (((cd)\underbrace{(ba)}_{1})((eg)f)) \rightarrow (\underbrace{((cd)(ab))}_{1}((eg)f)) \rightarrow$$

$$(\underbrace{((ab)(cd))}_{2}((eg)f)) \rightarrow (((a(b(cd)))\underbrace{((eg)f)}_{2}) \rightarrow ((a(b(cd)))(e\underbrace{(gf)}_{1})) \rightarrow$$

$$\underbrace{((a(b(cd)))(e(fg)))}_{2} \rightarrow (a\underbrace{((b(cd))(e(fg)))}_{2}) \rightarrow (a(b\underbrace{((cd)(e(fg)))}_{2}))) \rightarrow$$

$$(a(b(c(d(e(fg)))))).$$

---

(1) For the order relation, the alphabetic order (neglecting parentheses) can be chosen.

In this example the step number is 9.

3. THREE SCHEMES OF PARALLEL COMPUTATION.

If we neglect the restrictions given in the preceding section for the choice of the substring to be transformed and/or of the transformation rule to be applied we can obtain three schemes of parallel computation:

1) *Parallel-serial*. If we apply the rules in the given order to any reducible substring, we obtain the parallel application of the same rule to different substrings. The ambidexterous man points here at every step only with the left hand to the rule to be applied. This type of parallelism is implemented on the computer ILLIAC IV.[10]

   Example: One parallel-serial computation of the previous terminating generalized Markov algorithm for the same initial string is the following:

   (((dc)(ba))((eg)f)) → (((cd)(ab))((eg)f)) →
   1 1 1

   (((ab)(cd))((eg)f)) → ((a(b(cd)))(e(gf))) →
   2 2 1

   ((a(b(cd)))(e(fg))) → (a((b(cd))(e(fg)))) →
   2 2

   (a(b((cd)(e(fg))))) → (a(b(c(d(e(fg)))))).
   2

   The number of steps in this case is reduced to 6.

2) *Non-deterministic*. On the contrary if we neglect the order between the rules, but we decide to apply them to the leftmost-innermost reducible substring, several rules may be applied at the same time, because any substring may satisfy more than one string structure. The ambidexterous man points here at every step only with the right hand to the substring to be transformed.

   Example: With the same assumptions as in the preceding examples we have:

   (((dc)(ba))((eg)f)) → (((cd)(ba))((eg)f)) →
   1 1

   (((cd)(ab))((eg)f))
   1↙ ↘2

   (((ab)(cd))((eg)f)) ((c(d(ab)))((eg)f))
   2 ↓ 1 ↓

```
((a(b(cd)))((eg)f))          ((c((ab)d))((eg)f))
            2                      1
       ↓                                ↓
((a(b(cd)))(e(gf)))          ((((ab)d)c)((eg)f))
            1                      2
       ↓                                ↓
                                            etc.
```

3) *Strong-parallel*. This type of parallelism is obtained by the union of the two preceding cases: that is, we can choose freely both the rule and the substring. In this way different rules can be applied simultaneously to different substrings. The ambidexterous man is here superfluous.
Example: one strong-parallel computation, upon the same initial conditions of the previous examples, which terminates in only three steps is exhibited by:

```
(((dc)(ba))((eg)f)) → (((ba)(dc)(e(gf))) →
     1        2           1    1    1
                                2

((ab)((cd)(e(fg)))) → (a(b(c(d(e(fg)))))).
           2
       2
```

4. THE CHURCH-ROSSER PROPERTY.

At first we shall define the Church-Rosser property for a digraph.
Def. 6. A (possible infinite) *digraph is Church-Rosser* iff it is a connected digraph with at most one node with out-degree (1) zero and exactly one node with in-degree (1) zero.
That is, walking on the digraph at every branch node you can choose any direction whatsoever because there must exist eventually a join node. Some examples of Church-Rosser digraphs are given in Fig.1.
If the digraph is infinite the fact that it is Church-Rosser obviously cannot be verified by means of its structure, but must be proved by its (possibly recursive) definition.

---

(1) The out-degree (in-degree) of a node is the number of arrows leaving (entering) that node.[11]

We can represent an immediate transformation of a string into another string by means of an arrow connecting the former with the latter. In this way a digraph is naturally associated to every sequence of transformations of strings into strings, i.e. to every set of computations. A set of computations for a given initial string, which has a Church-Rosser digraph, from the above, possesses at most one irreducible string. This assures that the result of all these computations, if it exists, is unique.

The digraph of the computation of a generalized Markov algorithm is obviously Church-Rosser, because it is a chain. We will investigate when the digraph of a set of parallel computations is Church-Rosser. To this aim we need the following definitions:

Def. 7. A *transformation rule has the Church-Rosser property with respect to itself*, ,for a given input string, iff the graph of all possible transformations of this string, according to that rule, is Church-Rosser.

Def. 8. *Two transformation rules mutually possess the Church-Rosser property*, for a given input string, iff the graph of all possible tranformations of this string, obtained by choosing always the leftmost-innermost reducible substring according to either of the two rules, is

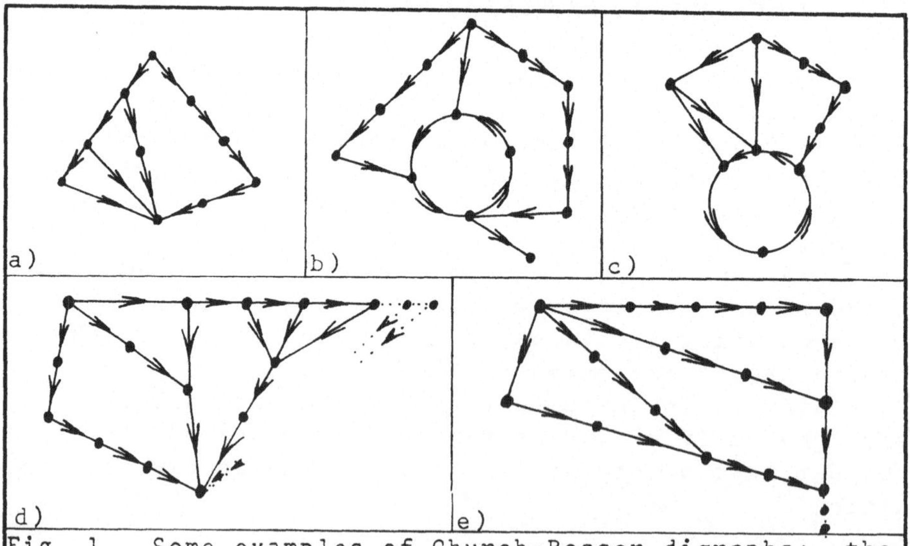

Fig. 1. Some examples of Church-Rosser digraphs: the digraphs a), b), c), are finite while the digraphs d) and e) are infinite.

Church-Rosser.

Def. 9. *A set of transformation rules has the Church-Rosser property,* for a given input string , iff the digraph of all possible transformations of this string according to the rules of the set is Church-Rosser.

One may even doubt that there exists a rule which does not have the Church-Rosser property with respect to itself. An example of such a rule is the following: $(c(\alpha_1\alpha_2))\to\alpha_1$ if we choose as initial string (c(c(ab))). In fact, applying this rule to two different substrings of (c(c(ab))) we obtain two different irreducible strings: (c(c(ab))) → (ca) (c(c(ab))) → c.

Another example of a rule not commuting with itself is given elsewhere.[12]

5. A DECISION METHOD FOR TESTING IF A GIVEN TERMINATING GENERALIZED MARKOV ALGORITHM CAN BE PARALLELIZED.

The preceding definitions of rules having the Church-Rosser property were limited to a given initial string.

Knuth[3] gives us a semi-decision method for testing if a given set of rules has the Church-Rosser property for every input string. The basic idea of this method is to limit the infinite set of strings to a finite set of "critical" string structures depending only on the l.h.s. of the rules. For every critical string structure σ the method gives a branch, i.e. two different string structures σ' and σ" obtained from σ by means of two different transformations. At this point we must find a join string structure ρ, such that there exist two sequences of transformations, one starting from σ' and the other from σ",both terminating with ρ. A necessary and sufficient condition for the set of rules to have the Church-Rosser property is that,for every critical σ the corresponding ρ exists. The search for a join string structure ρ may never terminate because starting from σ' and/or σ" an infinite sequence of transformations may be produced.We can assure an answer only for those sets of rules whose number of applications to every string is finite(1).

At this point we replace the definitions 7,8, and 9 with more general definitions.

Def. 7' (8'/ 9'). *A transformation rule has with respect to itself (two transformation rules mutually possess/ a set of transformation rules has) the Church-Rosser property* iff, for every critical string structure σ which can be transformed, by the application of this rule to two different substring, (of these two rules to the leftmost-innermost reducible substring / of the rules of the set) into two different string structures σ' and σ" there exists a join string structure ρ . The relation connecting definitions 7', 8', 9' is exhibited by:

Statement 1. *A set of transformation rules has the Church-Rosser property if:*

1) each rule of the set has the Church-Rosser property with respect to itself and

2) any pair of rules of the set mutually possesses the Church-Rosser property and

3) for every critical string structure σ which can be transformed, by the application of any two rules of the set to two different substrings of σ into two different string structures σ' and σ", there exists a join string structure ρ .

This statement is clear if we recall that the critical string structures determined by Knuth's method for a set of rules are of two types:

---

(1) A well-known method[13] to assure this condition is to define a well-order on the string structures in such a way that for every rule the l.h.s. be greater than the r.h.s. Let us note that an infinite sequence of transformations not necessarily corresponds to an infinite digraph, but it may correspond also to a digraph with a cycle. For example, infinite sequences of transformations are shown in digraphs b) and c) of Fig.1.
A total ordering relation would only assure the absence of cycles, whereas a well-ordering relation assures, moreover, the absence of infinite chains.

a) the critical string structures relative to each rule of the set

b) the critical string structures relative to a pair of rules of the set.

The existence of a join string structure ρ for critical string structures of type a) is assured by the Church-Rosser property of each rule of the set with respect to itself (condition 1). The critical string structures of type b) may be classified according to whether the two rules are applied to the same substring or not. In the first case the existence of a join string structure is assured by the Church-Rosser property of any two rules of the set (condition 2), because the transformed substring may always be the leftmost-innermost reducible. In the second case the existence of the join string is clearly assured by condition 3.

Conditions 1, 2, 3 are then sufficient for a set of rules to have the Church-Rosser property. They are not necessary because a set of rules not having the Church-Rosser property may acquire it by having adjoined to it another (suitable) rule.[3]

We can solve now the problem of determining whether a given terminating (1) generalized Markov algorithm can be computed according to one or more of the above three schemes of parallel computation, obtaining the same final string. Our decision method is based of the following three statements:

Statement 2. *Every terminating generalized Markov algorithm can be computed according to the parallel-serial computation scheme iff every rule belonging to its list has the Church-Rosser property with respect to itself.*

Statement 3. *Every terminating generalized Markov algorithm can be computed according to the non-deterministic computation scheme iff any two rules belonging to its list have mutually the Church-Rosser property.*

---

(1) A terminating generalized Markov algorithm is defined as an algorithm whose corresponding set of rules can be applied only a finite number of times to every string.

Statement 4. *Every terminating generalized Markov algorithm can be computed according to the strong-parallel computation scheme iff its set of rules has the Church-Rosser property.*

The proof of these statements follows immediately from the definition of Church-Rosser property. As an example, we give the demonstration of the first statement. "If" part: the Church-Rosser property assures that, no matter which substring we apply the rule to, after a finite number of transformations, we shall obtain the same string. "Only-if" part: if the rule does not have the Church-Rosser property there exists at least one critical string σ for which a corresponding ρ does not exist. This means that, starting from a string belonging to the structure σ, we can obtain (at least two) different irreducible strings, depending on the order of applications of the rule to the substrings. At most one of these final strings can coincide with the result obtained from the computation of the given terminating generalized Markov algorithm.

The preceding statement suggest a method for testing when and how a given terminating generalized Markov algorithm can be parallelized. If the set of rules has the Church-Rosser property, it may be applied according to the strong-parallel computation scheme. Otherwise if the rules of the set have the Church-Rosser property with respect to themselves (or any two rules of the set mutually possess the Church-Rosser property) the set may be applied according to the parallel-serial (non-deterministic) computation scheme. Finally, if no Church-Rosser property is valid, the given algorithm (without changing the rules) cannot be parallelized.

Examples: The transformation rules of the previous examples:

1. $(\alpha_1\alpha_2)\rightarrow(\alpha_2\alpha_1)$ if $\alpha_1>\alpha_2$
2. $((\alpha_1\alpha_2)\alpha_3)\rightarrow(\alpha_1(\alpha_2\alpha_3))$

have the Church-Rosser property with respect to themselves. Proof: for rule 1. the unique critical string structure is $\sigma_1=((\alpha_1\alpha_2)\alpha_3)$ where $\alpha_1>\alpha_2$ and $\alpha_1>\alpha_3$, with $\sigma_1'=(\alpha_3(\alpha_1\alpha_2))$

and $\sigma_1'=((\alpha_2\alpha_1)\alpha_3)$. In this case the join string structure is obviously $\rho_1=(\alpha_i(\alpha_j\alpha_1))$, where $\{i,j\}=\{1,2\}$ and $\alpha_j\geq\alpha_i$.
For rule 2. we have: $\sigma_2=(((\alpha_1\alpha_2)\alpha_3)\alpha_4)$ with $\sigma_2'=$ $=((\alpha_1(\alpha_2\alpha_3))\alpha_4)$ and $\sigma_2''=((\alpha_1\alpha_2)(\alpha_3\alpha_4))$. The join string structure is $\rho_2=(\alpha_1(\alpha_2(\alpha_3\alpha_4)))$.
The transformation rules 1. and 2. mutually possess the Church-Rosser property.
Proof: the unique critical string structure to which both the rules are applicable is: $\sigma_3=((\alpha_1\alpha_2)\alpha_3)$ where $\alpha_1>\alpha_3$ with $\sigma_3'=(\alpha_1(\alpha_2\alpha_3))$ and $\sigma_3''=(\alpha_3(\alpha_1\alpha_2))$. In this case we find: $\rho_3=(\alpha_i(\alpha_j\alpha_k))$ where $\{i,j,k\}=\{1,2,3\}$,and $\alpha_k\geq\alpha_j\geq\alpha_i$.
The above set of transformation rules has the Church-Rosser property.
Proof: we need only to test condition 3) of statement 1 because conditions 1) and 2) have been just now verified. The unique critical string structure such that rules 1) and 2) are applicable to different substrings is: $\sigma_4=((\alpha_1\alpha_2)\alpha_3)$ where $\alpha_1>\alpha_2$ with $\sigma_4'=((\alpha_2\alpha_1)\alpha_3)$, $\sigma_4''=(\alpha_1(\alpha_2\alpha_3))$. The join string structure in this case is: $\rho_4=(\alpha_i(\alpha_j\alpha_k))$ where $\{i,j,k\}=\{1,2,3\}$ and $\alpha_k\geq\alpha_j\geq\alpha_i$.

6. THE "NECESSITY" OF PARALLEL COMPUTATION RULES FOR RECURSIVELY DEFINED PARTIAL FUNCTIONS.

In the recursive definition of a partial function $f(x)\rightarrow \tau[f(x)]$ (1) the problem seems different because we cannot limit ourselves to finite computations (2).

---

(1) We write $f(x)\rightarrow\tau[f(x)]$ instead of the usual $f(x)\leftarrow \tau[f(x)]$[4,5,6] to be consistent with our notation for the transformation rules.
(2) In fact there can be functionals generating infinite computations; trivially, if $\tau=I$ the rule $f(x)\rightarrow f(x)$ is a cycle of length 1.

There are essentially two types of Church-Rosser transformation digraphs (for a given string) corresponding to infinite computations:

1) digraphs with the point (irreducible string) of out-degree zero. For example digraphs b) and d) of Fig.1 are of this type.
2) digraphs without the point of out-degree zero. For example, digraphs c) and e) of Fig. 1 are of this type.

Since the rule $f(x)\to\tau[f(x)]$ has the Church-Rosser property[2] the replacement of $f(x)$ with $\tau[f(x)]$ can be made in every substring preserving the possibility of finding a join string. But the problem is that the computation might never terminate, even if there exists the irreducible string, because of the choice, at some branchpoint, of a wrong transformation.

Different methods of computation of partial recursive functions haven been studied by Manna[4], Vuillemin[6], De Michelis.[5] According to Kleene's theorem,[14] they call the unique point with outdegree zero the value of the least fixedpoint function. Kleene himself[14] gives a computation scheme which assures that this value, assuming it exists, is obtained. This scheme substitutes simultaneously, when no other possible transformation rules defining τ can be applied, $\tau[f(x)]$ for every occurrence of $f(x)$.

The possibility of determinining which occurrences of $f(x)$ can be, step by step, replaced without losing the desired value depends essentially from the transformation rules used in the construction of τ. Manna et al.[4] consider that the preceding rules are restricted to:

a) the elementary operations on integers
b) the predicate of equality for integers
c) the definitions of the conditional statements:

   <u>if</u> <u>true</u> <u>then</u> A <u>else</u> $B\to A$

   <u>if</u> <u>false</u> <u>then</u> A <u>else</u> $B\to B$

This set of rules has the Church-Rosser property.[2] Manna et al. show that, besides the Kleene s computation scheme, there exists in this case another not losing the desired value: the leftmost-outermost computation scheme

(1). If we adjoin to the preceding set of transformation rules the following two:

$$0*\alpha\to 0$$
$$\alpha*0\to 0$$

we still obtain a set having the Church-Rosser property. Vuillemin,[6] shows that in this case there is a partial function, defined recursively, for which the leftmost-outermost computation scheme loses the value of the least fixedpoint function. It can, however, be reached by means of the parallel-outermost computation scheme (2).
We can show that, if we adjoin the transformation rule:

1. $\alpha/\alpha\to 1$

which preserve the Church-Rosser property for the set of rules, there exists a function for which even the parallel-outermost computation rule loses the desired value. In this case only the parallel substitution of $\tau[f(x)]$ for all the occurrence of f(x) (Kleene's computation scheme) can reach the value of the least fixed-point function. The mentioned recursive function is defined by:

2. $F(x,y,z)\to \underline{\text{if}}\ z=0\ \underline{\text{then}}\ F(y,x,-1)/F(x,y,-1)\ \underline{\text{else}}$
$F(F(x,y,z-1),y,z-1).$

We shall give the sequences of transformations of this function for x=y=z=1 using the parallel-outermost and the Kleene's computation scheme, omitting the applications of the rules different from 1. and 2.
Using the parallel-outermost computation scheme we have:

$$\underbrace{F(1,1,1)}_{2}\to\underbrace{F(F(1,1,0),1,0)}_{2}\to\underbrace{F(1,F(1,1,0),-1)}_{2}/\underbrace{F(F(1,1,0),1,-1)}_{2}\to$$
$$\underbrace{F(F(1,F(1,1,0),-2),F(1,1,0),-2)}_{2}/\underbrace{F(F(F(1,1,0),1,-2),1,-2)}_{2}\to\ldots$$

and so on. This sequence is infinite.

---

(1) This computation scheme replaces at every step only the leftmost-outermost occurrence of f(x), when none of the other rules is applicable.

(2) This computation scheme replaces at every step all the occurrences of f(x) which do not occur in any argument of another f(x), when none of the other rules is applicable.

On the contrary, using Kleene's computation scheme we have:

$$\underbrace{F(1,1,1)}_{2}\rightarrow F(\underbrace{F(1,1,0)}_{2},1,0)\rightarrow F(1,\underbrace{F(1,1,-1)/F(1,1,-1)}_{1},-1)/$$

$$\underbrace{/F(\underbrace{F(1,1,-1)/F(1,1,-1)}_{1},1,-1)}_{2}\rightarrow \underbrace{F(1,1,-1)/F(1,1,-1)}_{1}\rightarrow 1$$

which is the value of the least fixedpoint function.

This last example shows that a "parallel" computation may be necessary for recursively defined partial functions.

CONCLUSION

This paper stresses explicitly the relation between parallel computation and the Church-Rosser property. The interest of this connection is to illuminate semantic features of programs; the transformation presented here of a sequential algorithm into a parallel one (if it is possible), is essentially semantic. In fact, the algorithm so obtained is weakly equivalent to the original, i.e. only the input-output relation, not the computation, is conserved.

The connection between parallel computation and Church-Rosser property was observed by Rosen,[2] but the authors think that this paper is a further step in this direction and it is their intention to apply these methods to the solution of other computer problems.

REFERENCES

1.McCarty J., Recursive functions of symbolic expressions and their computation by machine. Part I, *Comm. ACM*, 3, 1960, 184.

2.Rosen B.K., Tree-Manipulating Systems and Church-Rosser Theorems, *ACM Journal*, 20, 1973, 160.

3.Knuth D.E. and Bendix P.E., Simple word problems in universal algebras, in *Computational problems in Abstract Algebra*, Leech J., Pergamon Press, Braunschweig, 1970, 263.

4.Manna Z., Ness S. and Vuillemin J., Inductive methods for proving properties of programs, in *Proceedings of an ACM Conference on Proving Assertions about Programs*, Las Cruces, 1972, 27.

5.De Michelis G., Recursive functions not dependent on the computation rules,in *Proceedings of International Computing Symposium 1973*, Davos, 1973.

6.Vuillemin J., Correct and optimal implementations of recursion in a simple programming language, in *Proceedings of fifth annual ACM Symposium on Theory of Computing*, Austin, 1973, 224.

7.Caracciolo di Forino A., Generalized Markov algorithms and automata,in *Proc. of the International Summer School on Automata Theory*, Caianiello E. Academic Press, New York, 1966, 115.

8.Cohen K. and Wegstein J.H., AXLE: An axiomatic language for string transformation, ACM *Comm.*, 8, 1965, 657.

9.Knuth D.E., Examples of formal semantics, in *Symposium on Semantics of Algorithmic Languages*, 188, Engeler E., Springer-Verlag, Berlin, 1970, 212.

10.Kogge P.H. and Stone H.S., A parallel algorithm for the efficient solution of a general class of recurrence equations, *Technical Report n.25 of the Standford University, 1972.*

11.Berge G., *Graphes et hypergraphes,* Dunod, Paris, 1970, 6.

12.Böhm C. and Dezani-Ciancaglini M., Listing the functional digraph structures, in *Proc. International Computing Symposium 1973,* Davos, 1973.

13.Floyd R.W., Assigning Meanings to programs, Proc. *Symposium Appl. Math.,* AMS 19, 1967, 19.

14.Kleene S.C., *Introduction to metamathematics,* D. Von Nostrand Company Inc., Netherlands, 1952, chap. XII.

ON THE SPEED GAINED IN PARALLEL METHODS

S. Winograd

IBM T. J. Watson Research Center
Yorktown Heights, New York

INTRODUCTION

In recent years there has been a growing interest in parallel computation as an approach for achieving high performance machines. Parallel machines, like the ILLIAC IV and STAR, have been built. Various programming language schemes, such as FORK, JOIN, LOCK, have been proposed to facilitate the direct expression of parallelism in the program. Several theoretical models have been proposed for studying the sequencing and control aspects of parallel computations. Many parallel algorithms have been proposed for a variety of numerical calculations. For further details on theoretical models for sequencing and control, as well as an extensive bibliography, the reader is referred to the recent paper by R. E. Miller[1]. W. L. Miranker[2] presents a survey of parallelism in numerical analysis.

In this paper we survey some of the results concerning the ultimate gain in speed of parallel execution of several numerical problems. Our point of view is that of complexity of computation, namely the determination of the best possible gain rather than the analysis of various parallel algorithms to find out the gain they achieve. We will concentrate on the numerical aspect of the problem and neglect the control aspects; that is, we will concentrate on maximizing the number of numerical operations done concurrently assuming that the data is available when needed. We make this assumption knowing full well that in many actual situations it is not valid, and that the data movement and sequencing restriction may slow the execution down.

In the next section we will consider some arithmetic problems, and in Section III we will describe results in analytic problems.

PARALLEL ARITHMETIC COMPUTATIONS

One of the simplest arithmetic calculations, which is also very amenable to parallel computation, is that of repeated addition. Consider the computation

$$y = \sum_{i=1}^{n} x_i. \tag{1}$$

In the sequential case n-1 additions are required to compute y, and assuming that each addition takes one time unit, the time required to perform the computation is n-1. To analyze the time required in the parallel case, assume n=15, and that 4 adders are available. In the first unit of time we can compute

$$y_1 = x_1 + x_2;\ y_2 = x_3 + x_4;\ y_3 = x_5 + x_6;\ y_4 = x_7 + x_8. \tag{2}$$

In the second time unit:

$$y_5 = x_9 + x_{10};\ y_6 = x_{11} + x_{12};\ y_7 = x_{13} + x_{14};\ y_8 = x_{15} + y_1. \tag{3}$$

In the third time unit only three adders compute:

$$y_9 = y_2 + y_3;\ y_{10} = y_4 + y_5;\ y_{11} = y_6 + y_7. \tag{4}$$

In the fourth time unit only two adders compute:

$$y_{12} = y_8 + y_9;\ y_{13} = y_{10} + y_{11}. \tag{5}$$

And in the fifth time unit we use only one adder to obtain:

$$y = y_{12} + y_{13}. \tag{6}$$

In general, the same method yields that $\lceil \frac{n}{m} \rceil - 1 + \lceil \log_2 m \rceil$ units of time are required to evaluate (1) using m processors.

Similar results are obtained when we consider the parallel evaluation of an inner product

$$z = \sum_{i=1}^{n} x_i y_i. \tag{7}$$

Under the assumption that an addition and a multiplication take a unit of time, m processors can compute (7) in $\lceil \frac{2n}{m} \rceil - 1 + \lceil \log_2 m \rceil$ units of time.

How good are these parallel methods? Is there a way for speeding up the computation? It is clear that if a sequential algorithm requires k arithmetic operations, then using m processors at least $\lceil \frac{k}{m} \rceil$ units of time are required. This is so because we can "simulate" the operations of m processors by one processor; put differently: the number of operations required in the sequential case is also the total number of operations required to solve the problem, and since at every unit of time at most m operations can be carried out, it follows that at least $\lceil \frac{k}{m} \rceil$ units of time are required. It was shown in [3] that n-1 additions are required to compute (1), and that n-1 additions and n multiplications are required to compute (7), and therefore the algorithms described above differ from the trivial lower bound by $\log_2 m$. What caused this "waste" of time? An examination of the algorithm for computing (1) reveals that in the third unit of time we used only three of the four adders, in the fourth unit of time we used only two of the four adders, and in the fifth unit of time we used only one. Therefore the only way for "speeding" the computation is to find a scheme in which every processor is used at every stage of the computation. No such scheme exists. Since the computation of (1) results in only one output, it is clear that in the last stage of the computation only one processor is required. Since the operations we consider are binary, then only two processors are required in the step before the last of the computation. Munro and Paterson[4] carried out this reasoning and obtained:

<u>Theorem 1</u>: Suppose the computation of a single quantity Q requires q binary operations, then the shortest parallel computation of Q requires at least $\lceil \frac{q}{m} \rceil - 1 + \lceil \log_2 \min(m,q+1) \rceil$ units of time.

In view of this result we see that the obvious method for computing (1) or (7) is optimal.

In the examples (1) and (7) the method for parallel evaluation was the obvious one, all that was required is the use of the associative law. Clearly many examples exist in which more sophisticated identities are necessary for efficient parallel evaluation. Consider the expression

$$y = a_1(a_2+a_3(a_4+a_5(a_6+a_7(a_8+a_9(a_{10}+a_{11}(a_{12}+a_{13}(a_{14}+a_{15}a_{16}))))))). \quad (8)$$

This expression requires 15 arithmetic operations, and it looks very sequential. In order to evaluate it the product $a_{15}a_{16}$ has to be found, and only then can we add a_{14} and so on. Can any speed-up be obtained if more than one processor are used in parallel?

One way of using two processors to evaluate (8) in 10 units of time is illustrated below. (We again assume that each arithmetic operation requires one unit of time.) This method can be described by the following table

Processor 1	Processor 2
$x_1 = a_1 a_3$	$y_1 = a_{15} a_{16}$
$x_2 = x_1 a_5$	$y_2 = a_{14} + y_1$
$x_3 = x_2 a_7$	$y_3 = a_{13} y_2$
$x_4 = a_5 a_6$	$y_4 = a_{12} + y_3$
$x_5 = a_4 + x_4$	$y_5 = a_{11} y_4$
$x_6 = x_1 x_5$	$y_6 = a_{10} + y_5$
$x_7 = a_1 a_2$	$y_7 = a_9 y_6$
$x_8 = x_7 + x_6$	$y_8 = a_8 + y_7$
$x_9 = x_3 + x_8$	
$y = x_8 + x_9$	

An examination of this method shows that it is based on writing (8) as

$$y = A + BX \qquad (9)$$

where

$$X = a_8 + a_9(a_{10} + a_{11}(a_{12} + a_{13}(a_{14} + a_{15}a_{16}))) \qquad (10)$$

$$A = a_1a_2 + a_1a_3(a_4 + a_5a_6)$$

$$B = a_1a_3a_5a_7$$

and using one processor to evaluate A and B, the second to evaluate X and finally computing (9).

Richard Brent[5] extended this method, and showed that it is applicable to any arithmetic expression in which every variable appears only once. He has shown that:

<u>Theorem 2</u>: Let E be an arithmetic expression with operations "+" and "·" in which every variable appears only once, and let $|E|$ be the number of variables in E. Then E can be evaluated in time $4 \log_2|E|$ with $|E| - 1$ processors. If division is also used then the number of processors may be as high as $3(|E|- 1)$.

The following result, also due to Brent[5] shows how to convert results about time of evaluation using large numbers of processors to fewer processors. The proof of this result, even though simple, will be omitted.

<u>Theorem 3</u>: If expressions $E_1,\ldots,E_m$ can be evaluated in time t with q operations, then they can be evaluated in $t + \frac{q-t}{p}$ using only p processors.

Combining Theorems 2 and 3 we obtain:

<u>Theorem 4</u>: Let E be as in Theorem 2, then using p processors E can be evaluated in time $4 \log_2|E| + \frac{2(|E|-1)}{p}$ if E does not involve division and in time $4 \log_2|E| + \frac{10(|E|-1)}{p}$ if it does.

Comparing Theorems 1 and 4, we see that the lower bound on the time to evaluate E with p processors is $\log_2|E| + \frac{|E|-1}{p}$ and that this lower bound can be reached within a factor of 2 (if no divisions are involved) or a factor of 10 (if division is involved). All this is, of course, in the special case that each variable in E appears only once. What about the more general case that some variables appear more than once? What gains can be achieved by parallel execution of a polynomial $\sum_{i=0}^{n} a_i x^i$ in which x appears more than once, or even in the simpler case of evaluating x^n?

Consider the evaluation of x^7. It is easy to verify that a single processor requires at least 4 units of time to compute x^7. Two processors working in parallel can reduce this time to 3 units of time: During the first unit of time they compute x^2, during the second unit of time the first processor computes x^3 while the second computes x^4;.during the third unit of time the first processor can compute x^7. The gain in speed using two processors was clearly minimal. Is this because of the particular parallel way we chose to compute x^7, or is it an inherent limitation of the problem? Let $f_p(n)$ denote the minimum number of time units required to compute x^n using p processors, and let $f_\infty(n)$ denote the time using unbounded number of processors. Thus $f_1(n)$ denotes the time to compute x^n in the sequential case.

<u>Theorem 5</u>: If $p_1 > p_2$ then

1) $\log_2 n \le f_\infty(n) \le f_{p_2}(n) \le f_{p_1}(n)$

2) $\lim_{n \to \infty} \frac{f_1(n)}{\log_2 n} = 1$.

We thus see that the problem of computing x^n is very sequential. Even if the number of processors at our disposal is unlimited, at least $\log_2 n$ units of time are required, which even one processor can compute x^n in time $\log_2 n + o(\log n)$. This is contrasted with the evaluations of arithmetic expressions in which every variable appears only once - where the gain is essentially linear with the number of processors.

We now turn our attention to the problem of polynomial evaluation. It was shown in [3] that at least n multiplications and n additions are required to evaluate $P_n(x) = \sum_{i=0}^{n} a_i x^i$, and thus 2n units of time are required in the sequential case. As an application of Theorem 1 we obtain that at least $\frac{2n}{p} - 1 + \log_2 p$ units of time are required to evaluate $P_n(x)$ using p processors, but can we achieve this lower bound?

Consider, for example, the evaluation of $P_{22}(x) = \sum_{i=0}^{22} a_i x^i$. In the sequential case we require 44 units of time. Assume we have two processors at our disposal. Dorn[6] proposed the following method: The polynomial $P_{22}(x)$ can be written as

$$P_{22}(x) = (a_0+a_2x^2+a_4x^4+\ldots+a_{22}x^{22}) + x(a_1+a_3x^2+\ldots+a_{21}x^{20} \tag{11}$$

Thus at the first unit of time one processor can compute x^2, at the next 22 units of time one processor computes $a_0+a_2x^2+\ldots+a_{22}x^{22}$ while the other computes $a_1+a_3x^2+\ldots+a_{21}x^{20}$ (which require only 20 units of time) and $x(a_1+a_3x^2+\ldots+a_{21}x^{20})$, finally we use the next unit of time to compute $P_{22}(x)$ using (11). Altogether this method requires 24 units of time.

In the case of three processors, a similar procedure is proposed. In this case we write $P_{22}(x)$ as

$$P_{22}(x) = (a_0+a_3x^3+\ldots+a_{21}x^{21})+ x(a_1+a_4x^3+\ldots+a_{22}x^{21}) + x^2(a_2+a_5x^3+\ldots+ a_{20}x^{18}). \tag{12}$$

This time we use the first two units of time to compute x^2 and x^3, the next 14 units of time to compute $a_0+ a_3x^3+ \ldots + a_{21}x^{21}$ by the first processor, $a_1+ a_4x^3+ \ldots+ a_{22}x^{21}$ by the second processor, and $x^2(a_2+ a_5x^3+ \ldots + a_{20}x^{18})$ by the third processor. The next unit of time is used to compute $x(a_1+ a_4x^3+ \ldots+ a_{22}x^{21})$ by one processor, and $(a_0+ a_3x^3+\ldots+a_{21}x^{21}) + x^2(a_2+ a_5x^3+\ldots+ a_{20}x^{18})$ by another. And finally we require one more unit of time to compute $P_{22}(x)$. Altogether 18 units of time were required.

Another method for computing $P_{22}(x)$ using three processors is based on the following identity:

$$P_{22}(x) = \sum_{i=0}^{8} a_i x^i + x^9 \left(\sum_{i=0}^{7} a_{i+9} x^i + x^8 \left(\sum_{i=0}^{5} a_{i+17} x^i \right) \right). \quad (13)$$

During the first 14 units of time the first processor computes $\sum_{i=0}^{5} a_{i+17} x^i$ (10 units of time), x^2 x^4, x^8, and $x^8 (\sum_{i=0}^{5} a_{i+17} x^i)$ which require additional 4 units of time. During the same 14 time units, the second processor computes $\sum_{i=0}^{7} a_{i+9} x^i$. During time unit 15, the first processor computes x^9 while the second computes $\sum_{i=0}^{7} a_{i+9} x^i + x^8 (\sum_{i=0}^{5} a_{i+17} x^i)$ and during those 16 time units the third processor computes $\sum_{i=0}^{8} a_i x^i$. Using the identity (13), we see that only 17 time units are necessary to compute $P_{22}(x)$ using 3 processors.

Munro and Paterson[4] generalize the second method, and showed that an nth degree polynomial can be evaluated in $\frac{2n}{p} \log_2 p + o(\log p)$ using p processors. In light of Theorem 1 this result is optimal up to a term which grows slower than log p.

PARALLEL SOLUTIONS OF ANALYTIC PROBLEMS

Considering the problem of parallel integration of differential equations, very little is known about the potential gain due to parallelism. The reader is referred to [2] for a description of various parallel methods for the integrations of ordinary differential equations. It is not known whether these methods are close to optimal. As a matter of fact very little is known even about optimal sequential methods for integrating differential equations.

Numerical solutions of partial differential equations lend themselves more readily to parallel execution. Consider the one dimensional heat equation $u_t(x,t) = u_{xx}(x,t)$, $u(x,0) = g(x)$. Standard discretization of the problem leads to

$$u(x,t+\Delta t) = u(x,t) + \frac{\Delta t}{(\Delta x)^2} (u(x+\Delta x,t) - 2\, u(x,t) + u(x-\Delta x,t)).$$

thus, if the values of $u(k\Delta x, \ell\Delta t)$ are known for all k, it is possible to compute in parallel the values of $u(k\Delta x,(\ell+1)\Delta t)$ for all k. We see therefore that it is possible to "parallelize" the numerical solution of this equation in such a way that the time required by p processors is about $\frac{1}{p}$ the time required by one processor. This situation is typical.

It was shown in [11] that every method for explicit solution of a partial differential equation can be so organized that p-fold reduction in execution time is obtained by the use of p processors.

This p-fold gain in execution time is predicated on the assumption that when a previously computed value of the solution is needed it is readily available. One attractive way for guaranteeing that is by having a small fast memory. Results are first stored in this fast memory, until they are no longer needed as inputs for future computation, at which time they are transferred to the large (and slower) memory. One of the results reported in [11] is that no such scheme can exist (except in some trivial exceptions). No matter how large is the temporary fast memory, it is bound to overflow, and therefore some of the results will have to be stored in the slower memory even though they are still needed for future calculations.

This result indicates that there exists a certain amount of "bookkeeping" and "data movement" overhead for the problem, and that parallel execution by itself cannot reduce this part of the computation. One can, of course, envision parallel data paths which will alleviate this problem, but this whole area still requires further investigation.

As a further example of possible gain in parallel operations, we consider the problem of finding the root of a function of a single variable by iterative method.

Let $f(x)$ be a function of one variable, and let r be a simple root, i.e., $f(r) = 0$ while $f'(r) \neq 0$. There are various iterative methods for determining r, for example, Newton's method and the Secant method. Each iterative method is a procedure for finding the next approximation for the root r given the previous ones. If we use x_n to denote the nth approximation then Newton's method is:

$$x_{n+1} = x_n - \frac{f(x_n)}{f'(x_n)} \tag{14}$$

and the Secant method is

$$x_{n+1} = x_n - \frac{x_n - x_{n-1}}{f(x_n) - f(x_{n-1})} f(x_n). \tag{15}$$

The various methods differ in the amount of information which they use. For example, Newton's method uses the value of the function and its first derivative at the previous approximation; while the Secant method requires just the value of the function, but at the previous two approximations. Various methods also have different rates at which the approximations converge to the root. If we use $e_n = |x_n - r|$ to denote the error after the nth iteration, then for Newton's method $e_{n+1} \sim C\, e_n^2$, and thus Newton's method is said to converge quadratically; while for the Secant method $e_{n+1} \sim C' e_n^{\lambda}$ where $\lambda = \frac{1+\sqrt{5}}{2} \sim 1.618$, and thus this method has a power of convergence λ.

The power of convergence of an iterative scheme is a convenient parameter to measure the number of iterations which are needed. If an iterative scheme has a power of convergence p, then the number of iterations needed to find the root with error ε grows as $\frac{\log \log \frac{1}{\varepsilon}}{\log p}$.

In the parallel case, when more than one processor is available, it is possible to use the additional computational power in two ways. We can use the extra computational power to speed up the time required for the execution of one iteration step; either by parallel evaluation of the value of the function and its derivative, or by parallel computation of the iteration step. The other alternative is to use the k processors in order to compute k new approximations to the root. It is difficult to assess the possible gains by parallel operations using the first alternative. Without more detailed knowledge of the function f, whose root is to be found, it is impossible to determine whether parallelism can be helpful in evaluating it or its derivative. The results described in the previous section suggest that parallel execution of each iteration step (which is in many cases an evaluation of an algebraic expression) can be sped up considerably by parallel computation. We will, however, concentrate on the second alternative, on the use of the k processors to compute at each step k new approximations of the root.

In order to determine the possible gain from parallel operations, we have to find out first the fastest way of finding the root in the sequential case. A good discussion of various sequential iterative methods can be found in [7]. It is shown there that there exists a family of iterative methods whose power of convergence increases indefinitely. However, in order to achieve this higher power of convergence they require the evaluations of higher and higher derivatives of the function f. Our problem is therefore to find out the highest power of convergence of iterative methods using no higher than the dth derivative of f. This problem was investigated in [8] and the results are summarized in:

<u>Theorem 5</u>: 1) No iterative scheme using up to d^{th} derivative and up to k processors can have a power of convergence higher than $k(d+1) + 1$. 2) For every k and d, and for every ε there

exists an iterative method using k processors and up to the dth derivative, whose power of convergence exceeds $k(d+1) + 1 - \varepsilon$. If the initial approximation to the root is within error e_1, and it is desired to find the root within error e_2, then about $\log_{k(d+1)+1} \frac{\log e_2}{\log e_1}$ iteration steps are required. It is clear, therefore, that using k processors reduces the computation time by a factor of log k. This small gain should be contrasted with the results about parallel evaluation of algebraic expressions, where the gain was linear with k (at least for small values of k).

Similar results were obtained by Karp and Miranker[9] in their study of another iterative process. Let f be a function defined on [0,1] which has a single maxima. It is desired to find an interval of length ε which includes the point where f assumes its maximum value. Since no smoothness conditions are imposed on f, the only method for determining this interval is by successively evaluating f at various points, and by comparing the value of f at these points. Kiefer[10] investigated the sequential case, and showed that in order to guarantee an interval of length ε containing the maximum, about $\log \frac{1}{\varepsilon}$ $\lambda = \frac{1+\sqrt{5}}{2}$ steps have to be taken, each step consists of an evaluation of f at a point. Karp and Miranker[9] showed that in the case of parallel operations, when each step is the evaluation of f at k points, about $\log_{\frac{k+2}{2}} \frac{1}{\varepsilon}$ steps have to be taken to guarantee the determination of an interval of length ε. So in this problem too the speed-up due to parallelism is only logarithmic with the number of processors.

CONCLUSIONS

This paper reported on some of the known results in the area of parallel computations. We saw that in the case of parallel evaluation of arithmetic expression the gain is essentially linear with the number of processors, except for a small "penalty" of about $\log_2 p$. The area of iterative methods is much more inherently sequential. There, the gain grew roughly as the logarithm of the number of processors, so that many more processors are needed to achieve a certain gain in speed.

The whole discussion in this paper centered around the arithmetic unit. An implicit assumption underlying the whole discussion was that the arithmetic unit can easily obtain the data it requires. This assumption is not always easily implementable in practice. In one specific case, that of numerical integration of partial differential equations, the existence of "inherent" overhead was shown. In general, much more work is needed to find out the flow of data required to achieve the maximum utilization of parallel processing.

Another criticism of the results reported in this paper is that the problem whose parallelism was investigated was too "small." Typically, the task of polynomial evaluation or finding the root of functions is only part of a larger problem whose solution is sought. It is not at all clear that the best organization of parallel computation is obtained by "parallelizing" these subtasks. It is conceivable that each processor will be assigned a job like polynomial evaluation, and that greater gains can be achieved by breaking the problem into larger subtasks to be executed in parallel. Very little is known about this possibility, but it definitely merits further research.

REFERENCES

1. Miller, R. E., A Comparison of Some Theoretical Models of Parallel Computations, IEEE Trans. on Computers, C-22, 710 Aug. 1973.

2. Miranker, W. L., A Survey of Parallelism in Numerical Analysis, SIAM Rev. 13, 524, Oct. 1971.

3. Winograd, S., On the Algebraic Complexity of Functions, Actes du Congres International des Mathematiciens, 1970; 3, 283, 1971.

4. Munro, I. and Paterson, M., Optimal Algorithms for Parallel Polynomial Evaluation, JCSS, 7, 189, April 1973.

5. Brent, R. P., The Parallel Evaluation of Arithmetic Expressions in Logarithmic Time, Proc. of the Carnegie-Mellon Conf. on Complexity of Sequential and Parallel Numerical Algorithms, Academic Press, 1973.

6. Dorn, W. S., Generalizations of Horner's Rule for Polynomial Evaluation, IBM J. of Res. and Dev., 6, 239, 1962.

7. Traub, J. F., Iteration Methods for Solution of Equations, Prentice Hall, 1964.

8. Winograd, S., Parallel Iterative Methods, Complexity of Computer Computations, Plenum Press, 1972.

9. Karp, R. M. and Miranker, W. L., Parallel Minimax Search for a Maximum, J. Comb. Theory, 4, 19, 1968.

10. Kiefer, J., Sequential Minimax Search for a Maximum, Proc. Amer. Math. Soc., 4, 502, 1953.

11. Karp, R. M., Miller, R. E. and Winograd, S., The Organization of Computations for Uniform Recurrence Equations, JACM, 14, 563, July 1967.

ALGORITHMES SERIELS, ALGORITHMES PARALLELES (1)

L. NOLIN

Université de Paris VII

Dans son exposé, C. Böhm a défini les notions de calcul en série et de calcul en parallèle ; il a également précisé la condition que doit remplir un formalisme servant à décrire des calculs pour qu'un parallélisme soit possible.

Nous ne faisons rien d'autre ici que d'illustrer ces notions en esquissant une théorie des algorithmes conçue tout spécialement pour les besoins de l'Informatique.

Dans une première partie, nous donnons les définitions essentielles et quelques résultats importants.

Nous définissons dans la seconde partie quelques algorithmes universels dont certains ont une parenté évidente avec des objets de la Logique Combinatoire.

Dans la troisième partie, enfin, nous montrons sur un exemple comment on peut repérer de façon effective les algorithmes applicables simultanément.

(1) Cette étude a été poursuivie avec l'aide du Centre National de la Recherche Scientifique (A.T.P. 7110).

I. ESQUISSE D'UNE THEORIE DES ALGORITHMES

Dans tout ce qui suit, T et $\mathcal{A}$ sont des collections qui ont successivement les propriétés H_1, H_1 et H_2, H_1 et H_2 et H_3, H_1 et H_2 et H_3 et H_4.

H_1 : T est une collection non vide, $\mathcal{A}$ une collection de parties de T, contenant $\emptyset$ et T comme éléments et qui est close pour l'intersection (infinie).

Les lettres X, Y, Z, V, W , affectées au besoin d'indices, désignent des éléments de $\mathcal{A}$.

Définition 1 :

$$\bar{\bigcup_{i\in \mathfrak{I}\neq\emptyset}} X_i$$

est le plus petit Z tel que

$$X_i \subset Z \text{ pour tout } i \in \mathfrak{I} \neq \emptyset \quad .$$

Définition 2 :
X est atomique ssi X n'a pas de partition propre (i.e. d'au moins deux éléments) dans $\mathcal{A}$.

Définition 3 :
$\mathfrak{F}$ est la collection de toutes les applications $f : \mathcal{A} \to \mathcal{A}$ qui sont normales (i.e. telles que, si

$$\bigcup_{i\in \mathfrak{I}\neq\emptyset} X_i \quad \text{et} \quad X_i \ (i \in \mathfrak{I}) \in \mathcal{A}$$

alors

$$f\left(\bigcup_{i\in\mathfrak{I}} X_i\right) = \bar{\bigcup_{i\in\mathfrak{I}}} f(X_i) \quad) .$$

Définition 4 :
F X Y est, par définition, si Y = T alors T, sinon la collection de toutes les $f \in \mathfrak{F}$ telles que $f(X) \subset Y$.

Définition 5 :
$\mathcal{A}_F$ est la plus petite collection qui contient tous les

$$F\,X\,Y \neq T$$

et qui est close pour l'intersection (infinie).

Les lettres X', Y', Z', V', W' affectées au besoin d'indices, désignent des éléments de :

$$\mathcal{A} \cup \mathcal{A}_F .$$

<u>Définition 6</u> :
Si

$$X' \in \mathcal{A} \cup \mathcal{A}_F \text{ et si } Y \in \mathcal{A} ,$$

<u>l'image de Y par X'</u>, en abrégé $X' [Y]$, est

$$\cap \{Z \ / \ X' \subset F\ Y\ Z\} .$$

<u>Théorème 1</u> :
Si

$Z_j \ (j \in \mathcal{J} \neq \emptyset,\ Z_j \in \mathcal{A})$ est atomique, alors

$$\Big(\bigcap_{i \in \mathfrak{I} \neq \emptyset} F\ X_i\ Y_i \Big) \Big[\bigcup_{j \in \mathcal{J}} Z_j \Big] = \bar{\cup} \{ \cap \{ Y_i \ / \ Z_j \subset X_i ,\ i \in \mathfrak{I} \} / j \in \mathcal{J} \} .$$

La démonstration est fondée sur les sept lemmes suivants :

<u>Lemme 1</u> :

$$F\ X\ Y \subset F\ X_1\ Y_1$$

ssi ou bien $Y_1 = T$ ou bien $X_1 \subset X$ et $Y \subset Y_1$;

car, si $Y_1 = T$, la chose est évidente ; si $Y_1 \neq T$,

l'implication de droite à gauche est évidente, et la réciproque vient du fait que la fonction f , telle que

$$f\ (Z) = Y \quad \text{pour tout} \quad Z \subset X, \quad \text{et} \quad f\ (Z) = T$$

sinon, appartient à $F\ X\ Y$.

<u>Lemme 2</u> :

$$\bigcap_{i \in \mathfrak{I} \neq \emptyset} F\ X\ Y_i = F\ X\ \Big(\bigcap_i Y_i \Big) \quad ;$$

si $\bigcup_i X_i \in \mathcal{A}\ (i \in \mathfrak{I} \neq \emptyset)$ alors $\bigcap_i (F\ X_i\ Y) = F\ \Big(\bigcup_i X_i \Big)\ Y$;

car 1°) $f \in \bigcap_i F\ X\ Y_i$

ssi $\forall i \,/\, f \in F\,X\,Y_i$, ssi $\forall i \,/\, f\,(X) \subset Y_i$,

ssi $f\,(X) \subset \bigcap_i Y_i$, ssi $f \in F\,X\,(\bigcap_i Y_i)$;

et 2°) $f \in \bigcap_i F\,X_i\,Y$

ssi $\forall i \,/\, f \in F\,X_i\,Y$, ssi $\forall i \,/\, f\,(X_i) \subset Y$,

ssi $\overline{\bigcup_i}\, f\,(X_i) \subset Y$, ssi (normalité) $f\,(\bigcup_i X_i) \subset Y$,

ssi $f \in F\,(\bigcup_i X_i)\,Y$.

<u>Lemme 3</u> :

$X'\,[Y] \subset Z$ ssi $X' \subset F\,Y\,Z$;

$(F\,X\,Y)\,[Z] = (\text{si } Z \subset X \text{ alors } Y \text{ sinon } T)$.

car 1°)

$\bigcap \{F\,Y\,V \,/\, X' \subset F\,Y\,V\}$ = (lemme 2)

$F\,Y\,(\bigcap \{V \,/\, X' \subset F\,Y\,V\})$ = (définition 6) $F\,Y\,(X'\,[Y])$;

ainsi $X' \subset F\,Y\,Z$

ssi $F\,Y\,Z \in \{F\,Y\,V \,/\, X' \subset F\,Y\,V\}$, ssi $F\,Y\,(X'\,[Y]) \subset F\,Y\,Z$,

ssi (lemme 1) $Z = T$ ou $X'\,[Y] \subset Z$, ssi $X'\,[Y] \subset Z$;

et 2°)

$(F\,X\,Y)\,[Z] \subset V$

ssi (précédent) $F\,X\,Y \subset F\,Z\,V$,

ssi (lemme 1) $V = T$ ou $Z \subset X$ et $Y \subset V$;

$(F\,X\,Y)\,[Z]$ est le plus petit de ces V (définition 5) donc

$(F\,X\,Y)\,[Z] = (\text{si } Z \subset X \text{ alors } Y \text{ sinon } T)$.

<u>Lemme 4</u> :

$X' \subset X'_1$ et $Y \subset Y_1$ entraine $X'\,[Y] \subset X'_1\,[Y_1]$;

en particulier, les éléments de $\mathcal{A} \cup \mathcal{A}_F$ sont <u>monotones</u> pour

l'inclusion.
Car

$$X' \not\subset X'_1 \quad \text{et} \quad Y \subset Y_1 \quad \text{et} \quad X'_1 [Y_1] \subset V$$

entraine (lemme 3) $X'_1 \subset F\, Y_1\, V$, donc (lemme 1) $X' \subset F\, Y\, V$,

donc (lemme 3) $X' [Y] \subset V$; et le corollaire est évident.

Lemme 5 :

Si $\bigcup_{i \in \mathfrak{I} \neq \emptyset} Y_i \in \mathcal{A}$ alors $X' [\bigcup_i Y_i] = \bar{\bigcup_i} (X' [Y_i])$;

car $\bar{\bigcup_i} (X' [Y_i]) \subset V$ ssi $\forall i \,/\, X' [Y_i] \subset V$,

ssi (lemme 3) $\forall i \,/\, X' \subset F\, Y_i\, V$, ssi $X' \subset \bigcap_i F\, Y_i\, V$,

ssi (lemme 2) $X' \subset F (\bigcup_i Y_i)\, V$, ssi (lemme 3) $X' [\bigcup_i Y_i] \subset V$.

Lemme 6 :

$$\bigcap_{i \in \mathfrak{I} \neq \emptyset} F\, X_i\, Y_i \subset F (\bigcap_i X_i) (\bigcap_i Y_i) \quad .$$

car (lemme 1) $\forall i \,/\, F X_i\, Y_i \subset F (\bigcap_i X_i)\, Y_i$,

donc $\bigcap_i F\, X_i\, Y_i \subset \bigcap_i (F (\bigcap_i X_i)\, Y_i)$

donc (lemme 2) $\bigcap_i F\, X_i\, Y_i \subset F (\bigcap_i X_i) (\bigcap_i Y_i)$.

Lemme 7 :

Si V est atomique, alors $\bigcap_{i \in \mathfrak{I} \neq \emptyset} F\, X_i\, Y_i \subset F\, V\, W$

ssi $W = T$ ou $\exists \mathfrak{J} \neq \emptyset , \mathfrak{J} \subset \mathfrak{I} \,/\, V \subset \bigcap_{j \in \mathfrak{J}} X_j$ et $\bigcap_{j \in \mathfrak{J}} Y_i \subset W$;

corollaire :

$$(\bigcap_{i \in \mathfrak{I}} F\, X_i\, Y_i) [V] = \bigcap \{Y_i \,/\, V \subset X_i\} \quad ;$$

car $f \in \bigcap_{i \in \mathfrak{I}} F\, X_i\, Y_i$ ssi $\forall i \,/\, f\, (X_i) \subset Y_i$;

si donc V est atomique :

ou bien non existe $\mathfrak{J} \neq \emptyset$ tel que $V \subset X_j \; (j \in \mathfrak{J})$

et alors $f\, (V) = T$;

ou bien existe un tel $\mathcal{J}$ et pour tout $j \in \mathcal{J}$, $f\ (V) \subset Y_j$,

donc $f\ (V) \subset \bigcap_{j\in\mathcal{J}} Y_j$, donc $f \in F\ V\ (\bigcap_{j\in\mathcal{J}} Y_j)$,

donc $f \in F\ V\ W$ si $\bigcap_{j\in\mathcal{J}} Y_j \subset W$;

la réciproque est évidente (lemmes 3 et 6) ; et le corollaire en découle par le lemme 3 .

Démonstration du théorème 1 :

$$(\bigcap_{i\in\mathfrak{I}} F\ X_i\ Y_i)\ [\bigcup_{j\in\mathcal{J}} Z_j] = \text{(lemme 5)}\ \overline{\bigcup}_{j\in\mathcal{J}} ((\bigcap_{i\in\mathfrak{I}} (F\ X_i\ Y_i))[Z_j])$$

$$= \text{(lemme 7)}\ \overline{\bigcup}_{j\in\mathcal{J}} (\bigcap_{i\in\mathfrak{I}} ((F\ X_i\ Y_i)\ [Z_i]))$$

$$= \text{(lemme 3)}\ \overline{\bigcup}\ \{\bigcap \{Y_i\ /\ Z_j \subset X_i\ ,\ i \in \mathfrak{I}\}\ /\ j \in \mathcal{J}\}\ .$$

Ce résultat permet de calculer l'image de Y par X' dans nombre de cas intéressants, et nous pouvons maintenant illustrer aisément les définitions données plus haut.

<u>Exemple 1</u> :

Prenons pour $\mathcal{A}$ une collection plus grande que l'ensemble $\mathcal{B}$ dont les éléments sont : $\mathbb{N}$ (l'ensemble des entiers naturels), ses sous-ensembles récursivement énumérables, et aussi l'ensemble des valeurs de vérité, {VRAI,FAUX} et ses sous-ensembles.

Parmi les éléments de $\mathcal{B}$, seuls sont atomiques l'ensemble vide et les singletons : {0} , {1} , ..., {VRAI} , {FAUX} ; {0,1} n'est pas atomique car $\{0,1\} = \{0\} \cup \{1\}$.

Comme il est entendu que si $X \in \mathcal{B}$

alors $X \subset F\ Y\ Z$ ssi $Z = T$,

on a $X\ [Y] = T$ pour tout Y ;

en particulier,

$$\{0\}\ [Y] = \{0,1\}\ [Y] = T = T\ [Y]\ ;$$

T joue ainsi le rôle de l'indéfini.

<u>suc</u> = $\bigcap_{i\in\mathbb{N}} F\ \{i\}\ \{i+1\} \in \mathcal{A}_F$

traduit, dans notre théorie, la fonction successeur et toutes celles qui prennent la même valeur qu'elle sur $\mathbb{N}$. Le théorème 1 nous permet de calculer les images que fournit suc de différents ensembles :

$$\underline{suc}\,[\{0\}] = \{1\}\ ,\ \underline{suc}\,[\{1\}] = \{2\}\ ,\ \ldots$$

mais aussi,

$$\underline{suc}\,[\{0,1\}] = \{1,2\}\ ,\ \underline{suc}\,[\{2i\ /\ i \in \mathbb{N}\}] = \{2\,i+1\ /\ i \in \mathbb{N}\},$$

$\underline{suc}\,[\mathbb{N}] = \mathbb{N}^+$ (l'ensemble des entiers $\neq 0$), ...

$$\underline{suc}\,[\{VRAI\}\,] = T\ .$$

Autre exemple, le vecteur booléen

V = <VRAI , FAUX , VRAI>

est traduit par

$$V' = <_1 \{VRAI\}, \{FAUX\}, \{VRAI\} >_3$$

égal par définition à

$$F\ \{1\}\ \{VRAI\} \cap F\ \{2\}\ \{FAUX\} \cap F\ \{3\}\ \{VRAI\}\ ;$$

alors V' [{i}]

traduit la ième composante de V si $i \in \{1,2,3\}$;

$$V'\,[\{1,2\}] = \{VRAI\}\ \bar{\cup}\ \{FAUX\} = \{VRAI\ ,\ FAUX\}\ ;$$

$$V\,[X] = T \quad si \quad X \not\subset \{1,2,3\}\ .$$

Ce formalisme très simple permet déjà de traduire certaines instructions et procédures des langages de programmation : nous en verrons des exemples dans la troisième partie ; il permet aussi d'en traduire certaines déclarations. Ainsi ,

BOOLEEN TABLEAU B [1:3]

est traduit par :

$$B' = F\ \{1,2,3\}\ \{VRAI\ ,\ FAUX\}\ (\in \mathcal{A}_F)$$

c'est à dire, en vertu du lemme 2 et d'une remarque précédente :

$$B' = <_1 \{VRAI\ ,\ FAUX\}\ ,\ \{VRAI\ ,\ FAUX\}\ ,\ \{VRAI\ ,\ FAUX\} >_3\ ;$$

quant aux <u>définitions</u>, elles ne posent pas plus de problèmes ; ainsi :

BOOLEEN TABLEAU B [1:3] = < VRAI , FAUX , VRAI >

est traduit tout simplement par l'élément V' de $\mathcal{A}_F$ défini plus haut.

<u>Fin de l'exemple 1</u>.

Pour pouvoir traduire commodément les procédures à plusieurs paramètres, nous allons supposer que $\mathcal{A}$ possède quelques propriétés supplémentaires.

H2 : <u>Soit $\mathcal{B}$ une collection de parties de T parmi lesquelles $\emptyset$ et T ; alors $\mathcal{A}$ est la plus petite collection de parties de T contenant $\mathcal{B}$ qui est close pour l'intersection (infinie) et pour l'opération</u>:

$$X, Y \mapsto FXY \quad .$$

La condition H1 étant satisfaite, les résultats précédents s'appliquent intégralement.

<u>Exemple 2</u> :

Soit $\mathcal{B}$ comme dans l'exemple 1 , $\mathcal{A}$ remplissant la condition H2.

Nous pouvons maintenant traduire des procédures qui calculent des fonctions de plusieurs variables, des relations, des fonctionnelles, ... Ainsi :

$$\leqslant_{\mathbb{N}} = (\bigcap_{\substack{i,j\in\mathbb{N}\\ i\leqslant j}} F\,\{i\}\ \ (F\,\{j\}\ \ \{\text{VRAI}\})) \cap (\bigcap_{\substack{i,j\in\mathbb{N}\\ i>j}} F\{i\}\ (F\{j\}\{\text{FAUX}\}))$$

entraine :

$$\leqslant_N [\{2\}] = (\bigcap_{\substack{j\in\mathbb{N}\\ 2\leqslant j}} F\,\{j\}\ \{\text{VRAI}\}\ ) \cap \{\bigcap_{\substack{j\in\mathbb{N}\\ 2>j}} F\,\{j\}\ \{\text{FAUX}\}\ \}$$

d'où :

$$(\leqslant_N [\{2\}])\ [\{1\}] = \{\text{FAUX}\}\quad ,\quad (\leqslant_N [\{2\}])\ \ [\{3\}] = \{\text{VRAI}\}, \ \ldots$$

(en notations usuelles : $2 \not\leqslant 1$, $2 \leqslant 3$, ...)

La matrice $W = \begin{bmatrix} 0 & 1 & 0 \\ 1 & 0 & 0 \\ 0 & 0 & 1 \end{bmatrix}$ est traduite par l'élément de $\mathcal{A}$, W' =

$$F\{1\}\,(F\{1\}\{0\}) \cap F\{1\}\,(F\{2\}\{1\}) \cap F\{1\}\,(F\{3\}\{0\})$$

$$\cap\, F\{2\}\,(F\{1\}\{1\}) \cap F\{2\}\,(F\{2\}\{0\}) \cap F\{2\}\,(F\{3\}\{0\})$$

$$\cap\, F\{3\}\,(F\{1\}\{0\}) \cap F\{3\}\,(F\{2\}\{0\}) \cap F\{3\}\,(F\{3\}\{1\}) \quad ,$$

qui est égal (lemme 2) à :

$$F\{1\}\,X_1 \cap F\{2\}\,X_2 \cap F\{3\}\,X_3 \quad ,$$

avec

$$X_1 = F\{1\}\{0\} \cap F\{2\}\{1\} \cap F\{3\}\{0\} \quad ,$$

$$X_2 = F\{1\}\{1\} \cap F\{2\}\{0\} \cap F\{3\}\{0\} \quad ,$$

$$X_3 = F\{1\}\{0\} \cap F\{2\}\{0\} \cap F\{3\}\{1\} \quad ;$$

$X_i = W'\,[\{i\}]$ traduit la $i^{ème}$ ligne de la matrice si $i \in \{1,2,3\}$;

$(W'\,[\{2\}])\,[\{3\}]$ traduit l'élément $W_{2,3}$;

$\bigcap_{\substack{i=1,2\\ j=1,2}} F\{i\}\,(F\{j\}\,((W'\,[\{i\}])\,[\{j\}]))$ traduit la sous-matrice $\begin{bmatrix} 0 & 1 \\ 1 & 0 \end{bmatrix}$.

La déclaration

ENTIER NATUREL TABLEAU E [1 : 3 , 1 : 3]

est traduite par l'élément de $\mathcal{A}$, E' =

$$F\{1,2,3\}\,(F\{1,2,3\}\,\mathbb{N})$$

qui est égal (lemme 2) à :

$$\bigcap_{\substack{i=1,2,3\\ j=1,2,3}} F\{i\}\,(F\{j\}\,\mathbb{N}) \quad .$$

La définition

ENTIER NATUREL TABLEAU W [1:3 , 1:3] =

< <0,1,0> , <1,0,0> , <0,0,1> >

est traduite par W' défini ci-dessus.

Fin de l'Exemple 2.

Les éléments de $\mathcal{A}$ examinés dans les exemples précédents possèdent quelques propriétés que nous avons utilisées, consciemment ou non ; comme on les retrouve dans toutes les applications et qu'elles ne découlent pas de H2 , nous allons imposer à $\mathcal{A}$ deux nouvelles conditions.

Définition 7 :

$\mathcal{A}_\emptyset = \{\emptyset\}$, $\mathcal{A}_T = \{T\}$, $\mathcal{A}_B$ = la fermeture de $\mathcal{B}$ par l'intersection (infinie), amputée de $\emptyset$.

Il est clair que $\mathcal{A} = \mathcal{A}_\emptyset \cup \mathcal{A}_T \cup \mathcal{A}_B \cup \mathcal{A}_F$.

H3 : Les éléments de $\mathcal{A}_B$ et $\mathcal{A}_F$ sont incomparables (par l'inclusion).

Corollaire 1 :

$\{\mathcal{A}_\emptyset , \mathcal{A}_T , \mathcal{A}_B , \mathcal{A}_F\}$ est une partition de $\mathcal{A}$;

$X \in \mathcal{A}_B$ et $X \subset F\ Y\ Z \Rightarrow Z = T$;

$X[Y] = T$ pour tout Y ssi $X \in \mathcal{A}_T \cup \mathcal{A}_B$;

si $X , Y \in \mathcal{A}_F$ alors $X \subset Y$ ssi $\forall Z \,/\, X[Z] \subset Y[Z]$.

Théorème 2 :

Tout singleton (appartenant à $\mathcal{A}_B$), $\emptyset$, tout élément de $\mathcal{A}_F$, est atomique.

H4 : T est atomique et tout élément non atomique de $\mathcal{A}$ a une partition en éléments atomiques.

Corollaire 2 :

Tout élément de $\mathcal{A}$, sauf $\emptyset$, a une partition unique en éléments atomiques ; le théorème 1 permet de calculer l'image de Y par X pour tous $X , Y \in \mathcal{A}$.

Nous appellerons désormais algorithmes les éléments d'une collection $\mathcal{A}$ qui remplit les conditions H_2 , H_3 et H_4 (donc H_1), ensembles de base les éléments de $\mathcal{A}_B$, algorithmes propres ceux de $\mathcal{A}_F$.

La condition donnée par C. Böhm pour qu'on puisse effectuer des calculs en parallèle se déduit de la remarque suivante :

Lemme 8 :

$$X\,[Y[Z]] = \cap\,\{W \,/\, \{V \,/\, X \subset F\ V\ W\} \cap \{V \,/\, Y \subset F\ Z\ V\} \neq \emptyset\}\ ;$$

$$(X\,[Y])\,[Z] = \cap\,\{W \,/\, \{V \,/\, X \subset F\ Y\ V\} \cap \{V \,/\, V \subset F\ Z\ W\} = 0\}.$$

Ce résultat, qu'on déduit essentiellement de H2 et des lemmes 1 et 3 met en évidence le fait que, dans le calcul de $X\,[Y\,[Z]]$ comme dans celui de $(X\,[Y])\,[Z]$, l'ordre des deux opérations V_1, $W_1 \rightsquigarrow V_1\,[W_1]$ est indifférent.

2 . ALGORITHMES UNIVERSELS.

Certains algorithmes peuvent être définis et utilisés quels que soient les ensembles de base choisis pour définir $\mathcal{A}$: ce sont les algorithmes universels ; nous ne citons ici que ceux dont nous nous servirons plus loin.

Pour les noter commodément, nous utilisons deux abréviations :

$$F_n\ X_1\ X_2\ \ldots\ X_n\ Y = F\ X_1\ (F\ X_2\ (\ldots\ (F\ X_n\ Y)\ \ldots))\quad ,$$

$$V\,[W_1,\ W_2,\ \ldots,\ W_n] = (\ldots\ ((V\,[W_1])\,[W_2])\ \ldots)\,[W_n]\quad ,$$

pour tout $n \in \mathbb{N}^+$.

Ainsi, dans l'Exemple 2 :

$$\leq_{\mathbb{N}} = (\bigcap_{\substack{i,j\in\mathbb{N}\\ i\leq j}} F_2\ \{i\}\ \{j\}\ \{\mathrm{VRAI}\}) \cap (\bigcap_{\substack{i,j\in\mathbb{N}\\ i>j}} F_2\ \{i\}\ \{j\}\ \{\mathrm{FAUX}\}),$$

$$\leq_{\mathbb{N}}\ [\{2\}\ ,\ \{1\}] = (\ \leq_{\mathbb{N}}\ [\{2\}])\ [\{1\}]\quad ,$$

$$W'\ [\{2\}\ ,\ \{3\}] = (W'\ [\{2\}])\ [\{3\}]\quad .$$

2.1. Algorithmes de composition.

Pour tous $m,\ n \in \mathbb{N}^+$, Comp_m^n =

$$\cap\ \{F_{m+n+1}\ V_0\ V_1\ \ldots\ V_m\ W_1\ \ldots\ W_n\ \ (V_0\ [V_1\ [W_1,\ldots,W_n],\ \ldots,\ V_m\ [W_1,\ldots,W_n]) \,/\, V_0,\ V_1,\ \ldots,\ V_m,\ W_1,\ldots,\ W_n \in \mathcal{A}\}.$$

(H4, en particulier, entraîne que $\mathrm{Comp}_m^n =$

$$\cap \{\dot{F}_{m+n+1} (F_m \quad Y_1 \ldots Y_m\, Z)\, (F_n\, X_1^1 \ldots X_n^1\, Y_1) \ldots (F_n\, X_1^m \ldots X_n^m\, Y_m)\, (X_1^1 \cap \ldots \cap X_1^m)\, (X_n^1 \cap \ldots \cap X_n^m)\, Z \;/\; X_i^{i'},\, Y_j^{j'},\, Z \in \mathcal{A}\})\ .$$

On montre que, si $W_1, \ldots, W_n$ sont atomiques, alors :

$$\mathrm{Comp}_m^n\, [V_0, V_1, \ldots, V_m, W_1, \ldots, W_n] =$$

$$V_0\, [V_1\, [W_1, \ldots, W_n], \ldots, V_m\, [W_1, \ldots, W_n]\,].$$

En particulier :

$$\mathrm{Comp}_1^1\, [V_0, V_1, W_1] = V_0\, [V_1\, [W_1]\,]\ ,$$

$$\mathrm{Comp}_2^1 [V_0, V_1, V_2, W_1] = V_0\, [V_1\, [W_1], V_2\, [W_1]\,]\ ,$$

$$\mathrm{Comp}_2^2\, [V_0, V_1, V_2, W_1, W_2] = V_0\, [V_1\, [W_1, W_2], V_2\, [W_1, W_2]\,]\ .$$

Ainsi,

Comp_1^1 (noté parfois <u>B</u>) est l'algorithme usuel de composition ;

Comp_2^1 est l'algorithme qui permet de traduire l'extension d'une opération scalaire en une opération vectorielle; par exemple, si + est l'algorithme de l'addition des entiers naturels

$$(+ = \bigcap_{i,j \in \mathbb{N}} F_2\ \{i\}\ \{j\}\ \{i+j\})$$

et si V_1 et V_2 traduisent des vecteurs d'entiers naturels (voir l'Exemple I), alors

$\mathrm{Comp}_2^1\ [+, V_1, V_2]$ est tel que

$$V_3\, [\{i\}] = (\mathrm{Comp}_2^1\ [+, V_1, V_2])\, [\{i\}] = +\, [V_1\, [\{i\}], V_2[\{i\}]]$$

pour tout $i \in \mathbb{N}$;

Comp_2^2 effectue la même extension, mais cette fois pour des matrices. Nous en verrons plus loin quelques applications.

2.2. Algorithmes de <u>projection</u>.

$$P_2^1 = \bigcap_X F_2\ X\ T\ X \quad ; \qquad P_2^2 = \bigcap_x F_2\ T\ X\ X \quad ;$$

et l'on a, pour tous X, Y :

$$P_2^1 [X, Y] = X \quad ; \quad P_2^2 [X, Y] = Y \quad .$$

Remarquons au passage que $P_2^1 [X]$ est l'algorithme qui donne, de tout Y, l'image X :

$$(P_2^1 [X]) [Y] = P_2^1 [X, Y] = X \quad .$$

2.3. Algorithme de <u>transposition</u>.

$$\underline{C} = \cap \{F_3 \; V \; X \; Y \; (V [Y, X]) \; / \; V, X, Y \in \mathcal{A}\} \cdot$$

On montre qu'on a :

$$\underline{C} = \cap \{F_3 \; (F_2 \; Y \; X \; Z) \; X \; Y \; Z \; / \; X, Y, Z \in \mathcal{A}\},$$

$$\underline{C} [V, X, Y] = V [Y, X] \quad \text{pour tous} \quad X, Y, V \; .$$

Ainsi, lorsque V traduit une matrice, $\underline{C} [V]$ en traduit la transposée, de sorte que $\underline{C} [V, \{i\}]$ traduit sa $i^{\text{ème}}$ colonne.

2.4. Algorithmes de <u>choix</u>.

Pour X_1 et X_2 donnés, $\Delta_{X_1,X_2} = F \; X_1 \; P_2^1 \cap F \; X_2 \; P_2^2$.

Ainsi,

$$\Delta_{X_1,X_2} [X, Y, Z] = (F \; X_1 \; P_2^1 \cap F \; X_2 \; P_2^2) [X, Y, Z] \quad ;$$

on montre que, si $X_1 \cap X_2 = \emptyset$ et si X est atomique et $\neq \emptyset$, on a :

$$\Delta_{X_1,X_2} [X, Y, Z] = \begin{array}{l} \text{si} \;\; X \subset X_1 \;\; \text{alors} \;\; Y \text{ , sinon} \\ \text{si} \;\; X \subset X_2 \;\; \text{alors} \;\; Z \text{ , sinon} \;\; T \; . \end{array}$$

Ces algorithmes permettent de traduire les tests usuels des langages de programmation ; par exemple, pour tout singleton $\{i\}$:

$$\Delta_{\mathbb{N}^+,\{0\}} [\{i\}, Y, Z] = \begin{array}{l} \text{si} \;\; \{i\} \subset \mathbb{N}^+ \;\text{alors} \;\; Y \text{ , sinon} \\ \text{si} \;\; \{i\} \subset \{0\} \;\text{alors} \;\; Z \text{ , sinon} \;\; T \; . \end{array}$$

2.5. Algorithmes d'<u>itération</u>.

Nous avons remarqué plus haut que tout algorithme est monotone dans le treillis (complet) $\mathcal{A}$, ordonné par l'inclusion ; l'opération $X, Y \mapsto X [Y]$ est suffisamment proche de l'application d'une fonction à son argument pour qu'on puisse transposer aisément des résultats connus.

En premier lieu, tout algorithme a un point fixe : pour tout Z , existe Y tel que $Z[Y] = Y$.

Ensuite, le plus grand point fixe de X (celui qui est le plus proche de l'indéfini T) est :

$$\bigcap_{n \in \mathbb{N}} Z^n [T] \quad \text{si et seulement si :}$$

(α) existe $m \in \mathbb{N}$ tel que $Z^m [T] \subset Z^{m+1} [T]$

(d'où l'égalité en vertu de la monotonie).

Ainsi, lorsque l'algorithme

$$P = \cap \{F_4\ Y\ V\ W\ X\ (\Delta_{X_1,X_2}\ [V[X],\ Y[V,\ W,\ W[X]],\ X])\ /\ Y,\ V,\ W,\ X \in \mathcal{A}\}$$

remplit la condition (α) , son plus grand point fixe est :

$\bigcap_{n \in \mathbb{N}} P^n [T]$, que nous désignons par $T_{q_{X_1,X_2}}$;

on a alors, car P est atomique, $T_{q_{X_1,X_2}} =$

$$P[T_{q_{X_1,X_2}}] = \cap \{F_3\ V\ W\ X\ (\Delta_{X_1,X_2}\ [V[X],\ T_{q_{X_1,X_2}}[V,W,W[X]],X])\}.$$

Supposons encore que $X_1 \cap X_2 = \emptyset$ et que $V[W^n[X]]$ est atomique pour tout $n \in \mathbb{N}$ et différent de $\emptyset$; alors :

$T_{q_{X_1,X_2}}[V,W,X]$ = si $V[X] \subset X_2$ alors X , sinon
si $V[X] \subset X_1$ alors $T_{q_{X_1,X_2}}[V,W,W[X]]$,
sinon T.

= si $V[X] \subset X_2$ alors X , sinon
si $V[X] \subset X_1$ alors :

(si $V[W[X]] \subset X_2$ alors $W[X]$, sinon
si $V[W[X]] \subset X_1$ alors
$T_{q_{X_1,X_2}}[V,\ W,\ W^2[X]]$, sinon T),

sinon T,
etc.

L'affectation «$X := \ldots$» mise à part, cela traduit parfaitement l'instruction de boucle des langages de programmation :

Tant que $V[X] \subset X_1$ faire $X := W[X]$
(et dès que $V[X] \not\subset X_2$ faire $X := X$) .

3. ALGORITHMES SERIELS, ALGORITHMES PARALLELES.

Soit M_0 une matrice booléenne carrée d'ordre N qu'on considère comme définissant une relation binaire sur l'ensemble $\{1, \ldots, N\}$ (ou encore un graphe G dont l'ensemble des sommets est $\{1, \ldots, N\}$). Le problème qu'on se pose est de trouver la fermeture réflexive et transitive de cette relation (ou encore le graphe des chemins, peut être vides, dans G) .

Un des algorithmes qu'on peut utiliser est le suivant, qu'on attribue à Warshall et à Roy .

1 - ajouter la matrice unité à M_0 ;

2 - prendre la 1ère colonne de la nouvelle matrice et sa 1ère ligne ; en faire le produit, soit la matrice M_1 ;

3 - ajouter M_1 à la matrice précédente ;

4 - refaire les étapes 2 et 3 en prenant successivement la 2ème colonne et la 2ème ligne, ..., la $N^{ème}$ colonne et la $N^{ème}$ ligne de la matrice obtenue à la fin de l'étape 3 précédente; la matrice obtenue au $N^{ème}$ coup est la matrice cherchée.

(Dans tout cela, l'addition et la multiplication des scalaires sont, respectivement, les opérations logiques <u>ou</u> et <u>et</u>; on les étend aux vecteurs et aux matrices comme on le fait dans l'algèbre linéaire).

Cette description de l'algorithme met assez bien en évidence les étapes du calcul qui peuvent être scindées en opérations exécutables simultanément et celles qui doivent se suivre dans un ordre déterminé. Mais elle a un grand défaut : elle s'adresse à un lecteur intelligent qui doit comprendre les allusions -exprimées de surcroît en français- et rétablir les indications passées sous silence parce que considérées comme évidentes.

Nous allons donc donner, dans notre théorie, une autre version de l'algorithme, laquelle peut être interprétée par un programme à défaut d'être parfaitement claire au lecteur non averti.

Pour aider cependant celui-ci autant que faire se peut nous donnons tout d'abord une première version rédigée à la manière des langages de programmation dits évolués.

Soit M_0 une matrice booléenne d'ordre N ;

P_1. $M_0 \longleftarrow \mathrm{Comp}_2^2$ [ou, M_0, I_N]

(où I_N est la matrice "unité" d'ordre N) ;

P_4. Pour K = 1, 2, ..., N , successivement faire :

P_2. $M_1 \longleftarrow \mathrm{Comp}_2^2$ [et, B [P_2^1, (C [M_0]) [K]], P_2^1 [M_0 [K]]

P_3. puis $M_0 \longleftarrow \mathrm{Comp}_2^2$[ou, M_0, M_1];

C'est bien ce qu'on veut faire ; en effet, pour tous I, J atomiques,

- par l'instruction P_1 on obtient une nouvelle matrice $M_0^{(0)}$ telle que

$M_0^{(0)}$[I,J] =(Comp_2^2 [ou, M_0, I_N]) [I,J]

= ou[M_0 [I,J], I_N [I,J]] ;

- par l'instruction P_2, on obtient une matrice M_1 telle que

M_1 [I,J] =

(Comp_2^2 [et, B [P_2^1, (C [M_0]) [K]], P_2^1 [M_0[K]]]) [I,J]

= et [(B [P_2^1, (C [M_0]) [K]]) [I,J], (P_2^1 [M_0 [K]]) [I,J]]

= et [P_2^1 [((C [M_0]) [K]) [I], J], (P_2^1 [M_0[K], I]) [J]]

= et [C [M_0, K, J], (M_0 [K]) [J]]

= et [M_0 [J, K], M_0 [K, J]] ;

- par l'instruction P_3, on obtient enfin une nouvelle version M_0' de M_0 telle que

M_0' [I, J] =

(Comp_2^2 [ou, M_0, M_1]) [I, J]

= ou [M_0 [I, J], M_1 [I, J]] .

Pour obtenir la version définitive, il suffit de considérer que l'algorithme opère constamment sur un quadruplet

M = <M_0, M_1, N, K>

abréviation de $\underset{\circ}{<} M_0, M_1, N, K >_3$ égal lui-même à

$$F\,\{0\}\,M_0 \cap F\,\{1\}\,M_1 \cap F\,\{2\}\,N \cap F\,\{3\}\,K$$

(voir l'Exemple 1).

Supposons donc qu'on possède les algorithmes $\underline{et}$ et $\underline{ou}$ (sur les scalaires) et la matrice unité I_N et posons les définitions suivantes (pour alléger l'écriture, nous remplaçons {0} par $\underline{0}$, {3} par $\underline{3}$, etc ...) :

$$\mathcal{A} = \langle Comp_2^2\,[\underline{et}, M\,[\underline{0}], M[\underline{1}]], M\,[\underline{1}], M\,[\underline{2}], M\,[\underline{3}]]\rangle;$$

$$\mathcal{B} = \langle M\,[\underline{0}], Comp_2^2\,[\underline{et}, \underline{B}\,[P_2^1, (\underline{C}\,[M\,[\underline{0}]])\,[M\,[\underline{3}]]], P_2^1\,[(M\,[\underline{0}])\,[M\,[\underline{3}]]]], M\,[\underline{2}], \underline{suc}\,[M\,[\underline{3}]] \rangle ;$$

$$\mathcal{C} = \langle Comp_2^2\,[\underline{ou}, M\,[\underline{0}], M\,[\underline{1}]], M\,[\underline{1}], M\,[\underline{2}], M\,[\underline{3}] \rangle ;$$

(où $\underline{suc} = \bigcap_{i\in\mathbb{N}} \underline{Fi}\ \ \underline{i+1}$).

$\mathcal{A}$, $\mathcal{B}$ et $\mathcal{C}$ sont donc les résultats obtenus en appliquant à un quadruplet M les algorithmes qui traduisent les étapes 1, 2 et 3 ; ces algorithmes sont donc, respectivement ,

$$\mathcal{A}' = \bigcap_M F\,M\,\mathcal{A}, \quad \mathcal{B}' = \bigcap_M F\,M\,\mathcal{B}, \quad \mathcal{C}' = F\,M\,\mathcal{C}.$$

L'algorithme correspondant à l'étape 4 est celui qui fait appliquer $\mathcal{B}'$ puis $\mathcal{C}'$, c'est à dire $(\underline{C}\,[\underline{B}])\,[\mathcal{B}', \mathcal{C}']$, tant que l'image $\mathcal{D} = \dot{-}\,[M\,[\underline{3}], M\,[\underline{2}]]$ que donne de M l'algorithme $\mathcal{D}' = \bigcap_M F\,M\,\mathcal{D}$ est incluse dans {0}
(en notations usuelles $M_3 \dot{-} M_2 = 0$, i.e. $K \dot{-} N = 0$; $i \dot{-} j$ étant 0 ssi $i \leq j$).
L'image fournie d'un quadruplet M par l'algorithme correspondant à l'étape 4 est donc

$$\mathcal{E} = T_{q_{\underline{0},\mathbb{N}^+}}\,[\mathcal{D}', (\underline{C}\,[\underline{B}])\,[\mathcal{B}', \mathcal{C}'], M\,] ;$$

l'algorithme lui-même est $\mathcal{E}' = \bigcap_M F\,M\,\mathcal{E}$.

Le composé de $\mathcal{A}'$ et de $\mathcal{E}'$, $(\underline{C}\,[\underline{B}])\,[\mathcal{A}', \mathcal{E}']$, fournit du quadruplet initial $\langle X, I_Z, Z, 1\rangle$ l'image

$$\mathcal{F} = (\underline{C}\,[\underline{B}])\,[\mathcal{A}', \mathcal{E}', \langle X, I_Z, Z, 1 \rangle] ;$$

l'algorithme au grand complet est donc :

$\mathcal{F}' = \bigcap_{X,Z} F_2\, X\, Z\, \mathcal{F}$; l'image qu'il fournit d'une matrice M_0 d'ordre N est $\mathcal{F}'[M_0,\ N]$.

Enfin, le "programme" destiné à résoudre le problème initial est le couple formé par l'ensemble d'algorithmes

$\{\underline{suc},\ \underline{et},\ \underline{ou},\ I_N,$ (pour tout $N \in \mathbb{N}$), $\mathcal{A},\ \mathcal{A}',\ \mathcal{B},\ \mathcal{B}',\ \mathcal{C},\ \mathcal{C}',\ \mathcal{D},$

$\mathcal{D}',\ \mathcal{E},\ \mathcal{E}',\ \mathcal{F},\ \mathcal{F}'\}$ et l'algorithme distingué $\mathcal{F}'$ (le premier qu'on doit appliquer).

L'éxécu tion du calcul $\mathcal{F}'[M_0,\ \underline{4}]$ pour M_0 traduisant $\begin{bmatrix} FFFV \\ VFFF \\ FFVF \\ FVVF \end{bmatrix}$, par exemple, est parfaitement décrit par <u>une suite de transformations de l'expression</u> $\mathcal{F}'[M_0,\ \underline{4}]$:

$\mathcal{F}'[M_0,\ \underline{4}]$

$= (\underline{C}[\underline{B}])\,[\mathcal{A}',\ \mathcal{E}',\ <M_0,\ I_4,\ \underline{4},\ \underline{1}>]$

(résultat de la substitution de M_0 et $\underline{4}$ à X et Z dans $\mathcal{F}$)

$= \mathcal{E}'[\mathcal{A}'[<M_0,\ I_4,\ \underline{4},\ \underline{1}>]]$

(α) $= \mathcal{E}'[<M_0^{(1)},\ I_4,\ \underline{4},\ \underline{1}>]$

$= T_{\underline{q_0},\mathbb{N}^+}[\mathcal{D}',\ (\underline{C}[\underline{B}])\,[\mathcal{B}',\mathcal{C}'],\ <M_0^{(1)},\ I_4,\ \underline{4},\ \underline{1}>]$

$= T_{\underline{q_0},\mathbb{N}^+}[\mathcal{D}',\ (\underline{C}[\underline{B}])\,[\mathcal{B}',\mathcal{C}'],\ ((\underline{C}[\underline{B}])\,[\mathcal{B}',\mathcal{C}'])$
$[<M_0^{(1)},\ I_4,\ \underline{4},\ \underline{1}>]]$

(puisque $\mathcal{D}'[<M_0^{(1)},\ I_4,\ \underline{4},\ \underline{1}>] = \dot{-}[1\ ,\ 4] = \underline{0} \subset \underline{0}$) ;

(L'algorithme $((\underline{C}[\underline{B}])\,[\mathcal{B}',\ \mathcal{C}'])\,[<M_0^{(1)},\ I_4,\ \underline{4},\ \underline{1}>]$

$= \mathcal{C}'[\mathcal{B}'[<M_0^{(1)},\ I_4,\ \underline{4},\ \underline{1}>]]$

(β_1) $= \mathcal{C}'[<M_0^{(1)},\ M_1^{(1)},\ \underline{4},\ \underline{2}>]$

$=<Comp_2^2\ [\underline{et},\ M_0^{(1)},\ M_1^{(1)}],\ M_1^{(1)},\ \underline{4},\ \underline{2}>$

(γ_1) $= <M_0^{(2)},\ M_1^{(1)},\ \underline{4},\ \underline{2}>$)

et on retombe sur

$t_{\underline{q_0},\mathbb{N}^+}[\mathcal{D}',\ (\underline{C}[\underline{B}])\,[\mathcal{B}',\ \mathcal{C}'],\ <M_0^{(2)}\ ,\ M_1^{(1)},\ \underline{4},\ \underline{2}>]$

qui se transforme de la même façon en

$$T_{\underline{q_0}, \aleph^+} [\mathcal{D}', (\underline{C}\ [\underline{B}])\ [\mathcal{B}', \mathcal{C}'], < M_0^{(3)}, M_1^{(2)}, \underline{4}, \underline{3} >]$$
$$= T_{\underline{q_0}, \aleph^+} [\mathcal{D}', (\underline{C}\ [\underline{B}])\ [\mathcal{B}', \mathcal{C}'], < M_0^{(4)}, M_1^{(3)}, \underline{4}, \underline{4} >]$$
$$= T_{\underline{q_0}, \aleph^+} [\mathcal{D}', (\underline{C}\ [B])\ [\mathcal{B}', \mathcal{C}'], < M_0^{(5)}, M_1^{(4)}, \underline{4}, \underline{5} >]$$
$$= < M_0^{(5)}, M_1^{(4)}, \underline{4}, \underline{5} >$$

(puisque $\mathcal{D}' [< M_0^{(5)}, M_1^{(4)}, \underline{4}, \underline{5} >] = \dot{-}[\underline{5}, \underline{4}] = \underline{1} \subset \aleph^+$).

La matrice cherchée, $M_0^{(5)}$ est donc :

$$< M_0^{(5)}, M_1^{(4)}, \underline{4}, \underline{5} > [\underline{0}] .$$

Il est clair que chacun des passages d'une expression à l'autre doit être exécuté dans l'ordre même où il apparait ici ; la suite de ces transformations est donc l'image d'un calcul en séquence. Nombre d'entre elles sont des substitutions de définissant à défini qui traduisent des appels de procédures ; naturellement, comme dans le calcul des prédicats, il s'agit de substitutions simultanées d'arguments aux places libres correspondantes.

Restent les transformations qui fournissent α, β_1, γ_1, ..., β_4, γ_4. Considérons par exemple celle qui fournit β_1. L'algorithme $\mathcal{B}'$ y décrit le calcul qui transforme le quadruplet $< M_0^{(1)}, I_4, \underline{4}, \underline{1} >$ dans le quadruplet $< M_0^{(1)}, M_1^{(1)}, \underline{4}, \underline{2} >$.

Tout d'abord, il est possible d'effectuer simultanément la transformation de I_4 en $M_1^{(1)}$ et celle de $\underline{1}$ en $\underline{2}$; ensuite, la première de ces opérations peut être éclatée en opérations élémentaires simultanées. En effet elle équivaut à

$$\mathrm{Comp}_2^2\ [\underline{et}, \underline{B}\ [P_2^1, (\underline{C}\ [M\ [\underline{0}]])\ [M\ [\underline{3}]]], P_2^1\ [(M\ [\underline{0}])\ [M\ [\underline{3}]]]] ,$$

en l'occurrence :

$$\mathrm{Comp}_2^2\ [\underline{et}, \underline{B}\ [P_2^1, (\underline{C}\ [M_0^1])\ [\underline{1}]], P_2^1\ [M_0^1\ [\underline{1}]]] ,$$

donc, comme on l'a vu plus haut

$$\mathrm{Comp}_2^2\ [\underline{et}, P, Q]$$

où P (resp$^{\underline{t}}$ Q) est la matrice dont toutes les colonnes (resp$^{\underline{t}}$ lignes) sont identiques à la première colonne (respt ligne) de $M_0^{(1)}$. Ces deux constructions peuvent être effectuées dans un ordre quelconque ; celles-ci faites, rien n'empêche qu'on ne décompose l'algorithme Comp_2^2 en N^2 opérations effectuées simultanément sur les éléments de P et de Q , ou encore en une suite de N opérations globales sur les lignes ou les colonnes de P et de Q.

L'Informatique considère aujourd'hui que l'éxécutant auquel s'adresse l'usager est un software s'appuyant sur un hardware. Alors que la machine sait seulement effectuer quelques opérations triviales et les enchainer, sur ordre, de façon simple, l'éxécutant peut appliquer, suivant le programme qu'on lui fournit, des consignes beaucoup plus élaborées. Il peut par exemple décider à tout moment, compte tenu des ressources dont il dispose dans l'instant, qu'un calcul complexe sera décomposé en une suite d'opérations ou qu'il sera réalisé par un ensemble d'opérations simultanées.

Encore faut-il qu'on lui fournisse le moyen de faire ce choix. La théorie que nous venons de présenter le permet puisqu'elle bannit jusqu'aux affectations qui, dans les langages de programmation, fournissent l'occasion d'empiéter sur les attributions de l'éxécutant. Et il suffit de choisir convenablement les ensembles de base pour qu'elle permette de traduire, non seulement des calculs, comme nous l'avons montré, mais n'importe quel comportement algorithmique.

PARALLEL PROCESSING AND COMPUTER SEMANTIC MEMORY

F. Sirovich

Istituto di Elaborazione della Informazione(+)

The interest for computer semantic memory arises in many fields. Semantic memory is needed for

i) Data representation for Question Answering systems (see Shapiro[1]), where one needs
   - most flexible data structures to have maximum flexibility in answering questions,
   - deductive capabilities to get answers not explicitly contained in memory.

ii) Data representation for natural language comprehension. (See Quillian[2]).

iii) Subject matter storage for Computer Aided Instruction. (See Charbonnel[3]).

iv) Data base for Artificial Intelligence systems, where one needs to represent the webb of interrelations occurring among the objects of a given problem domain, which is generally quite limited in size but still richly structured.

All approaches to such representation problems are centered around the notion of "concept", and accordingly the problem of semantic memory can be defined as consisting of

1) Designing an efficient way to encode concepts in computer memory;

(+) Consiglio Nazionale delle Ricerche, Via S. Maria, 46, 56100 Pisa, Italy.

2) Designing effective procedures to store and retrieve information in the computer memory, so that the procedures which have to operate on such concepts can operate effectively by beeing provided with all and only relevant information.

Retrieval processes constitute a crucial problem and they have to be efficient to respond to the following tasks.

- Associative retrieval. An item has to be retrieved by means of a partial description of it;
- Concept discrimination: The system should be able to identify a concept occurrence and discriminate it from other similar concepts;
- Providing a uniform representation for data and system's programs. This requirement is particularly important for applications to automatic problem solving, an area in which many interesting ideas (see for example Minsky[4]) are being investigated about knowledge representation (declarative vs. procedural) and system organization (hierarchy vs. heterarchy vs. production systems).

Production system is an interesting system organization that has been introduced and used by Newell[5], as a tool for modelling human problem solving behavior. I will try to show that it can be extended to provide a very general framework for semantic memory. Moreover, such an organization leads to a particular kind of parallel processing which it is interesting to comment on.

PRODUCTION SYSTEMS

Formally, a *production system*, or *combinatorial system*, Π consists of a single non empty word, called the *axiom* of Π, and a finite set of productions called the *productions of* Π. A production is a recursive binary predicate $R^{g\,h\,k}_{\bar{g}\,\bar{h}\,\bar{k}}(X,Y)$ which is true for given words X_0, Y_0 if and only if (possibly empty) words P_0, Q_0 exist such that

$$X_0 = g\,P_0\,h\,Q_0\,k$$
$$Y_0 = \bar{g}\,P_0\,\bar{h}\,Q_0\,\bar{k}$$

A production will be symbolized as

$gPhQk \rightarrow \bar{g}P\bar{h}Q\bar{k}$.

The alphabet A of Π consists of all letters that occur either in the axiom of Π or in the g, h, k, $\bar{g}$, $\bar{h}$, $\bar{k}$ that define the productions of Π. By word on Π we mean a word in which only letters from the alphabet of Π appear.

Combinatorial systems are essentially symbol manipulation systems. In fact, we can say that if $R(X_0, Y_0)$ is true, then Y_0 is a consequence of X_0 with respect to R. Immediately, we define as proof in a combinatorial system Π a finite sequence $X_1, X_2, \ldots, X_m$ of words such that X_1 is the axiom of Π and for each $i, 1 < i \leq m$, X_i is a consequence of X_{i-1} with respect to one of the productions of Π. X_m is then called a theorem of Π. Combinatorial systems then formalize the procedure to follow in order to obtain theorems starting from an axiom and by use of inference rules which are represented by the productions. An important point is that a particular kind of combinatorial system (a semi-Thue system) can imitate the behavior of a given simple Turing machine. Moreover, one of the characteristics of such a system is that it is monogenic, i.e., at most one production can be applied to any given theorem which therefore has at most one consequence in II. This is of course needed in order to imitate a sequential machine like the Turing machine. In fact, roughly speaking, the succession of theorems obtainable in such a system corresponds to the successive instantaneous descriptions of the imitated Turing machine.

Such a combinatorial system can therefore be seen as a program which operates on the axiom corresponding to the starting description of the Turing machine. The combinatorial system operates by discovering the production which applies to the axiom (a process which we will call matching), determines the words P and Q occurring in the left side (condition side) of the production, and finally rewrites the axiom as $\bar{g}P\bar{h}Q\bar{k}$, which is the first obtained theorem of II and on which the operation is repeated again.

The production systems used by Newell allow for a more complex matching to be specified in the condition side and more sophisticated symbol manipulations to be executed in the right side (action side) of the production. Essentially these extensions are introduced in order to achieve conciseness in expressing programs, and they consist in the following:

i) Variables are allowed in both sides of a production. When a variable is first found during the matching process, it is bound with the corresponding element of the matching theorem. Any subsequent occurence of the variable, in both sides of this production, is substituted by the bound element. Obviously, productions with variables are a sort of rule-of-inference scheme which is equivalent to the set of productions that can be obtained from it by making all the consistent substitutions for variables (i.e., it is an estension of the role of the P and Q words).

ii) Negative conditions are allowed that again yield to inference schemes. In fact, negative conditions define the class of strings accepted by the condition side as the complement, with respect to A^*, of a certain string (or set of strings).

iii) Multiple conditions are specified by means of the logic operators AND and OR (and the already mentioned NOT).

iv) Ordering is imposed on Newell's production systems, in order to force the system to be monogenic (the first production whose condition is satisfied is executed). This shortcut again yields to a more compact system because the condition sides should otherwise be extended in order to assure that they are mutually exclusive in any circumstance.

v) A stack of strings (theorems) is introduced on which the production system is called to operate. During each match, the strings in the stack are searched starting from the top. When a condition is satisfied, the strings it matches are brought on the top of the stack before the action is executed.

vi) The action side specifies elementary operations (to be executed on the strings on the top of the stack) like modifying or deleting some elements of the strings or inserting new elements. Moreover, an action may require the insertion of one or more strings on the top of the stack, as well as the execution of elementary operators (e.g., simple operators we assume to be primitive for the system, or operators we are not interested to represent explicitly).

Newell has shown that this kind of organization can effectively be used to model human problem solving behavior and hence is a useful programming language. If we want to generalize and extend this method to be the support not only for control structures but also for all the knowledge available to the system, we have to tackle two related problems.

The first problem is to design an effective way to operate a production system. In fact, using the logical ordering of the productions as an access scheme is unfeasible even for systems of moderate size. If instead we could use a discrimination tree access, we would have an accessing time roughly proportional to the logarithm of the number of productions.

The second problem is to design a symbolic language to represent semantic information. Such a representation language must in a sense lead to the discrimination tree access and, at the same time, provide the flexibility and power which is needed for a sophisticated representation of any kind of knowledge.

A PROPOSED REPRESENTATION LANGUAGE

The representation language proposed in the sequel is centered around the notions of concept and concept token. A <u>concept</u> can be defined as a set, i.e., a collection of objects called <u>concept tokens.</u> Thus, the representation language will be seen as a language to define sets.

A simple solution to the problem of defining a set makes use of the well known <u>principle of abstraction</u>. A formula P(x) defines a set with the convention that the elements of the set are exactly those objects for which the formula is true.

The defining property P(x) can also be seen as a <u>membership criterion</u> by means of which an information processing system is able to decide whether a given object belongs to the concept or, more generally, to discriminate the occurrence of a concept token. The representation language has then to be structured in such a way that the interpretation of the set definition in procedural terms is simple.

The fact that an object has a certain property can be conveniently represented by its membership to a suitable concept (precisely, to the set of objects which have that property). As a trivial example consider the following.

example-1 :: = ((IS-A integer))

We have here the representation of a concept (incidentally called example-1) which is defined by stating that its tokens belong to the set of the integers. The membership relation is here denoted by the symbol IS-A. Of course it is trivial to set up mechanisms to combine properties by use of the logic operators AND,OR and NOT.

An interesting kind of concept is a relation i.e., a set whose elements are (ordered) n-tuples of components. A typical way to define a relation consists of defining a cartesian product, in which the relation is contained, along with the defining property of the relation itself. Such a defining property will consist in specifying that some m-tuples of components of the relation belong to some other suitable relation. The relation components are identified, in the representation language, by means of local variables. For example,

```
part-of-named ::= ((HAS-AS-A-PART   x1)
                   (HAS-AS-A-PART   x2)
                   (HAS-AS-A-PART   x3)
                   (x1 HAS-AS-A-PART  x3)
                   (x3 IS-A  x2)
```

specifies that tokens of part-of-named are triplets (x1,x2,x3) such that x3 is a component of x1 and in addition x3 belongs to the concept x2.

In every day's language, relations are often used in a way we could call "functional", i.e., they are used to designate the image of a given element under the function extracted from the relation. In the proposed representation language, relations can be used in an analogous fashion. We can designate, for example, by the following expression

(part-of-named car engine)

the engine of a car (i.e., that component of a car which is a token of the concept engine).

It is obvious that the use of the notion of image of a function has the only goal of producing less awkward definitions of relations. As an example, consider the definition of the character E.

```
E character ::=((HAS-AS-A-PART x1)
                (x1 IS-A vertical-stroke)
                (HAS-AS-A-PART x2)
                (x2 IS-A horizontal-stroke)
                ((top x1) IS-SAME (left-ext x2))
                (HAS-AS-A-PART x3)
                (x3 IS-A horizontal-stroke)
                ((middle x1) IS-SAME (left-ext x3))
                (HAS-AS-A-PART x4)
                (x4 IS-A horizontal-stroke)
                ((bottom x1) IS-SAME (left-ext x4)))
```

The character E is here described as a relation (let us call it semantic relation to avoid confusion with the system relations like IS-A). Such a semantic relation has four coordinates x1, x2, x3 and x4, which are defined in succession, along with the constraints on them.

In conclusion, the defining property of a concept is expressed in terms of other concepts by use of few system relations. The defining property can be seen as a criterion which allows the discrimination of the occurrence of a concept token. What kind of processing is then needed to implement such a discrimination?

Firstly, we need to be able to check whether a specified system relation occurs between two given objects. For example, it is needed to check whether some property is present in the description of the given object, which of course will be described in the same representation language.

Secondly, we need to be able to "compute" the image of a function, i.e., to find the missing coordinates of a token of some relation. Such a computation may simply consist in an associative selection of the memory, or even in the invocation of a stimulus recognition process.

Whenever, the defining criterion is satisfied by the given object representation, we can say that the given object belongs to the concept. The defining criterion can therefore be seen as the condition side of a production. The action side of the production will encode the fact that the given object belongs to the concept.

We can see the semantic memory as a large production system whose conditions play the role of conditions under which particular symbol manipulations (called actions) occur, concept tokens are discriminated, or, more generally, information is gathered from the memory. The collection of action sides actually represents the knowledge stored in the memory, and the condition sides represent the way to access to such a knowledge according to contextual relevance.

We will now turn the attention to the processes needed to effectively retrieve information from such a semantic memory.

INFORMATION PROCESSING IN THE SEMANTIC MEMORY

The process to retrieve the relevant condition side is the crucial point of such a system because of the large number of productions. The problem can be sketchily posed in the following way.

As we have already said, we have a stack memory which contains the analogous of the last found theorem of a monogenic combinatorial system. The stack actually contains symbol structures which act as as information request to the semantic memory. The symbol structures are input to the retrieval process which has the task of locating the first production whose condition side matches some of the symbol structures contained in the stack. Not all the symbol structures in the stack are relevant to select the eventual production.

The retrieval process can be seen as carried on by a multiprocessor organized as a binary tree. Let me briefly describe such a processor by means of the example shown in Fig.1. The nodes of the tree represent processors that are able to perform a simple information processing. The edges of the tree represent channels on which information is sent from one processor to others. The processors can be thought of as transducers activated by the reception of an input.

Let us suppose that node N1 is the root of the discrimination tree. The processor then receives as input the symbol structures contained in the stack.

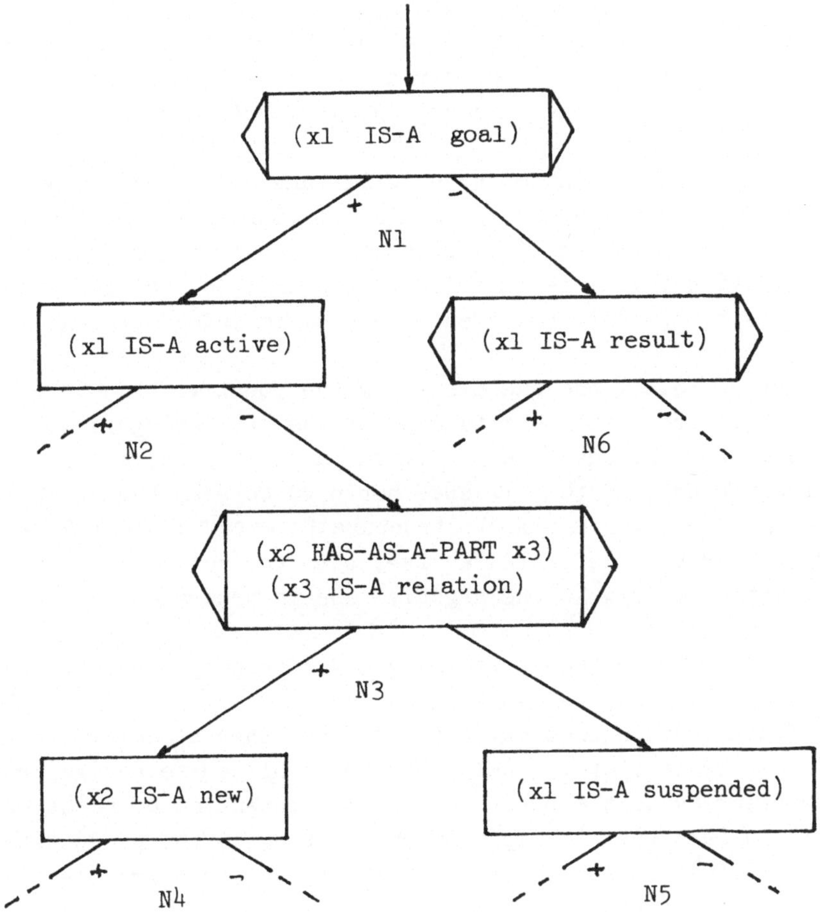

Fig. 1 - A typical semantic memory processor. Wedged nodes represent attention nodes as distinct from discrimination nodes. See the text for explanations.

The function performed by the processor consists in a simple pattern matching on the input. It has to scan the stack from the top to find a symbol structure which has the property of being a goal in its description. If such a symbol structure is present, then all productions which contain such a condition have to be tried and therefore N1 sends out along the positive branch the same symbol structures it received, plus the binding between the variable x1 and the just discovered symbol structure.

The subtree dominated by N2 will take care of discriminating among conditions which contain goals. Of course, it may very well happen that even if a goal is present in the stack, no condition side involving a goal is satisfied. This implies that the negative branch has to be activated even if a goal is indeed present.

Let us follow the example along the positive branch. N2 has again to perform a simple matching but the processing is quite different from the one done by N1. In fact, N2 receives the variable x1 already bound, so that it does not need to scan the stack and already knows on which symbol structure to execute the match. If x1 has the property of being active, then the positive branch only is activated, otherwise the negative one is activated.

N1 and N2 are then performing a different processing. The former will be called attention node and the latter discrimination node. The attention node may activate one other processor only (the negative one) or two new processors, of course assigning a higher priority to the positive one. The intention of an attention node is in fact that if imposing an ordering on the productions. The discrimination node instead activates only one processor according to the result of the test.

Let me now briefly comment on the accessing scheme I have sketchily described(+). A first interesting feature is the organization as a discrimination tree. The condition sides are factorized according to their common parts. Thus, we avoid the extensive repetitions of the same test which would occur in a serial scan of the productions. The access time can be expected to grow almost with the logarithm of the number of productions of the system, instead of linearly. The tree will certainly not be balanced especially because it has to implement the logical ordering on the productions, but the situation will tend to settle to a good compromise with the increase of the number of productions.

(+)The interested reader can find more details about the retrieval process and a description of the storage process in another paper by the author[6].

Secondly, the tree can fruitfully be seen as a multiprocessor system. It is profitable to think of the tree as a parallel processor for the following reasons.

Parallelism can be seen as a useful tool for describing (or programming) the access scheme. We need to describe the computation of the various processors and then to describe simply how they interact. Of course the tree could have been seen as a passive net, and we could have described the behavior of a single more complex processor which "travels" on the tree up and down, searching for the desired production. The control machinery which allows such a processor to carry on this search may come to be of substantial complexity and may constitute a difficulty both from the point of view of a formal description of the accessing scheme, and from the point of view of the programming of the system on a digital computer. There is a general feeling that concurrent processing may be of help as a programming tool for large, complex, multi-task systems. Many programming languages are being developed which allow the user of a serial digital computer to define and work in such a framework (see for example PASCAL and SAIL [7,8]).

CONCLUSIONS

The goal in designing the proposed semantic memory was not that of modelling nature, but that of performing a task which, as a coincidence, is also a task very well performed by natural beings. Yet it seems to me that the very nature of the task leads to an organization of the system which far closer to nature than the tools we are starting from. I find very encouraging that the present organization is similar to the model of the human memory sketched by Dr. Young, and to the mechanism proposed by Feigenbaum and Simon to model human learning of lists of words,[9] which is an almost perfect model explaining all experimental results about such a behavior.

REFERENCES

1. Shapiro, S.C., A net structure for semantic information storage, deduction and retrieval, Proc. Int. Jt. Conf. Art. Intel., 512, 1971.

2. Quillian, M.R., The Teachable Language Comprehender: a simulation program and a theory of language, Comm. ACM, 12,459,1969.

3. Charbonnel, J.F., AI in Cai: an Artificial Intelligence approach to Computer Assisted Instruction, IEEE Trans. on Man-Machine Systems, 11,190,1970.

4. Minsky,M. and Papert, S., Artificial Intelligence progress report, AI Memo No. 252, MIT, Cambridge, Mass., 1972.

5. Newell, A., A theoretical exploration of mechanisms for coding the stimulus, in Coding Processes in Human Memory, Melton and Martin Eds., Winston, 373, 1972.

6. Sirovich, F., Some ideas on semantic memory in automatic learning of heuristics,Atti Convegno AICA di Informatica Teorica, 413,1973.

7. Hansen, P.B., Concurrent programming concepts, invited paper to be published in ACM Computing Surveys, 1973.

8. Feldman, J.A., Low J.R., Swineheart D.C. and Taylor R.H., Recent developments in SAIL - An ALGOL-based language for artificial intelligence, Proc. Fall Jt. Comp. Conf., 1193,1972.

9. Feigenbaum, E.A., and Simon, H.A., Generalization of an elementary perceiving and memorizing machine, Proc. of IFIP Congr., 401, 1962.

LARGE SCALE PARALLEL PROCESSING IN EXPLOITATION SYSTEMS

Shuhei AIDA

University of Electro-Communications
Chofu-Shi, Tokyo 182, Japan

1. Fundamental Concepts

In this paper, the author would like to discuss some of the fundamental points concerning the environmental issues as these relate to the field of technological development, that is Eco-Technology (*).

It goes without saying that our awareness of the environmental issue is stimulated by the rapid growth in technology. It is often pointed out that if the technological innovations continue to the extent they do today, the year 2000 will be the year of technological control. It goes beyond my imagination what existence will be like under these conditions.

An understanding of technology and what it means to man and his environment is needed. We must understand how man plans to develop technology and how he will apply it to his environment.

Suppose we take the virtues of the computer as one of the driving forces in the rapid advances seen in technology. It is clear that we owe a great deal to its capacity to perform operations. The speed of the computer allows the establishment of new relationships that we have found in physics and chemistry. Computer advances provided the means for the conceptualization of new technology. Furthermore, new technologies seem to provide the impetus for change in the social system which moves toward greater harmony between nature, man, and technology.

Prior to this stage in man's development society seemed to operate in a mediun where nature and technology were in

(*) Eco-Technology is the abbreviation of Ecological Technology. "Informatics in Eco-Technology" Proceedings of NATO Advanced Institute on Information Sciences, University of Pittsburgh, August, 1972.

equilibrium. This equilibrium provided also a degree of diversity for human expression and existence. The further expansion of technology, however, threatens this balance with a destructive force—equivalent to the speed in which a computer functions.

The fibre of modern civilization rests on three forces, namely, matter, energy and information. Our existence rests on these forces. The author has represented these forces in Figure 1, and the author would like to refer to this as the "human development space". It's a conceptual tri-dimensional space, and a large scale parallel processing system from the points of information sciences.

Let me briefly expand on the major components of this configuration.

<u>Structure/ Matter-Energy Component</u> (M-E)

This represents the area of great utility and upon which man built his technology to capitalize on the energy resources available to him. The greatest expansion has occurred in this area, and where industrial growth is manifested. This area provides us with the GNP index as well as the pollutants which we must control.

<u>Structure/ Matter-Information Component</u> (M-I)

This represents the utilization of industrial resoures through information systems which make known the power available from the environment. Information systems of greater power and versatility further increase our capability to capture our resouces. The capability has been significantly extended through the computer.

The author should point out here that the M-E component depends on the development of hardware while the M-I component is highly dependent on advances in software. In either case, however, both are bound by the natural resources that are available for their development.

<u>Structure/ Energy-Information Component</u> (E-I)

This is the unexplored component. This component marries the power available in nature with the inherent capability of man to control intellectually this power. The author refers to this as eco-energy or intelligenetics, the ability of man to aply his own intellect to control the environmental forces about him. The E-I component provides the basis for what is necessary to the development of those characteristics which typify human function, e.g. artificial intelligence.

As seen, ecology is highly related to the components that we have discussed, namely, energy, matter, and information. Knowledge about these components is the foundation for ecology. The energy-information component provides the central basis for the development of a symbiotic relation between the three dimentional space. Admittingly, energy-information is limited to matter, but only to the extent that the media through which energy-information finds expression. Ultimately, the limits will be obscured. New methodologies will be uncovered where organismic and machine systems will arrive at a true co-existence. Ecological balance and its reality in every realm of life would be the prevailing reality.

Here the author would like to extend the previous formulations to the matter of systems engineering.

In Figure 1, the author has shown the three primary components. The author would like to refer these as matter, energy and information for the purpose of the ensuring discussion.

Structure is a conceptual space which consists of matter, energy and information. The structure includes the social arrangements (society) that make the physical substance of the environment function, as well as the biological and mechanical counterparts that enable the organisms to react and mold the environment.

By energy the author means the raw substance and power required for the organism, institution, or technology to functoon.

By information the author refers to the means through which the organism relates and creates aspects of his own nature as well as that of the environment.

The dimensional space indicated in Figure 1 is the total environment that man lives in. Each environment is characterized by a particular climate by the energy flow through it, and by the function of information which applies and utilizes this energy.

The concept of tri-dimensional space was initially proposed by Leonardo da Vinci in his descriptions of the anatomy of the human body. Da Vinci saw the human body as a spatial entity— a three dimensional spatial entity. Within this concept, the flow of information is conceived as existing and operating within an inner space, interacting with and influencing the total organismic structure. Physiology emerged as a science with this conceptual view of the human organism. Da Vinci's farsighted view of the organism perceived the anatomical, physiological as well as the ecological charactaristics of the human body.

The three factors which the author has proposed are needed if we are to explore the interdisciplinary nature of the subject we are here to discuss. Energy cannot be considered separate from its environmental/informational counterparts. Energy provides all of man's needs and goes beyond these when we consider the forces of the physical environment which are highly dependent on it.

Energy is all pervasive, perhaps more than most of us realize. Beyond the fact that energy causes all matter, energy finds its expression in different kinds of forms—physically and psychologically. It is in the psychological area where the greatest challenge exists and where our knowledge is most deficient. As a matter of fact, it is in this area in modern science where there is great despair and challenge.

Figure 2 shows the transition of the Society and the Basic Science in this concept. At this point let us extend our

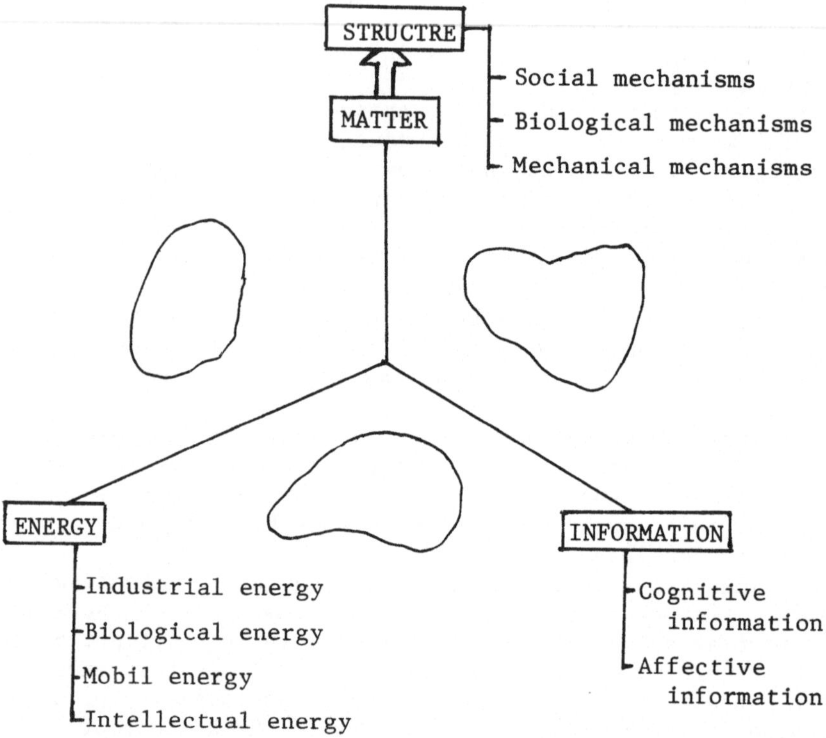

Fig. 1) Development space by the structure, energy and information (SEI space)

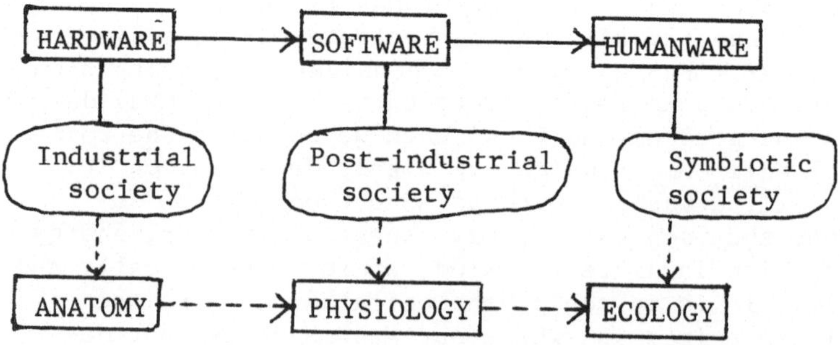

Fig. 2) Transition of society and basic science

discussion by relating the concepts of structure, energy, and information so some specific fields.

2. Urban Information System

Urban information system is for the urbanization with the concepts of Eco-Technology. In this example, the author would like to refer to a Plan for Remodeling the Japanese Archipelago. Some of the areas of the Archipelago are studied with cities and other areas that are relatively unexplored and retain their natural settings. When the Japanese Archipelago is studied in the light of its physical structure it helps us understand its geographical states and the various functions which identify those states. For example, the geographical classification covers Hokkaido, Tohoku, Kantoh, ect. while the functional classification includes manufacturing, agriculture, forestry, fishery, etc. Add to this the traffic and communication networks which bind the Japanese Archipelago as an environment.

We know that energy and information must be part of the underlying structure of the Japanese Archipelago. When these are included, we obtain a clear picture of the three dimensional aspects of the archipelago, particularly, in its interactive, dynamic character.

Basically, that approach would asuume that sufficient data concerning the present situation were available. Another approach related to the foregoing would be to emphasize the creation of road networks and new communication lines around the industrial belt of a city. But it would appear that before such action could be undertaken that a more fundamental understanding of the migration of people must be available. In other words, what are the factors that instigate the movement of people, particularly in respect to the raw resources available to them and the intellectual activiry that such resources could instigate? Perhaps in this manner we would achieve an understanding of the relationship between different priorities which exist in the environment for man—in dynamic and not static terms. If we were to conclude for example, that industrial and agricultural locations be dictated solely by geographical, physical condition, it is quite possible for us to obtain a biased snese of the importance of these factors to the dynamics of the Japanese Archipelago.

The author does not want to restrict this view to physical factors alone. These statements, for example, can be extended to legalistic as well as economic considerations underlying city life—the legislative and taxation systems are systems which come to mind.

Now what all this amounts to is that a remodeling of the city requires a much more extensive look at the forces that play on a city than are readily apparent and what the author is arguing as a tridimentional look at the problem space, one that does not solely emphasize the physical dimensions, but one that incorporates the structural, energy factors within people's sets of values.

Energy is indepensable to social activity, that is, energy is an essential force underlying the social environment.

Specifically, regarding energy, we can say that when we try to work on something, there is an automatic flow of energy. The author submits that this is followed by the "information effect", a change in the energy flow followed by a transformation of energy. Energy may be classified as natural and as artificial energy. Energy is found in the green forests, in the biological organisms of the animal. Yet, we also refer to the spiritual energy of man in the process of his intellectual thinking and activity.

Now in this paper concerning a plan for remodeling the Japanese Archipelago, it is important to have a precise idea of the amount of energy stored in natural itself. It will provide the basic data as to artificial energy which can be obtained from natural energy. Further, it can provide the basis for a more precise understanding of the dynamic nature of the Japanese Archipelago. It can further indicate the role and importance of information to the remodeling process, and also provide the basis for integrating the citizen in the total process. Remodeled structure should be designed by the interrelated area of Energy-Information as shown in Figure 3.

At this point , the author should mentioned some of the rationale and logic underlying the remodeling plan. Because of the war, Japan has moved from a largely stable, non-flexible economy, where natural resources were not a problem, to what may be called a "scrap and build" economy, where resources are scarece. Today, scrapping is often necessary as a pre-condigion for construction work. Thus, the cost of scrapping is increasing year after year. One can readily appreciate that if the remodeling of the Archipelago were to proceed on a "non-scrap" policy, many conflicts would a arise from environ-mentalists who would see a direct threat to the environment and nature in such a policy.

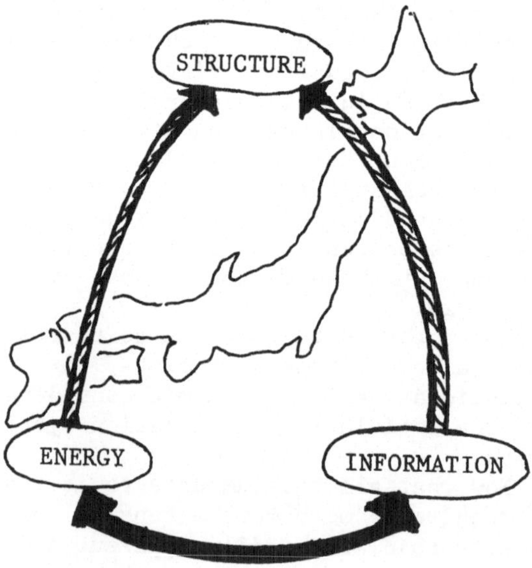

Fig, 3)

As such, one can appreciate the importance of an analysis of Japan's climate and national geographical characteristics prior to any remodeling of the Archipelago. It is our ability to see the interrelationship between the environment, the energy within the environment, and our understanding of the relationship between the two that will ensure a credible argument for the remodeling of the Japanese Archipelogo.

Urban information system study on the interrelationship of information and energy should give shape to the structure of eco-urbanization as it should stand. In this case, eco-urbanization covers both the regional and metropolitan development. On the other hand, the clear understanding of the present structure should help to set up a definite plan on the emergency information system. Figure 3 is the observation on a plan for remodeling the Japanese Archipelago in this regard. It indicates that the ideal structure of the Japanese Archipelago can be proposed only after the interrelationship of information and energy, that is urban information system, is made clear at many different levels.

All this is the process of eco-urbanization in which systems like EMISARI (Emergency Management Information System And Reference Index) system are designed. Figure 4 is an example for the emergency information system designe by SEI approaches.

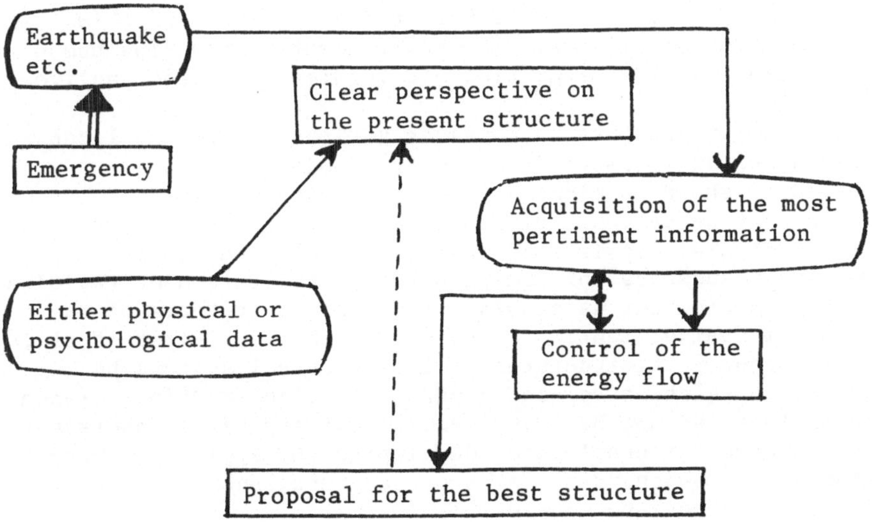

Fig. 4) Emergency information system by SEI approches (Urban information systrm)

The followings are the elaboration of the Figure 4.

i) the clear perspective on the present structure.

ii) the selection of the information by the degree of emergency

iii) the control of the energy flow by the information

<u>Structure</u> : highway, energy station, energy path, park, school, information network, etc.

<u>Energy</u> : traffic movement, movement of people, flow of energy, etc.

<u>Information</u> : degree of emergency, kind of emergency, damage, crowdiness, announce effect, etc.

This concept is the process of identification, estimation and control in the control systems.

3. Exploitaion System in Oceanospace

There are three essential factors underlying the conception of a development system: Structure, Energy and Information as discussed before. What is involved, in other words, is an effort to delineate the field to be developed along with the technological system to be developed in connection with it in the space constituted by the three fundamental factors.

A concrete spatial conception would be the environment in which human beings live. One structural constituent of this space is weather or climate. The environment surrounding human life is formed by the energy flow and information effect active within it.

It is important that we consider various systems applicable to oceanospace development deriving from this fundamental conception as shown in Figure 1. Figure 5 provides such a schematic representation applied to oceanospace development.

The next step consists of development of technological systems in oceanospace for the purpose of utilization or exploitation. Here again, let us consider the process in terms of the three fundamental elements of development theory. First of all, it is important that structures of technological systems be considered in relation to oceanospace, to which existing science and technology has not been applied. It is especially necessary that considerable technological development be applied to hard structures, for which pressure, ocean current, buoyancy, etc., must also be taken into account.

The second of the three factors, energy, refers to energy resource required for the structure which will be operating in the oceanospace. Underwater operations involving resistance water pressure will pose problems with respect to air pressure, oil pressure and electricity as well as to methods energy supply, depending on a variety of conditions.

In order to ensure the smooth operation of all technological systems for ocean development in accordance with the objectives

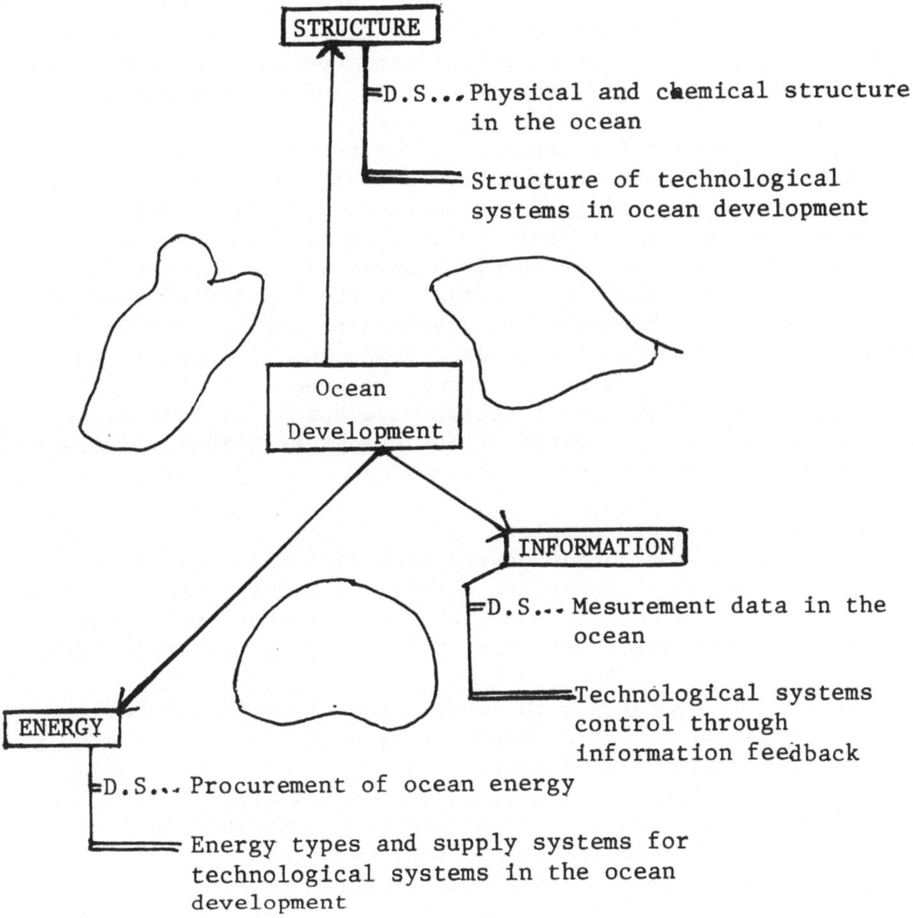

Fig. 5) Oceanospace development by SEI space.

in view, it is indispensable that there be adequate control of the equipment itself through the feedback of information. Such an arrangement would function as a nervous system able to detect subtle changes in oceanospace,adjust the flow of energy supplied to the equipment systems (structures), and operate the development systems for sea-bottom resource excavation and for sea-bottom construction.

Most important in this regard is the recognition that ocean space is for all intents and purposes the ultimate source energy for lifeand world of ecology. Cosequently, the mesurement of development systems, which tends to be essentially oriented toward the physical science, requires above all an understanding of sea-bittom space from the viewpoint of the biological sciences. An understanding of the ocean and development systems are precisely what are needed for ocean development to progress as expected. The establishment of dynamic mesurement systems will lead to the materialization of better development systems as well ascontribute to the improvement of mesurement techniques, thereby making it possible to create new technology.

3.1 Manned or Unmanned Systems

For the success of ocean development systems, it will be necessary to overcome the servere conditions peculiar to the ocean environment,e.g., turbulent waves, the corrosive effect of the sea, extremely high water pressure, the absence of air and light, etc.

Some ways of combatting the above might involve the following:

1) The use of different kinds of vessels.
2) Divers using diving gear.
3) The use of submersibles, diving bells, etc.
4) Diving using a sea-bottom observation tower, etc.
5) The use of a marine operational base, etc.
6) The use of remote control equipment.

Given the diverse development objectives and goals for which the above will be developed, it is difficult to apply any overall orientation here; indeed, it maybe expected that these methods will be utilized in combination when appropriate. Nonetheless, these techniques should be continually refined in tandem with the development of advanced (looking ahead 5 - 10 years) fundamental technology, the support afforded by this accumulated technology, the consolidation of experimental facilities, and research on corrosion and pressure resistance capacities; thesetechniques are fu fundamental to so many areas of ocean development that they maybe considered indispensable. In particular, as far as sea-bottom excavation is concerned, 2) and 6) above are especially significant inasmuch as they relate to the vital question of which course development systems will follow—that of manned or unmanned systems.

The manned systems implied by 2) above involve not only

problems which must be solved, e.g., human safety, the removal of obstacles, and diving limits etc., but also several problems which are essentially irresoluable. The type of unmanned systems suggested in 6) above, on the other hand, must confront technological barriers which exceed by far current technological levels, e.g., pattern recognition, position control, etc.

Any selection and assigning of priority with regard to these two alternatives must ultimately involve a consideration of such given condition as goals, ocean depth, environmental condition, expenses, etc. It is thus difficult to say generally which of the two systems, the manned or unmanned, should be adopted, although it is probably safe to view the trend in marine operations in ocean developments as in the direction of the latter.

3.2 Settling up Information Control Systems

Operations in the sea, needless to say, must be performed under extremely special conditions. Consequently, it is necessary from both the safety and maintenance standpoint, to make every possible effort to ensure the satisfactory performance of the information control system. The primary steps to be taken in this respect are:

1) Measurement and control of the operational environment
2) Scheduling of the operations to be performed
3) Emergency information control and procedures
4) Lighting and power systems management
5) Designing operational equipment systems for installation in capsules
6) Efforts at developing unmanned systems

On the basis of the problems discussed above in connection with development systems, the following items are proposed for cosideration.

3.2.1 Improving overall reliability

As we become better aware of the special nature of operations in the ocean, it becomes urgent that we try to improve the reliability of these operations. Especially at the present time, when so great a part of these operations is dependent on human efforts, it is essential (as well as important in terms of improving overall reliability) that as information control system be developed which can be used in stuations by the individual development systems.

3.2.2 Materials and Machinery Resaerch

Research of materials and machinery in the oceanospace environment are important for creating the special ocean development technology which has not been known to use until now. Above all, the development of materials especially suitable use in the ocean and of a machinery science based on such materials is of vital importance for the proper operation of the various subsystem incorporated in the development system.

3.2.3 Consolidation of Legal Systems

It is anticipated that the legal aspects of development will become increasingly complicated in view of its close link with environmental problems. Accordingly, it is desirable that an effort be made to consolidate presently existing legal systems in order resolve the problems of ocean development and the environment in accord with climatic and natural considerations: I.e., concretely, there should be established an overall water area control system integrating agriculture, forestry, and fishery industries and coastal development, etc.

The ocean is a particularly pertinent area of interest because it illustrates the ecological problem which concerns us at present. We all know that modern science and technology have the potential to lead all of the ocean-living creatures to extinction. The awareness of the potential certainly makes us sensitive to the need for exploring new aspects of science and technology which would reduce this theat. On the other hand, the author wants to stress that science and technology should not be restricted in their attempts to develop new mechanisms for coping with our environment. Now that the ocean may be a new area for resource, we should examine it from an ecological point of view. The author contends that we should pay more attention to the biological environment. This ecological approach would serve as the initial step to harmony between the ocean and technology, especially in the range of the monsoon zone.

In Japan there are many plans now being carried out, for example OSPER project proposed by the author, and its modified system to be described on Appendix. We should seek on this occasion an outline of the fundamental issues and a methodology to solve them from the ecological point of view, however, it's not a little hard works.

In the old days, man must have learned a great deal from nature in order to accomplish the engineering that he undertook. And it must have been in the process of construction work that the theory of civil engineering was brought forth.

Finally, development of theory and technology are analoguous should be in harmony with ecology. Most fundamental in our approach to the environmental system is our understanding of nature and our ability to be responsive to nature in our ability to learn from it.

APPENDIX

OCEAN SPACE ROBOT-MEASUREMENT(*)

1. Needs of the development and its purpose:

To carry out the ocean development project, the pre-requisites are to collect and grasp detail information and data on the ocean space. However, the survey area is a three dimentional ocean — an enormous space. Besides its huge area, the condition and phenomena of ocean is changing each second. To obtain a data and complete data on this dynamic object, on the changes taking place in accordance with time and space in the ocean, much labour and expenses are required.

A survey of the ocean is carried out by research ships, submarines and buoys; and also for the purpose of discovering more efficient surveying methods, this research is being positively promoted.
Numerous studies and books on new surveying methods are introduced , but most of these methods are now in an experimental stage. The conventional method for the ocean survey requires much labour and time, therefore it is impossible to say whether the method can efficiently measure and collect data focusing on a spacific item whose space distribution and change in time are surveyed. The object of this project is to develope an ocean measuring system mainly by using un-manned robots which can measure and analyse multiple research items in a certain ocean area and can grasp the space distribution and change in time of the ocean condition of the area.

The measuring system by the un-manned robot can be applied to various kinds of researchs which are connected to ocean development, for example,

a) The basic and academic research of ocean.
b) A survey of distribution of ocean minerals.
c) A detail and long-team survey on ocean phenomena, topographic situation and the nature of soil for space use of ocean oil storage system and construction in some spacific ocean area.
d) The current issue is a survey of water quality to determine the present situation for cuntermeasures and monitoring of ocean especially in pollution offshore.

2. Outline of system (Figure 6)

This system comprises of six sub-systems:

1) Vertical type robot
2) Horizontal type robot

(*) This project is being carried out by OSR project Team, (Leader Professor S. Watanabe) in Machine and Systems Development Center of Japan Sociefy for the Promotion of Machine Industry.

Fig. 6] Schematic outline of OSR system

3) Mooring buoy
4) Mothership
5) Sea-bottom base
6) Land base

Each of six sub-systems is devided into three sub-systems, namely measurement, data processing and communication. Besides these other sub-systems, each sub-systems is divided into smaller sub-systems. The six sub-systems described above 1) to 6) plus the corresponding three sub-systems will be named hereafter as the SYSTEMS, and the system of which is constructed by these smaller systems will be named hereafter as the TOTAL SYSTEM.

2.1. Verticl Robot System

The purpose of this un-manned robot is to measure and analyse the vertical distribution of ocean components. The robot comprises of several similar type vertical robots. These robots measure their locations automatically by means of the supersonic signal which is sent from the sea-bottom base, and also control the location and posture with the self-contained equipment; power source, gyro and propulsion equipment. By this control system, the un-manned robot submerges vertically in the ocean space. During its movement, the robot surveys the ocean components with the use of various sensors.

The collected data of the ocean is recorded in the magnetic tape of the robot. When the robot emerges, the data is sent to the land base. The robot has an underwater communication system which enables it to send emergency signals such as underwater refuge signal due to rough weather, the emergency floating signal for the robot, malfunctioning and accident of the robot and/or some other required assistance and emergency information to the land base and to the mothership and also the ability to establish communication with them.

2.2 Horizintal Robot System

The purpose of this robot is to survey the horizontal distribution of ocean components. This system consists of several similar type horizontal robots. The robot measures its own location and posture with the self-contained posture controller such as gyro and depth meter and by supersonic signals sent by the sea-bottom base. The robot also travels horizontally according to a pre-determined path and surveys with various. The surveyed data is recorded in the self-contained magnetic tape and the pre-determined job is completed, the magnetic tapes are collected by the mothership. And also the emergency signals of abnormal travel such as emergency due to malfunctioning and accident of the robot are sent to the land base or the mothership by underwater communication.

2.3 Mooring buoy system

The un-manned robot also consists of same similar type

systems. The purpose of this robot is to survey the ocean differences, Besides this function, the mooring buoy acts as the relay station between land base and robot. The robot and mooring buoy are inter-communicable with surface electrical weve communication and underwater sound communication. The communication between mooring buoy, land base and mothership is attained through the electrical wave.

2.4 Mothership

The role of the mothership is launching and collecting of robot and mooring buoy system, checking and examining the SYSTEM and maintenance work to attain normal operation such as charging of batteries of robot. The required devices for these parposes must be equipped to the ship.

2.5 Sea-bottom Base

The role of the sea-bottom base is to inform the robot of its location by sending the supersonic signal. This also consists of three bases principally.

2.6. Land Base

The role of land base is to deliver and record measuring data collected by the robot and the mooring buoy system to the computer center at the base for processing and analysing the data according to measuring purpose. The analysis and then offered to users of this data. And also according to the results of the survey, a new survey plan is prepared by the staff of land base. They prepare the travelling schedule of the robot system and send the robot system operation control order to the mothership. Besides these jobs, the land base has the function of construction, check, exemination, repair and maintenance of each system.

2.7 Information Processing System

The role of the information processing system which networks the above mentioned systems, is to collect the surveyed data and to send the robot operation order. The functions of the information processing system is the survey control, data processing and emergency countermeasures.

3. Mission of the SYSTEM

To schieve the object of the system development, considering the importance of the method and factor for the system development, the mission of the TOTAL SYSTEM is as follows:

1) To control the submergence and emergence of the vertical robot with axis contol .
2) To control the direction of travel of the horizontal robot according to a pre-determined path.
3) To launch the mooring buoy and survey the ocean components.
4) To perform the total analysis of space and time differences

of oceanic data by combining the vertical robot, horizontal robot and mooring buoy.

5) To perform the real time processing of the OSR system.
6) To make use of the surveyed results to be demonstrated at The International Ocean Exposition, Okinawa, 1975.

4. Planning Requirements for the SYSTEM

4.1 The requirements for the surveyed ocean area are as follows:

Area of sea	1 to 10 Km
Depth	250 m below sea level
Maximum current speed	3 knots (1.5m/S)
Temperature	-4C to +30C
Wave (height)	10 m (maximum)
Wind (speed)	30 m (Maximum) per second

4.2 Requirements for measurement:

Items to be measured:

- Pressure of water
- Temperature of water
- Current Direction
- Speed of current
- Hydro conductivity
- pH (acid and alkali)
- Amount of soluble oxygen
- Transparency

Accuracy of position measurement:

Horizontal - within 5% derangement
Vertical - within 1% derangement

Measuring conditions for Vertical Robot and Horizontal Robot are as follows:

	Time Interval for Measurement	Space Interval for Measurement	Period
V. Robot	3 hours	1 m	5 days
H. Robot	30 seconds	45 m	50 hours
Mooring buoy	5 minutes	more than 5 levels	1 month

CELLULAR ORGANIZATION AND COMMUNICATION IN DICTYOSTELIUM DISCOIDEUM AND OTHER CELLULAR SLIME MOULDS*

Morrel H. Cohen** and Anthony Robertson††

Department of Biophysics and Theoretical Biology
University of Chicago

ABSTRACT

Metazoan development provides one of the best examples of the central role of intercellular communication in cellular organization. The relevant aspects of Metazoan development will be briefly reviewed. A case will be made for the existence of control systems operating at the multicellular level through intercellular communication during development. The development of Dictyostelium discoideum will be considered in some detail to illustrate the general points made. Its life cycle will be described, but the focus will be on interphase, the period of transition from the vegetative to the aggregative state, and on aggregation. The elements of the aggregative control system and their emergence during development will be discussed. The dynamics of the aggregation control system and of the resulting morphogenesis will be described mathematically as well as qualitatively.

INTRODUCTION

We have been concerned with Metazoan development and in particular with the transition from the fertilized egg to the

*Supported in part by the National Institutes of Health, the National Science Foundation, and the Sloan and Otho S. Sprague Foundations.
**NIH Special Fellow, 1972-73.
††Alfred P. Sloan Research Fellow, 1973-75.

adult organism. This transition entails profound and dramatic changes at all of the diverse levels of biological organization. We have viewed it from the cellular level and the lowest multicellular level, that of the field (Robertson and Cohen, 1972). From that viewpoint organized complexity (a multicellular organism) emerges from simplicity (a single cell) by only six unitary processes: growth, division, movement, contact formation, differentiation, and secretion. The final complexity results from frequent repetition of these few processes. Frequent repetition implies many opportunities for gross errors to arise at the cellular level. Development, however, is reliable. Reliability in the face of repeated opportunity for error implies the existence of control systems guiding the course of development.

Control can be of two extreme types, intracellular and global. Intracellular control can, for example, be simply the reading out of an inbuilt program. It is global control which guarantees the orderly emergence of spatial heterogeneity and with which we are primarily concerned. In very crude terms, we study maps and clocks in developing organisms.

Such field-wide control of cell behavior can occur only if there is communication between cells. Intercellular communication lies at the heart of any multicellular control system. In studying developmental control systems, therefore, we are inevitably led to the investigation of intercellular communication in embryos or other developing systems and of its role in the dynamics of the control systems.

The time dependence of developmental phenomena can reveal dynamical aspects of the control system and therefore of intercellular communication. Time dependence, crudely speaking, can be of two extreme kinds, steady or periodic. Periods of events likely to be important in the control of development are probably of order minutes (Goodwin and Cohen, 1969). The unitary process best suited to the detection of periods in the minute range is morphogenetic movement. Time-lapse microphotography or video tape recording becomes, in the present argument, the basic experimental tool.

One needs, in addition to an experimental technique, a well chosen model organism. We have worked primarily with the cellular slime mold, Dictyostelium discoideum, which was discovered by Raper (1935), who described its life cycle and pointed out its importance for developmental biology (Raper, 1940). In this organism, development is relatively simple and the unitary processes are well separated in space and in time, with morphogenetic movement playing a primary role. The literature on D. discoideum is reviewed by Bonner (1967) and specific aspects of its biology are described by many authors (e.g. Ashworth, 1971; Gerisch, 1968;

Newell, 1971; Robertson and Cohen, 1972; Shaffer, 1962). Accordingly, in the next section we give only a brief outline of its life cycle and then go on in subsequent sections to describe the early stages, interphase and aggregation, in more detail and with occasional use of mathematics. We finish with a section on other cellular slime molds containing a few general remarks on intercellular communication.

LIFE CYCLE OF D. DISCOIDEUM

D. discoideum amoebae hatch from elliptical spores about 6μ long which germinate in moist places in the soil and on decaying vegetation (Cotter and Raper, 1966, 1968 a and b). The amoebae feed on bacteria which they find by chemotaxis (Potts, 1902; Bonner, et al, 1970). As long as food is available they divide about every four hours. When the food is exhausted the amoebae enter a period of differentiation called interphase (Bonner, 1963) which lasts for eight hours (Gerisch, 1968; Cohen and Robertson, 1972a). During interphase the amoebae develop those competences which are required for aggregation. These include: the ability to respond to extracellular cyclic adenosine monophosphate (c-AMP) by chemotaxis towards a c-AMP source (Konijn, et al, 1967); the ability to relay a pulsatile signal of c-AMP when stimulated by a suprathreshold extracellular c-AMP concentration (Robertson, Drage, and Cohen, 1971); the ability to form EDTA stable intercellular contacts (DeHaan, 1959); and the ability, for some cells, to release periodic pulses of c-AMP autonomously (Cohen and Robertson, 1972a). Development of each of these competences must involve changes in many cellular properties, in particular in the cell membrane. Such changes are being investigated by many workers (see Newell, 1971 for a review). A detailed knowledge of molecular mechanisms is not necessary in the present context.

At the end of interphase some amoebae begin to release pulses of c-AMP autonomously. Their neighbours are competent to respond to autonomous signals by chemotaxis and by signal relaying. Each signal can only travel a limited distance because its amplitude is reduced by both diffusion and the action of an extracellular phosphodiesterase (PDE). As there is a threshold concentration for signal relaying, there is also a critical density of amoebae below which a signal cannot be propagated. This density corresponds to a distance of approximately 75μ between amoeba centers. After receiving a signal an amoeba becomes refractory to further stimulation (Shaffer, 1957; Gerisch, 1968; Cohen and Robertson, 1971a). The refractory period is a function of age, decreasing from more than six minutes at the beginning of aggregation, to approximately two minutes within four hours.

The refractory period ensures that signals propagate unisensally from a center and that a territorial boundary will form between neighbouring centers as colliding signals annihilate each other. These properties of the system lead to the outward propagation of periodic waves of c-AMP concentration from centers, marked by inward waves of cell movement (Gerisch, 1968; Cohen and Robertson, 1971a and b; Robertson, 1973).

Because each cell relays the signal, amoebae outside a center tend to move towards their nearest central neighbors, forming streams radiating from the center (Shaffer, 1957; Cohen and Robertson, 1972; Nanjundiah, 1973). An early aggregate therefore consists of a central mass of cells with randomly bifurcating streams leading away from it. During the later stages of aggregation the central cell mass develops a nipple-shaped tip (Raper, 1940; Robertson, 1972). Cells within the tip remain there during the rest of morphogenesis (Takeuchi, 1969). In late aggregation, cells in the aggregate secrete a mucopolysaccharide slime. This is liquid at the highest, most central region of the aggregate, but hardens as it flows downwards (Shaffer, 1965). As cells are still entering the aggregate within streams, and as the slime is only liquid at the center, pressure in the center can only be relieved by the erection of a cylindrical column, bearing the tip and covered in slime. This grows until it becomes unstable and falls over to form the slug, or migrating pseudoplasmodium. Cells within the slug are joined by antero-posterior contacts and by lateral membrane interdigitations. They migrate as a unit within the slime sheath, which remains stationary with respect to the substrate and collapses behind the slug. Raper showed that the tip controls slug migration (Raper, 1940) and that extra tips grafted into the side of a slug will take over cells posterior to the graft site. He concluded that the tips acted as though they were organisers, controlling cell movement and defining the developmental axis of the slug.

The slug migrates for a distance which depends on environmental conditions (Raper, 1940; Newell, Telser and Sussman, 1969). At the end of migration the cell mass rounds up and rotates so that the tip again moves to the top. At this stage the cells in the anterior third of the slug have become "pre-stalk" cells showing histological and biochemical changes, and these in the posterior two thirds are "pre-spore" cells, again with a characteristic histology (Bonner, Chiquoine and Kolderie, 1955; Gregg, 1965; Gregg and Badman, 1970; Hohl and Hamamoto, 1969; Maeda and Takeuchi, 1969) which has been developing since late aggregation. Their axial organisation is retained on rotation of the cell mass. The tip is therefore on top of a mass of pre-stalk cells, which is on top of the pre-spore cells. Formation of the fruiting body ensues. Raper and Fennell have described this in great detail (Raper and Fennell, 1952). A tube of cellulose fibrils is formed with its top at the base of the tip. It extends downwards until it makes contact with a group of cells at the base of the cell mass. These will form the base-plate of the mature fruiting body. At this stage the outer cells in the tip are organised radially, like a columnar epithelium, and it has been suggested that the stalk cellulose is secreted from their central faces (Bonner, 1967). Pre-stalk cells, which move inside the cellulose tube, vacuolate and increase in volume, as well as producing cellulose walls. This differentiation may be triggered by contact with cellulose (Farnsworth, 1973). Cells are continually added to the top of the stalk and the cellulose fibrils are continually laid down at the top of the cellulose cylinder. The stalk, therefore, elongates until the stock of pre-stalk cells is exhausted. We have noticed that the tip of the fruiting body shows periodic jerks, while elongation of the stalk itself by vacuolation of cells within it is continuous (Robertson, 1972). The period of tip jerks is close to five minutes, and the period distribution is remarkably similar to that of autonomous cells in early aggregation (Durston, 1973b). We have therefore speculated that cell movement to the top of the stalk is periodic and under the control of the same signalling system as is responsible for cell movement during aggregation and during slug migration, which also shows a five minute periodicity although it is less well defined (Robertson and Cohen, 1972).

This description of the life-cycle of D. discoideum shows that there is good reason to assume that the tip retains a special role throughout. Rubin and Robertson (Robertson, et al., 1972) therefore repeated Raper's grafting experiments and further made grafts of tips from all developmental stages into all other stages. They also assayed tip function by grafting tips into fields of amoebae at different stages of interphase. Their experiments showed that tips from all stages have qualitatively similar properties. The signal a tip supplies is apparently not a function of its developmental age, but the response it evokes is characteristic of the recipient structure. Tip function remains constant until all tip cells have been used up in the process of fruiting body formation. The presence of a tip in a pseudoplasmodium inhibits further tip formation. Removal of a tip leads to the determination of a new tip within 44 minutes, and its visible emergence within about 1 1/2 hours, confirming Farnsworth's observation for the conus (Farnsworth, 1972) and Bonner's film (Bonner, 1959). Other pseudoplasmodial portions cannot replicate tip function. In the case of slug midportions, tip functioned is mimicked when, after a significant delay, the midportion regulates to produce its own tip. The signal from a tip corresponds in all respects to a continuous c-AMP signal.

The analogy between the tip and the "organiser" of classical embryology (Spemann, 1938) first suggested for the slug tip by

Raper (Raper, 1940) is now well-founded (Robertson and Cohen, 1972; Robertson, et al., 1972; Farnsworth, 1973). The tip controls morphogenetic movement at all stages in the D. discoideum life-cycle from late aggregation onwards. The tip defines the direction of the developmental axis and controls a field of cells. It is able to do this by steadily secreting an attractant, presumably c-AMP, at concentrations above the threshold for signal relaying (Rubin and Robertson, unpublished results; Durston, 1973b).

This feature of its signal, fundamental to the tip's role as an organiser, allows it to "take over" and dominate centers of any type whose signals are initiated by autonomous cells or pulsing microelectrodes. Finally, a new tip or organiser can be produced by a regulative process when a tip is removed. This is in clear analogy with the regulative abilities of Metazoan embryos whose organisers have been removed, and with the properties of the hypostome of Hydra, also an organiser (cf. Wolpert, Hicklin and Hornbruch, 1972). A further implied, but not explicitly demonstrated, role of the tip is the control of differentiation. Regulation following tip renewal involves regulation of the proportions of pre-stalk and pre-spore cells (Raper, 1940; Bonner, 1967; Robertson, 1972). It is possible that the tip supplies positional information (Wolpert, 1969) by way of the gradient of signal, probably c-AMP, that it produces.

Thus the tip, as is the autonomous cell in early aggregation, is the center of communication in the developing organism.

INTERPHASE: THE AGGREGATION COMPETENCES

To repeat, interphase is a period of differentiation during which the four competences essential for normal aggregation emerge (Cohen and Robertson, 1972a).

1) The ability to respond to a suprathreshold extracellular concentration of c-AMP by chemotaxis toward a c-AMP source; 2) The ability to relay a pulsatile signal of c-AMP when stimulated by a suprathreshold extracellular concentration of c-AMP higher than that for chemotaxis; 3) The ability to release periodic pulses of c-AMP autonomously; and 4) The ability to form EDTA stable polar intercellular contacts. The first three of these provide the basis for morphogenetic movement and its control during aggregation. Knowledge of the rates at which these competences appear during interphase as each relevant stage of differentiation is completed is of basic importance. Without it, biochemical and genetic studies of differentiation for aggregation are severely hampered (Cohen and Robertson, 1972b) as are mathematical analyses of wave propagation and

aggregation.

Let $x_i(t)$ be the fraction of amoebae in a large uniform population possessing the ith competence (i = 1 for movement, i = 2 for relaying, i = 3 for autonomy) at an interval t after the beginning of interphase. We have formulated techniques for the measurement of each of these, but detailed quantitative results are available only for $x_3(t)$. Obtaining $x_1(t)$ requires counting cells in a given population responsive in a characteristic way to a standardized artificial c-AMP source. Obtaining $x_2(t)$ requires determining the time of onset of long-range wave propagation in response to an artificial signal for populations of different density (Gingle, unpublished). $x_3(t)$ has been obtained from about 11 hrs. onward (Hashimoto, Cohen, Raman and Robertson, unpublished) by the use of the Konijn-Raper small drop technique (Konijn and Raper, 1961). Our present knowledge of the x_i is shown schematically in Fig. 1 (Cohen and Robertson, 1972a).

In the small drop technique, a drop of amoebae is placed on hydrophobic agar. The amoebae do not move outside the confines of the drop. The total number η of amoebae and their number density N are readily measured. For a number M of drops having similar values of η and N, the fraction P not containing any autonomously signalling cells can be determined vs. time. If the differentiation of any cell into an autonomous cell were an internal event only and independent of the presence of all other cells, P would be given by

$$P = (1-x_3)^{\eta} \cong e^{-\eta\, x_3(t)} \tag{1}$$

with $x_3(t)$ independent of N. The dependence of P on η would then immediately give $x_3(t)$ as

$$\frac{\ell n\, P}{\eta} = x_3(t) \tag{2}$$

We have used Eq. (2), augmented by a suitable statistical analysis, to interpret our data and find that differentiation into autonomy is indeed a random event independent of the presence of other amoebae at sufficiently short times and small amoebae number and density. At later times and higher amoebae number, the rate of differentiation into autonomy appears to be enhanced by a diffusible molecule secreted by the amoebae. From 11 to 15 hrs. x_3 increases roughly linearly with t and saturates at about 10^{-2} after 18 hrs. Since aggregation is normally initiated at about 8 hrs., the gap from <8 to 11 hrs. will have to be filled in by another method employing larger values of η than the small drop technique.

WAVE PROPAGATION

The Signalling Range

The cellular event basic to the control of morphogenetic movement during aggregation and to the communication between cells in D. discoideum is the release of a burst of molecules of c-AMP into the aqueous film surrounding each amoebae on, e.g., agar. The c-AMP molecules diffuse into the agar and, concurrently, become converted into linear AMP by phosphodiesterase (PDE), both bound to the plasma membrane and released into the agar. We have worked out the resulting somewhat complicated mathematical problem of determining the c-AMP concentration $C(\underset{\sim}{r},t)$ at a distance $\underset{\sim}{r}$ on the agar surface from the point of release and a time t after release. The result is qualitatively the same as found earlier (Cohen and Robertson, 1971a) and is sketched in Fig. 2. The important points to note at present are that $C(\underset{\sim}{r},t)$ has a maximum value $C_m(\underset{\sim}{r})$ and that $C_m(\underset{\sim}{r})$ is a monotonically decreasing function of distance, as sketched in Fig. 3. Thus at distances larger than R, the signalling range (Cohen and Robertson, 1972a), the value of $C(\underset{\sim}{r},t)$ never rises above C*, the threshold for the stimulation of signal relaying,

$$C_m(R) = C^*. \qquad (3.1)$$

Because $C(\underset{\sim}{r},t)$ depends on amoeba density N through the action of the PDE, so does R. Determination of R vs. N can lead in principle to determination of the PDE activity both in membrane bound and free form.

High Density Fields

The significance of R is that an amoeba cannot stimulate another amoeba at a greater separation to relay its signal. R imposes a scale of distance on the mean amoeba separation. In the high-density case when there are many amoebae within the range of a single signal, on average, i.e., when

$$\pi R^2 N \gg 1, \qquad (4)$$

the field of amoebae may be regarded as a continuous medium for the propagation of signalling waves.

After 6 hrs. when $x_2(t)$ has reached unity, the field is a sensitive medium (Robertson and Cohen, 1972; Robertson, 1972; Durston, 1973a), in a condition to propagate waves in response to any suprathreshold signal of external origin. Such sensitive media are well understood (Wiener and Rosenblueth, 1946; Krinskii, 1968). Heart tissue (Ruch and Patton, 1966, p. 139) (ibid, p. 144), certain solutions of chemical reactants (Zaikin and Zhabotinsky, 1970; Winfree, 1972), multivibrator networks, and models of neural networks are known to be sensitive media. Wave propagation in such media is well understood (Krinskii, 1968; Goodwin and Cohen, 1969; Winfree, 1972). It follows the eikonal equation but in addition possesses a Huyghens construction without a superposition principle. Because of the existence of the refractory period, the boundary conditions are absorbing for propagation into a boundary. The eikonal equation admits the propagation of kinks in wave fronts, However, there are no shadows as occur in geometric optics because of the Huyghens construction, and beams spread out to fill the medium. A point source produces circular wave fronts. Spiral propagation occurs (Winfree, 1972, 1973) with the inner end of the most stable spiral describing a circle of perimeter vT_R, with v the propagation speed and T_R the refractory period.

What further remains to be understood is the propagation speed v. We have constructed a theory of v which relates it to the parameters of the signalling system: the number of c-AMP molecules released, η ; the diffusion coefficient of c-AMP in agar or H_2O, D; the threshold concentration for signal relaying, C*; the delay time between signalling and receiving a suprathreshold signal, T_D; and the free and membrane-bound PDE activity. We have measured D, and T_D can be obtained by other measurements (see below). The density dependence of v can therefore in principle give values for η/C^* and the PDE activities.

Low Density Fields

When condition (4) is not well met, we are dealing with a low density field. The field is patchy, density fluctuations are important, and the continuum treatment can no longer be used. When several amoebae in a wavefront signal together, the concentration produced near one of the signalling amoebae is larger than that produced by that amoebae alone. The range of a signal is enhanced by cooperative effects within a wave front. As the density is reduced, however, the randomness of the amoeba distribution breaks up the regular wavefronts so that cooperative effects are reduced. They are reduced also by the increasing separation of amoebae. Thus, as shown directly by Futrelle (unpublished) in computer simulations, simultaneous signalling by several amoebae does not have any significant quantitative effect on wave propagation at low densities. Under those circumstances, wave propagation can be considered a percolation process on a random medium (Shante and Kirkpatrick, 1971; Cohen and Robertson, 1972a). Long range propagation of signals cannot be sustained by the field unless the amoeba density exceeds a critical density N* given by

$$\pi R(N^*)^2 N^* = 4.5 \tag{5}$$

For N less than N*, signal propagation is restricted within isolated clusters of amoebae, the mean size of which falls rapidly with decreasing density. The criterion (5) for N* derives from percolation theory (Shante and Kirkpatrick, 1971), and, as noted above, is only approximately applicable to our problem. However, the form of (5) should remain correct, the only change being in the numerical value of the right hand side, which should decrease. Futrelle's simulations show that the decrease should be small, and we shall ignore it.

Our current best estimate of N* from the literature (Konijn and Raper, 1961) and from our own unpublished measurements (Hashimoto, Gingle, unpublished) is about $2.5 \times 10^4 cm^{-2}$, which yields a value of 75μ for R via (5).

For densities just above N*, wave propagation is restricted to narrow channels within the field (Futrelle, Robertson, unpublished). Indeed, some portions of these channels degenerate to strings of individual cells (Futrelle, unpublished), each one signalling only the next cell along the string as propagation proceeds. It is possible to identify signalling cells visually because they round up during the delay period between signalling and being signalled (Drage and Robertson, unpublished). The delay time, T_D, and the signalling range R(N) can thus be measured directly. Preliminary values of 15 sec. and 75μ are consistent with earlier results.

Propagation along strings occurs whenever the number of sensitive cells just exceeds the critical density N*. This occurs during interphase when $x_2(t)N$ increases above N*. Identification of the onset of the capability of long-range wave propagation during interphase thus permits determination of $x_2(t)$, of R(N), and of T_D. Similarly, the refractory period T_R has a fairly broad distribution of values at the time aggregation normally starts. Thus if a pulse is followed by another at an interval within the distribution of T_R such that the density of cells is again just above N*, propagation along strings occurs.

This permits the determination of the probability distribution of T_R, of R(N), and of T_D. Such experiments are under way (Gingle, unpublished).

PACEMAKERS

In the preceding section, we have sketched the properties of a field of amoeba as a continuum or discrete sensitive medium. Without a source of cyclic AMP to initiate wave propagation, however, wave propagation simply does not occur in such a medium. This is observed in our small drop experiments (Hashimoto, Cohen, Raman, and Robertson, unpublished). For example, a certain percentage of drops containing a small enough number of amoebae does not aggregate even though the density is well above N*. Sources of c-AMP can be external, e.g., a microelectrode (Robertson, Drage, and Cohen, 1972), an autonomous cell, or a more complex entity consisting of a group of cells. We call entities of this sort which elicit a periodic response from the field pacemakers. Durston (1973a,b) has made a detailed study of natural pacemakers in D. discoideum.

He finds two geometries for wave propagation away from pacemakers, concentric and spiral. The latter have been observed to be single or double. Spiral to concentric switches have been observed which are correlated with the emergence of the tip.

Histograms were constructed for the intervals between successive waves for all waves observed, for various geometries, and for various stages of the life cycle. Time courses of intervals were determined for individual centers.

Durston's observations and analyses taken together with other observations and theoretical analyses suggest the following conclusions:

Autonomous cells contain an autonomous oscillator having a fairly sharply defined 5' period which is independent of developmental age. This oscillator is linked to the c-AMP release mechanism but is independent of it. Differentiation into autonomy could therefore take place in two ways. First, the oscillator could be present in all cells, e.g., a metabolic oscillator such as the glycolytic oscillator in yeast, and only the link need be constructed. Second, both the oscillator and the link must be constructed.

The refractory period decreases from about 7 1/2' at the onset of wave propagation to about 2 1/2' after several hrs. Initially the refractory period is broadly distributed; the width of its distribution decreases markedly with time.

This rather broad initial probability distribution of values of refractory period means that the field is then spatially heterogeneous with regard to refractivity. Consequently, spirals are initiated in an otherwise homogeneous field at local maxima of refractivity by waves propagating from autonomous cells.

Finally, tips are steady sources of c-AMP.

Aggregation Dynamics

The preceding sections make clear that we have arrived at a solid though incomplete understanding at the cellular level of morphogenetic movement and its control by c-AMP waves via chemotaxis during aggregation. Accordingly, we have constructed a crude theory of aggregation dynamics and the resulting gross morphology of an aggregating field in the continuum density range.

The two centrally important features of the aggregation process are (1) the formation of stable aggregation centers and (2) the flow of amoebae into these. With regard to the first, let N_c be the density of stable aggregation centers. These can be autonomous cells, spirals, or more complex pacemakers. To be stable, they cannot either cease operation or be entrained by some neighboring faster pacemaker. The rate of emergence of such new stable centers must relate to the rate of emergence of autonomous cells, but we do not as yet know this relation. Accordingly, we have made the simplest possible assumption

$$\frac{dN_c}{dt} = \alpha \frac{dN_3}{dt} = \alpha \frac{dx_3}{dt} N_F = \beta N_F , \tag{6}$$

where N_3 is the density of autonomous cells within the unaggregated part of the field, which has mean density N_f. We expect β to be substantially less than unity because of entrainment.

With regard to the second, the flow of the amoebae towards a single center can be described by elementary hydrodynamics provided one ignores or averages over stream formation, or

$$\frac{\partial N}{\partial t} + \underline{\nabla} . (V_A \hat{r} N) = 0. \tag{7}$$

Here N is the density of unaggregated amoeba, V_A is the mean aggregation velocity,

$$V_A = u/T \tag{8}$$

with u the mean step length in response to a signal and T the mean signal period, and $\hat{r}$ is a unit vector pointing away from the aggregation center.

These two equations (6) and (7), if one ignores variations in aggregation territory size, lead immediately to a simple equation for the time dependence of the mean density N_F of amoebae left unaggregated

$$\frac{d^3f}{d\eta^3} + f^2 = o, \qquad f(o) = 1 \tag{9}$$

where

$$\left.\frac{df}{d\eta}\right|_{\eta = o} = \left.\frac{d^2f}{d\eta^2}\right|_{\eta = o} = o$$

$$f = \frac{N_F}{N_o} \tag{10}$$

$$\eta = \frac{t}{t_o} \tag{11}$$

$$t_o = (2\pi\beta v^2 N_o)^{-1/3} \tag{12}$$

Here N_o is the initial uniform amoeba density at the beginning of interphase, f is the fraction of amoeba remaining unaggregated, and η is the time measured in units of t_o, a parameter establishing a density dependence of $N_o^{-1/3}$ for the morphological time scale during aggregation.

Solution of (9) is straightforward and yields the following principle results. The time t_A required to aggregate is 1.88 t_o. At a time 40% of t_A, the field becomes patchy, with territories of individual aggregates marked by clear borders. The final aggregate density is

$$N_A = 0.748\ (\beta N_o/v)^{2/3}, \tag{13}$$

which has a remarkable 2/3 power dependence on initial amoeba density.

Over a broad range of densities, one aggregate produces one fruiting body so that (13) implies that the fruiting-body density is proportional to the 2/3 power of the amoeba density in the continuum range. This is shown to be the case for Hashimoto's data in the density range $6x10^5$ cm^{-2} to $6x10^6$ cm^{-2} (Hashimoto, Robertson, and Cohen, unpublished) in Fig. 4. Inserting the known value of V_A and an estimate of dx_3/dt into the corresponding value of 0.748 $(\beta/v)^{2/3}$ obtained from the data of Fig. 4 via Eq. (13) gives a value of α of about 10^{-2} which indicates that very few of the autonomous cells finally give rise to centers.

The principal shortcoming of this theory is the very crude way in which the generation of new centers and their survival is treated. Before it can be improved, knowledge of x_3 at the end of interphase and up to about 11 hrs. must be obtained.

Communication Systems in the Cellular Slime Moulds

We have described a communication system which controls aggregation in the cellular slime mould D. discoideum. A small fraction of the cells is competent to release pulses of an acrasin (most likely c-AMP) autonomously with a period of about 5 min. Essentially all of the cells are competent to relay a suprathreshold c-AMP signal. After releasing a signal itself a cell is refractory to further c-AMP signals for a period of about 7 1/2 to 2 1/2 depending on developmental history. Cells respond to a c-AMP signal above another threshold lower than the relaying threshold by a movement step towards the signal source. The step duration, about 100 sec., substantially exceeds the duration of the signal itself. We have visual evidence from time-lapse films that the movement step is preceded by an approximately 15 sec. interval during which the cell tends to become hemispherical, and all movement ceases, even the weak random movements common between successive movement steps. The agreement between the duration of the 15 second period of stationarity and rounding with that of the delay period between signalling and being signalled (Cohen and Robertson, 1971a) as well as its preceding of the movement step lead us to suggest that the two are coincident. This in turn implies a complementarity between morphogenetic movement and secretion in D. discoideum cells (Robertson, 1973).

Such a complementarity exists in all species of cellular slime moulds for which the information is available. Before discussing the point further, however, we present in Table I a classification of some of the cellular slime moulds according to the known features of their aggregation control system, i.e. their intercellular communication system (Cohen and Robertson, 1972a). The various species are classified according to two main characteristics: whether they do or do not possess founder cells, i.e. source cells specialized and morphologically distinct from the remaining cells in the field which act as centers of aggregation, and whether almost every cell in the field is a local source of acrasin (aggregative signal). The cells can be steady local sources, as in D. lacteum, or periodically relaying as in D. discoideum, and this is indicated in a third column.

What is remarkable is that the founder cells are hemispherical in shape and do not move (Bonner, 1967, plate 2), precisely as for D. discoideum cells during the stage we have tentatively identified as the period between signalling and being signalled. It thus appears that in all cases where we have information, cells have similar morphologies and remain stationary while signalling. This suggests that founder cells in P. violaceum and P. pallidum are steady sources of acrasin and that the observed periodicity in their signal propagation derives from the refractory period of the field. Experiments to test this possibility are under way.

It can be seen from the table that the diversity of known communication systems present in the cellular slime moulds derives from only three binary choices: founder cells or no founder cells, local signal sources in the field or no local signal sources, steady local sources or pulsatile local relaying. These observations have considerable relevance to evolutionary and genetic questions.

TABLE I. Communication Systems of Some Cellular Slime Moulds

Species	Founder Cell	Local Signal Production	Periodic Relaying
D. minutum	+	-	-
D. lacteum	-	+	-
P. violaceum	+	+	+
P. pallidum	+	+	+
D. rosarium	?	+	+
D. purpureum	-	+	+
D. mucoroides	-	+	+
D. discoideum	-	+	+

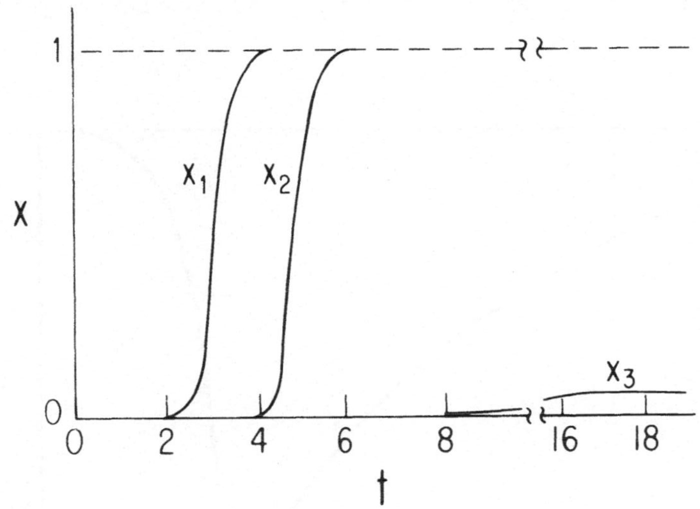

Fig. 1.-Emergence of competences during and after interphase.

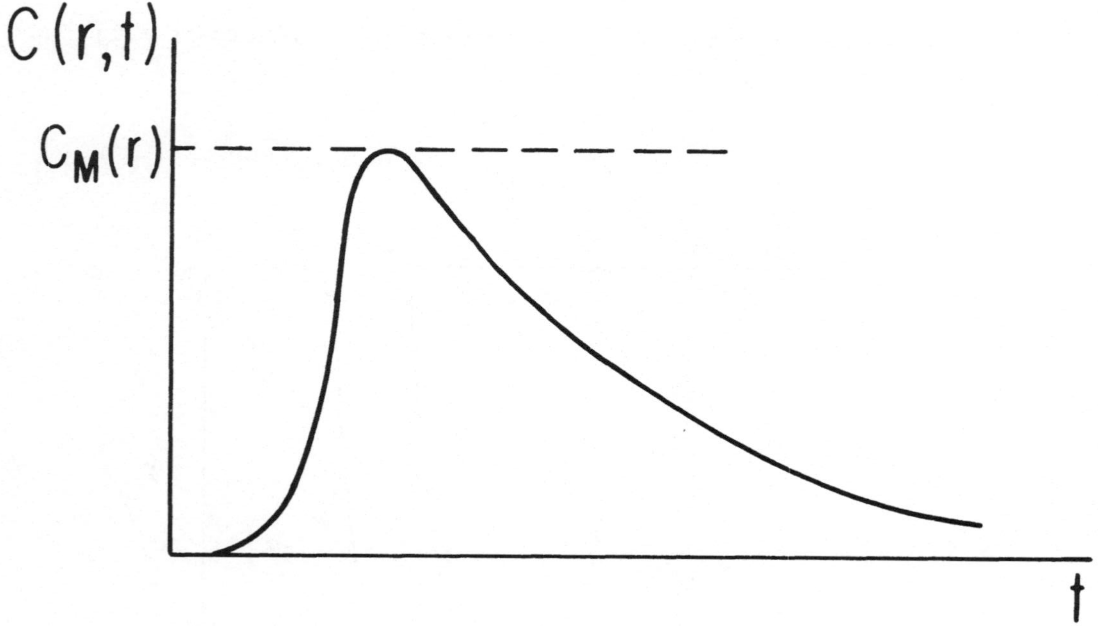

Fig. 2.-Time dependence of c-AMP concentration C(r,t) at the agar surface a distance r away from the point of signal release.

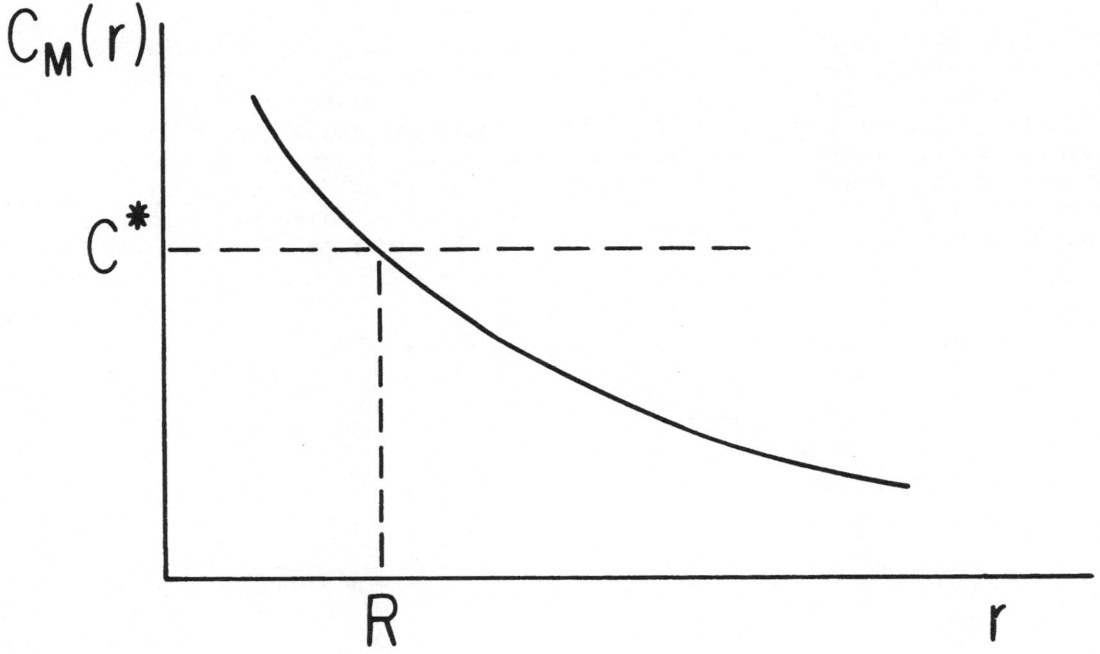

Fig. 3.-The dependence of maximum signal concentration $C_M(r)$ on distance from the signal source. For distance larger than R, C(r,t) never rises above C*, the threshold for relaying.

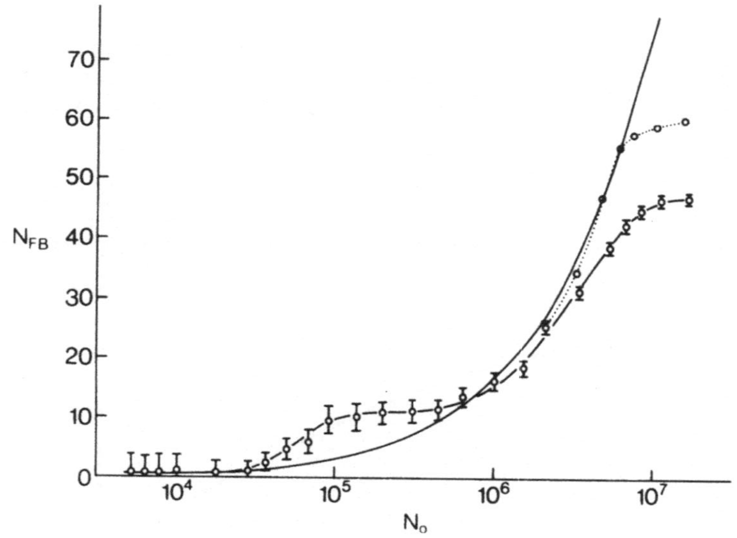

Fig.4.-Dependence of fruiting body density N_{FB} on cell density N_o at the beginning of interphase. Experimental points are shown by circles with error bars for fruiting body counts made 48 hrs. after the beginning of interphase and without error bars for 96 hrs. after interphase begins when the fruiting body count saturates (Hashimoto, unpublished). The theoretical result for the aggregate density, Eq. (13), is plotted as a solid line with the value of β chosen by a least squares fit to the data between $1.5x10^5$ and $1.5x10^6 cm^{-2}$ for N_o. The fit to the 96 hr. data is excellent between $5x10^5$ and $6x10^6 cm^{-2}$.

REFERENCES

Ashworth, J.M., Cell development in the cellular slime mould, Dictyostelium discoideum. Symp. Soc. Exptl. Biol. 25, 27, 1971.

Bonner, J.T., Fruiting in the cellular slime molds, Precision Films, Inc., New York, 1959.

Bonner, J.T., Epigenetic development in the cellular slime molds. Symp. Soc. Exptl. Biol. 17, 341, 1963.

Bonner, J.T., The Cellular Slime Molds, Princeton University Press, Princeton, 1967, 205 pp.

Bonner, J.T., Chiquoine, A.D. and Kolderie, M.Q., A histochemical study of differentiation in the cellular slime molds, J. Exptl. Zool., 130, 133, 1955.

Bonner, J.T., Hall, E.M., Sachsenmaier, W. and Walker, B.K., Evidence for a second chemotactic system in the cellular slime mold Dictyostelium discoideum, J. Bacteriol. 120, 682, 1970.

Cohen, M.H. and Robertson, A., Wave propagation in the early stages of aggregation of cellular slime molds, J. Theor. Biol. 31, 101, 1971a.

Cohen, M.H. and Robertson, A., Chemotaxis and the early stages of aggregation in cellular slime molds, J. Theor. Biol. 31, 119, 1971b.

Cohen, M.H. and Robertson, A., Differentiation for aggregation in the cellular slime molds, in Cell Differentiation, Harris, R., Allin, P., Viza, D., Eds., Munksgaard, Copenhagen, pp. 35-45.

Cohen, M.H. and Robertson, A., Control of development: the cellular slime molds, Dev. Biol. 27, 589, 1972b.

Cotter, D.A. and Raper, K.B., Spore germination in Dictyostelium discoideum, Proc. Natl. Acad. Sci. U.S.A. 56, 880, 1966.

Cotter, D.A. and Raper, K.B., Properties of germinating spores of Dictyostelium discoideum, J. Bacteriol. 96, 1680, 1968a.

Cotter, D.A. and Raper, K.B., Spore germination in strains of Dictyostelium discoideum and other members of the Dictyosteliaceae, J. Bacteriol. 96, 1690, 1968b.

DeHaan, R.L., The effects of the chelating agent ethylenediamine tetra-acetic acid on cell adhesion in the slime mould Dictyostelium discoideum, J. Embryol. Exptl. Morphol. 7, 335, 1959.

Durston, A.J., Pacemaker activity during aggregation in Dictyostelium discoideum, Dev. Biol., in press, 1973b.

Durston, A.J., Dictyostelium discoideum aggregation fields as excitable media, J. Theoret. Biol., in press, 1973a.

Farnsworth, P., Morphogenesis in the cellular slime mould Dictyostelium discoideum: the formation and regulation of aggregate tips and the specification of developmental axes, J. Embryol. Exptl. Morphol. 29, 253, 1973.

Gerisch, G., Cell aggregation and differentiation in Dictyostelium, Current Topics in Dev. Biol. 3, 157, 1968.

Goodwin, B.C. and Cohen, M.H., A phase shift model for the spatial and temporal organization of developing systems, J. Theoret. Biol. 25, 49, 1969.

Gregg, James H., Regulation in the cellular slime molds, Dev. Biol. 12:3, 377, 1965.

Gregg, J.H. and Badman, W.S., Morphogenesis and ultrastructure in Dictyostelium, Dev. Biol. 22:1, 96, 1970.

Hohl, H.R. and Hamamoto, S.T., Ultrastructure of spore differentiation in Dictyostelium: the prespore vacuole, J. Ultrastruc. Res. 26, 442, 1969.

Konijn, T.M., Barkley, D.S., Chang, Y.Y. and Bonner, J.T., Cyclic AMP: a naturally occurring acrasin in the cellular slime molds, Am. Naturalist, 102, 225, 1968.

Konijn, T.M. and Raper, K.B., Cell aggregation in Dictyostelium discoideum, Dev. Biol., 3, 725, 1961.

Konijn, T.M., van de Meene, J.G.C., Bonner, J.T., Barkley, D.S., The acrasin activity of adenosine-3',5'-cyclic phosphate, Proc. Nat. Aca. of Sci., 58, 3, 1967.

Krinskii, V., Fibrillation in excitable media, Probl. Kibern., 20, 59, 1968.

Maeda, Y. and Takeuchi, I., Cell differentiation and fine structures in development of cellular slime molds, Dev. Growth Diff. 11:3, 232, 1969.

Nanjundiah, V., Chemotaxis signal relaying and aggregation morphology, J. Theoret. Biol., in press, 1973.

Newell, P.C., The development of the cellular slime mould Dictyostelium discoideum: a model system for the study of cellular differentiation, Essays in Biochem., 1, 87, 1971.

Newell, P.C., Telser, A. and Sussman, M., Alternative developmental pathways determined by environmental conditions in the cell slime mold, J. Bacteriol., 100, 763, 1969.

Potts, G., Zur Physiologie des Dictyostelium mucoroides, Flora, 91, 281, 1902.

Raper, K.B., Dictyostelium discoideum, a new species of slime mold from decaying forest leaves, J. Agr. Res., 50, 135, 1935.

Raper, K.B., Pseudoplasmodium formation and organization in Dictyostelium discoideum, J. Elisha Mitchell Scient. Soc., 56, 241, 1940.

Raper, K.B. and Fennell, D.I., Stalk formation in Dictyostelium discoideum, Bull. Torrey Bot. Club, 79, 25, 1952.

Robertson, A., Quantative analysis of the development of cellular slime molds, in Some Mathematical Questions in Biology, Cowan, J.D., Ed., American Mathematical Society, Providence, R. I., 1972, 47.

Robertson, A., Information handling at the cellular level: intercellular communication in slime mould development, in The Biology of Brains, Academic Press, London, 1973.

Robertson, A. and Cohen, M.H., Control of developing fields, Ann. Rev. Biophys. Bioengineering, 1, 409, 1972.

Robertson, A., Cohen, M.H., Drage, D.J., Durston, A.J., Rubin, J., and Wonio, D., Cellular interactions in slime-mould aggregation, in Cell Interactions: Proceedings of the Third Lepetit Colloquium, Silvestri, L.G., Ed., North-Holland, Amsterdam, 1972, 299.

Robertson, A., Drage, D.J. and Cohen, M.H., Control of aggregation in Dictyostelium discoideum by an external periodic pulse of cyclic adenosine monophosphate, Science, 175, 333, 1972.

Ruch, T.C. and Patton, H.D., Physiology and Biophysics, Saunders, Philadelphia, 1966, 1244 pp.

Shaffer, B.M., Properties of slime mould amoebae of significance for aggregation, Quar. J. Micro. Sci., 98, 377, 1957.

Shaffer, B.M., The acrasina, part I, Adv. Morphogen., 2, 109, 1962.

Shaffer, B.M., Cell movement within aggregates of the slime mould dictyostelium discoideum revealed by surface, J. Embryol. Exp. Morph., 13:1, 97, 1965.

Shante, V.K.S. and Kirkpatrick, S., An introduction to percolation theory, Adv. in Phys., 20, 325, 1971.

Spemann, H., Embryonic Development and Induction, Yale University Press, New Haven, 1938, 401 pp.

Takeuchi, I., Establishment of polar organization during slime mold development, in Nucleic Acid Metabolism, Cell Differentiation and Cancer Growth, Cowdrey, E.V. and Seno, S., Eds., Pergamon Press, Oxford, 1969, 297.

Wiener, N. and Rosenblueth, A., The mathematical formulation of the problem of conduction of impulses in a network of connected excitable elements, specifically in cardiac muscle, Arch. Inst. Cardiologia de Mexico, 16, 105, 1946.

Winfree, A., Spiral waves of chemical activity, Science, 175, 634, 1972.

Winfree, A., Scroll-shaped waves of chemical activity in three dimensions, Science, 181, 937, 1973.

Wolpert, L., Positional information and the spatial pattern of cellular differentiation, J. Theoret. Biol., 25, 1, 1969.

Wolpert, L., Hicklin, J. and Hornbruch, A., Positional information and pattern regulation in regeneration of Hydra, Symp. Soc. Exptl. Biol., 25, 391, 1971.

Zaikin, A.N. and Zhabotinsky, A.M., Concentration wave propagation in two-dimensional liquid-phase self-oscillating system, Nature, 225, 535, 1970.

NEURAL AND LOGIC ASPECTS OF LANGUAGE PROCESSING I

Douglas A. Young

Department of Computer Science,
University of Manitoba,
Winnipeg, Manitoba R3T 2N2, Canada.

I suppose that most people would agree that language is basically comprised of symbols or words and the rules for their use; that words comprise groups of one or more special symbols called letters, and that these words can take either visible or audible form. However, when considerations of language extend to the relationship between language and thought, language and truth, and language and grammar some difficulties and differences of opinion arise. Here, as in so many other areas of human activity, the symbol has unhappily tended to obscure the meaning.

In the first of two lectures on symbol and natural language processing in the brain, I want to speak about the origins and nature of language, and of some of the implications that seem to arise. In my second lecture, I shall introduce some neuron networks that could help to explain some of our linguistic and cognitive capabilities.

For most of you, the capability of using language is taken as a matter of course; you probably have never thought very much about words, or about thought at all. Yet this capability, that most of us have in full measure, provides us with our principal means of communication with one another, of storing and recording "knowledge", and, some would argue, with the very means of thinking itself. These alone would form sufficient reason for us to investigate the nature of the language process but, since reasoning, speech initiation and language comprehension form the most advanced capability of brains that nature has produced, and since the brain is the greatest and most versatile parallel processor known to us, it is appropriate that this first lecture be devoted entirely to a consideration of words, language and concepts.

It seems likely that somewhere between one million and five hundred thousand years ago[1], man began to use simple visual and voiced symbols and signs in order to communicate within small groups. He already had sensory concepts of things he had heard, seen, smelt or felt, since, without such concepts, language would have been impossible. He also already had directional and locational concepts, such as would be needed, for example, to allow him to appreciate the variety of the different ways by which he might return to his cave or outwit an animal that he was hunting. With the use of symbols, certain parts of his brain would have begun to have been used in an unusual way. This new form of para-lingual use of the brain was to inevitably equip him the better for survival the more he used it. The processes of mutation and natural selection would have facilitated the growth and neurological development of those cortical areas that were to become genetically assigned to speech and language.

By what sort of process, then, did this primitive use of voiced symbols begin ? Little is known about the details, and the best one can do is to establish hypotheses as to its structure.

Let us suppose that you, as an early man, had enjoyed a certain new dish - a kind of lizard stew - and that you wanted your wife to produce another stew like that. The obvious way to do this would have been to have gone and gathered several of the necessary ingredients and to have started to munch and to draw the attention of your spouse so that, recognizing you to be hungry, if only because she was also, she put together the notion of what you were wanting. But the time would have come when you would have wanted to be able to demonstrate all this quickly to her, without going through the business of gathering all the ingredients each time. So you devised a visible sign - like  - resembling the shape of a lizard, and pointed to your mouth as you simulated the eating process.

It is of course very simple to suggest that "you devised a visible sign to resemble a lizard" but such devising must have involved a very considerable jump in mental processes, just to have thought of it - let alone to have implemented it. Indeed, it is difficult to imagine not having the understanding that enables one to appreciate the relationship of the words to objects and events. But it would not appear unreasonable to suggest that perhaps early man's consciousness of the possibility of symbols arose from some chance reflective awareness of the fact that an oft repeated stimulus "meant" an event or thing, following from the regularity of the response. If he had been aware of this "meaning" enough times and over a long enough period, there might well have been an increased tendency to think of miming a stimulus, so to speak, in order to produce the response of an awareness of its significance or meaning in the hearer or observer.

To revert to your menu, however, let us suppose that you wished to express a desire for some variety, like lizard with snake heads garnished with sunflower sauce. It would probably have taken a very long time to have worked out how to sign or express this to your spouse and then to actually get it across to her. But in the course of going through the motions, time after time, of drawing the lizard, the snake head and the crushed sunflower, together in front of the cooking pot, in order for her to understand what you meant by "altogether", it would likely have happened that you found yourself repeatedly making certain similar noises. You saw the effect ultimately, of your noisy demonstrations, and your wife became, as you did too, somewhat conditioned to associating this particular noise with the special dish of lizard with snake heads garnished with sunflower sauce rather than with any of these delicacies on their own.

From what has been said so far, it can be clear why it is often claimed that there are substantial grounds for the hypothesis that all language is ultimately derived from ostensive definitions and that the "meaning" of words and symbols is dependent upon the cognitive grasp of their symbolic function in relation to objects and events. In the end, most if not all words have their "meaning" and are explainable in and through activities. Complex thoughts, too, are, when all else fails, ultimately explainable in more simple terms which in turn are explainable only by activities or ostensively, using appropriate sequences of motor and sensory processes.

At this point, before introducing the next major factor in cerebral language processing, I would like to digress for a moment to computer processing of natural language.

In the late 1950s and early 60s, when the need for computers to handle words as opposed to numbers was bringing great pressure to bear on computer manufacturers and engineers, the only effective recall method for stored natural language statements was one of key-words. But key-words tend all too often to provide either too much or too little information.

If one wished to refer for instance to the term "encephalitis" in connection with some medical enquiry, one would likely obtain some information which, even where "encephalitis" is used in combination with two or three other key-words, would not be relevant (e.g. where "encephalitis" was a high frequency occurrence word, but did not represent the major topic of discussion). On the other hand one might miss much relevant information because the term "encephalitis" was too specific (e.g. where some papers had been published on "viral infections of the central nervous system", the actual term "encephalitis" itself might have seldom occurred throughout any of the papers, and hence retrieval would have been omitted).

Automatic translation was another area where great strides were demanded and made in the basic work of handling language by computer. But, because the meaning of a word or group of words in one language is often different from the meaning of an 'equivalent' word or words in another language, progress has been very slow.

Since the days when computerised key-word analysis and machine translation began, not much progress was made until 1971, when Winograd advanced the whole field of natural language processing to a new level of competence.[2] Concepts were no longer considered the realm of psychologists and philosophers alone, but were things that should form an essential part of any versatile natural language processing system. But although the emphasis was not turned toward systems that would in some ways "understand" language, Winograd still adhered to a belief in the importance of grammar and of parsing procedures and in the sequential (as opposed to the parallel) processing of statements, questions and commands. Only when we can produce computer-like systems that will process concepts in something like the way in which humans process concepts, will any really effective natural language processing become possible.

What then is a concept ? For many people, it is an "idea", or something vaguely "of the mind". But however any one concept is defined, it is usually explainable by means of words and or demonstrations. Some people in attempting to define concepts refer to certain specific ones in terms of "<u>the</u> concept of ....." (e.g. "the concept of goodness", of "blue" or of "electricity"), as if a concept is something which is in some way exactly the same for everyone in the world and which can be defined in terms universally agreed by all.

This kind of use and explanation of the word "concept" may be satisfactory for some, but for my present purposes I need something more accessible and specifiable. I define a concept as being a disposition possessed by a person (or animal) to respond to certain kinds of linguistic or sensory stimuli in (what is considered as) an appropriate way (Konorski, gnostic units; [3] Brain, schema. [4]). In describing a concept in this way, we have facilitated its translation into units of hardware. For clearly a system of potential response patterns is convertible into terms of neural networks or of electronic circuitry. (I have purposely omitted the nebulous and precarious aspect of the subjective experience of a concept, and therefore of consciousness, since I consider consciousness, in so far as it is a subjective experience, as being a logical construction).

It is important to note here that a concept is not simply an element of a linguistic system, but of the non-linguistic (sensory, motor and emotive) systems also. We have visual concepts (e.g. of a tree), auditory concepts (the sound of a bell), tactile concepts (the feel of a blanket), vestibular concepts (tilting backwards), emotive concepts (awareness of beauty). All forms of concepts go towards the make up of knowledge and depending on the topic, one or more sensory, proprioceptive, emotive or other concepts may be involved in the circuitry that facilitates the use of a particular term or phrase in a particular circumstance.

We will be discussing some specific circuits in the next paper, but having obtained a definition of a "concept" that enables us to convert to circuitry, what kind of process is envisaged in which concepts can be used to handle natural language ?

In discussing key-words earlier, we saw how some of these cannot give adequate coverage of the meaning which the user wishes to convey to the computer system. Our own knowledge and experience of the use of words, on the other hand, does enable us to appreciate and to use the whole meaning or any of the meaning of a word or phrase. Part of the difference between our own neurological language processing system and a computer system is that all meanings of a term or phrase are not only available to us almost immediately, but are brought to our attention according to the circumstances of the use of the term or phrase.

Meaning, as we have indicated previously, is not simply a matter of word equivalence; eventually words and sentences can no longer "explain" meaning. However, in so far as it is a matter of word equivalence (and I don't mean just on the basis of one-to-one equivalence or synonyms, but of verbal definitions) meaning comprises verbal concepts. Verbal concepts are contained within groups of specific interconnected word-initiating centres in the language areas of the brain. Hence if the term "butterfly" is linked with, among many other concept groups of word centres, the verbal concept "a flying insect", it is possible to represent this connection in computer or hardware terms. "Flying" and "insect" in the end have no further definition possible than ones which are ostensively given. A computer system could store this part description of a butterfly, but whereas a neural network or its electronic equivalent would access any or all of the verbal concepts involved in explaining or defining a butterfly, the computer would only very tediously and under specific instructions be able to search for any or all of the partial descriptions. Whereas the computer would do all its processing serially, the neural network or its equivalent could process the total activity in parallel. This is not, of course, to say that the spoken answer to the question, "What is a butterfly ?" is also in parallel, nor that the conscious awareness of all these partial descriptions of a butterfly are simultaneous, but rather that much of the initial answer to the question, "What is a butterfly ?" is switched in and is interassociating immediately and unconsciously, ready to be sequenced into the slower speech or writing output system.

In the example of the question, "What is a butterfly ?", we have confined ourselves to verbal concepts. But in explaining or giving the meaning of a word or phrase, many non-verbal concepts can be involved. Thus, in the process of describing a butterfly, a connection may be made between the word-centre for "butterfly" and the visual memory area and hand motor areas, such that you can "describe" the insect by remembering your usual vision of a butterfly flying and by raising your two hands together, back to back with thumbs linked, and flapping the two palms and sets of fingers up and down. Accompanying this demonstration might be the remark, "They fly like this".

Again, a complex experience concept, probably stored in the temporal lobe, may be recalled, in which you, as a small girl, were frightened by a butterfly which entangled in your hair. While you may not mention this specifically, experience of this recall would probably have some effect on the explanation you were giving of what a butterfly is.

While these types of connectivity undoubtedly occur in human language processing events, they would clearly be more difficult to include in any artificial concept-handling system. However, its difficulty would lie not in the language outcome of such recalls but in translating the emotive, tactile and other stored information of such events into symbolic or descriptive form. Such difficulties serve well to emphasise the importance of concept handling capabilities in any computer system that aspires to the kinds of language processing capability which many computer scientists and others would like to see produced.

REFERENCES

1. Young, J.Z., An introduction to the story of man; Oxford University Press; London; 1971; p.497.
2. Winograd, T., Procedures as a representation for data in a computer program for understanding natural language; Massachusetts Institute of Technology Project Mac; Cambridge, Mass.; 1971.
3. Konorski, J., Integrative Activity of the brain; University of Chicago Press; Chicago and London; 1967; Chapter II.
4. Brain, R., The neurology of language, Brain, 84,145-66, 1961.

NEURAL AND LOGIC ASPECTS OF LANGUAGE PROCESSING II

Douglas A. Young

Department of Computer Science,
University of Manitoba,
Winnipeg, Manitoba R3T 2N2, Canada.

Since it is known what hardware the brain has with which to process information, the problem is to find out how in principle such hardware, the neurons and synapses of all the different functional areas, can be linked together to provide the kind of language-handling system that we each possess. Words and concepts have been defined in the first paper; what is the neurophysiological and other evidence for such definitions ? Let us first consider some of the basic facts.

The principle areas of the brain of a right-handed person that are concerned with language (see Fig. I) are the anterior speech area (Broca), located in the lateral part of the left frontal lobe; the supplementary speech cortex, located in the superior part of the left hemisphere; the posterior speech cortex, comprising Wernicke's area and other associated areas; the writing centre, located between the supplementary speech cortex and Broca's area; the naming centre (Mills) located in the inferior posterior part of the temporal lobe, beneath Wernicke's area; and the pulvinar, located in the posterior section of the thalamus. That these areas are concerned with language is established by impairment or cessation of the affected part of their specific functions when a lesion or electrode interference is introduced.[1,2]

All the language areas are very extensively interconnected through the thalamus and there are also considerable direct interconnections. They are also linked both directly and indirectly through the thalamus and mid-brain, with all the sensory and many motor and allied areas.

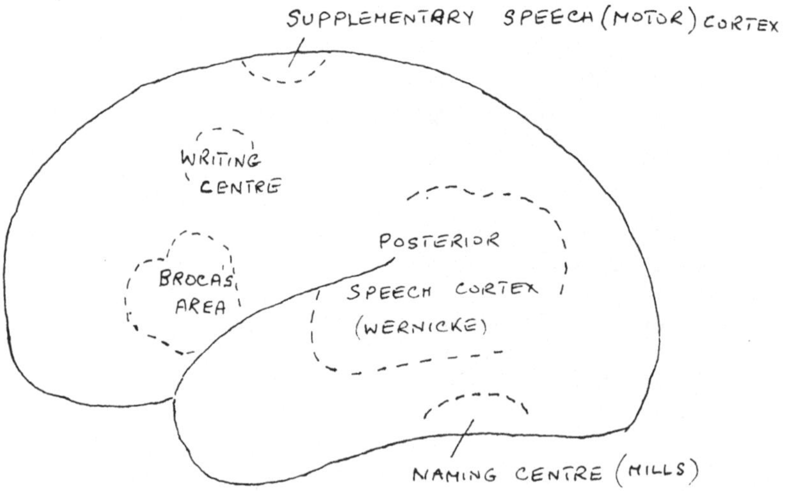

Fig. I. Principle language areas of the brain.

Given then that these are the centres of the brain, in which language (spoken, heard, written, read or "thought") is processed, and given that the brain's functions other than support systems are achieved through various types and sizes of neurons and by axons, dendrites and synapses, how could the language processing be achieved ?

The first point to make is that actual speech comprises the correct sequencing and spacing of groups of phonemes. In order to achieve this, a neuron or nucleus of neurons would need to be the initiation point for each phoneme selection and sequence. (Young children often sequence phonemes or other noises very consciously, in attempting to improve their speech sounds acquired through echolalia; but very soon, as for normal adults, the words come "automatically" without conscious effort).

A second point is that there is very considerable clinical neurophysiological evidence to show not only that each word in our vocabulary has a specific neuron or group of neurons through which it is initiated, but that each concept also has its ganglionic equivalent. (Penfield and Roberts op.cit., p.230). The clinical evidence is contained in the records of case histories in which electrodes have, with the patient's conscious approval, been introduced into parts of the language areas and individual words have been "forgotton" for just as long as the electrode was either present or was electrically stimulated. Among the large number of case histories that can be cited, I will just refer to two from Penfield and Roberts (op.cit., pp.143, 227).

An operation was conducted on a 33 year old right-handed man, who had been struck in the left posterior frontal region by a tool box at the age of 28, and who subsequently had begun to have recurrent seizures. An excision was made in the supplementary speech area. Electrode stimulation was used to guide the operating team and to keep them informed of the patient's progress. Several objects were presented to him during the course of these stimulations, and he was asked what they were. At one point in the procedure a flag was shown him and the patient repeated, "This is a ......" several times. After the removal of the electrode he was able to say, "A flag".

In the second case a patient had undergone an operation in which part of the left temporal lobe had to be removed. An electrode was inserted in the superior part of the anterior speech area, and the patient was shown for verbal identification a picture of a human foot. He said, "Oh I know what it is. That is what you put in your shoes". After the electrode was withdrawn he was able to say that it was a "foot". When the electrode was applied to a point in the supramarginal gyrus, and the patient was shown a picture of a tree, he said, "I know what it is" and was silent. When the electrode was withdrawn, he said at once, "tree". When the electrode was applied to a point in the posterior temporal region and he was shown a different picture, he was completely silent. A little time after the electrode was withdrawn, he exclaimed suddenly, "Now I can talk - butterfly" (which was correct). "I couldn't get that word 'butterfly', and then I tried to get the word 'moth'".

For the reasons given, it is logical to work on the hypothesis that each word has some kind of neuron grouping forming its means of initiation in speech. It may be argued that the hypothesis would not be acceptable since we are all losing by degeneration tens of thousands of brain cells every day of our adult lives. While such a loss would not necessarily involve more than something under 5% of the total number of cells, the proportion of degeneration in the speech and associated areas could indeed be considerable. However, apart from the fact that most of us retain our full linguistic faculties until very late if not for all our lives, there is very little substantial evidence for the claim that so many cells are in fact being lost each day. The two or three principle researchers in this area produced the data from some work with a very small part of the brain, and this work was done a considerable time ago and has not been verified by anyone since. Apart from these reasons for not accepting the argument based on degeneration losses, the question has to be asked if we are losing as many cells as this per day, what other sort of system than that proposed in the hypothesis could possibly handle the language processing that we achieve throughout our lives ?

If, then, we assume that in some sense each word in our vocabulary has a group of one or more neurons through which that word may be "used", what sort of logical arrangement could accomplish this ?

The argument of whether we think wholly, partly or not at all with words continues to rage, but the trend of the discussion is usually along psychological ground. Let us look at the question neurophysiologically. If one uses a system of word-initiating neurons or nuclei for speech, the logically previous system must also be concerned with words and with the sequencing of words. From the simple fact that in most everyday conversation one's spoken answer to a question or comment is usually far too quick for one to have had time to consider and decide exactly how one was going to say it, even if as one does occasionally spend a little longer deciding on the use of one particular word, it would seem reasonable to suppose that a network of interconnected word nuclei exists. If there is such a network, it would seem from these very actions of commenting on comment and of answering questions that the system that decodes the auditory system's message of a heard sentence or question is the same system, in terms of word nuclei, that initiates the reply.

That such a principle as here stated links two major parts of the language system is strongly indicative that the reading and writing subsystems also have much of their processing activity within this same central system. Verbal thinking also, if it is not anything other than ideational speech, would seem to be containable within this central system.

As has already been indicated, there is ample neuroanatomical evidence of the interconnectedness of the various parts of the language processing system, and of the connections between the language, sensory and motor systems. If, however, we use the "subjective" evidence, some additional evidence comes to light. Consider what happens when you are asked the way to the post office, assuming you know the way. Unless you have repeated the answer to this question so many times that you are used to it, you will probably have to think for a moment how you do in fact get from where you are to the post office. As you do so and as you start to describe how to get there, you might well be inclined to point in some initial direction, and then, by almost enacting some of the necessary turns and in feeling the orientation movements you went through the last time you went to the post office, and in remembering some of the things which you see on the way there, you describe the route the person should take.

Notice that you do not recall a set of statements or even of words to help describe how one gets to the post office. Rather does one rely upon sensory concepts, visual and proprioceptive primarily, with maybe some audio, tactile or olfactory concepts also. In addition one may recall some experience one had had en route once before. But while these sensory concepts may take as little time as usual to recall, there is some kind of direct connection between these concepts and the language area that initiates the spoken description of the route that you give. We do not experience a conscious selection of words; they just "seem to come". Also the "thinking" that you go through in considering the route you take will not be at all an ideational language experience; words wont usually come into the thinking except incidentally.

Consider what you do when you "think of" somebody. How much is your mind full of sentences and words describing aspects of that person ? Do you not rather re-experience your memories of that person, re-hear some things that he said, re-see what she looked like, or re-feel how you reacted to him or her ? And from these concepts you could describe the person, could you not ?

Again, when you are thinking how best to make something, you usually re-experience concepts you have of what you want to make, what it is for, what its sizes or colour should be, etc. Is not your final choice of design a primarily non-verbal event in which your last few decisions of choice "just come", however eventually ?

So there is considerable "subjective" evidence for there being not only direct links between some of the sensory and motor concept areas and the language areas, but also that much of our thinking is done on a non-verbal basis. (I do not mean by this that words do not "come to mind", in the course of the thinking, but that if they do they come incidentally and are not essential to the thinking processes to which I have referred).

From the evidence of the neuroanatomist, the neurophysiologist and one's everyday subjective experience of thinking about simple things, and from the hypotheses and definitions to which I have previously referred, it is possible now to propose some simple elements of neural logic circuitry for processing some simple language events.

Let us take the example of someone being asked, "What is a butterfly ?". In Fig. 2 there are laid out some of the words that could be introduced in answering that question. But note that, in addition to the word nuclei and connections, there are connections with the sensory, motor and other conceptual areas. The experiments of word association researchers have shown that the kinds of word associations that exist are such as to strongly support not only the hypothesis of word cells or nuclei, but also direct lines between word cells, some weighted more than others.

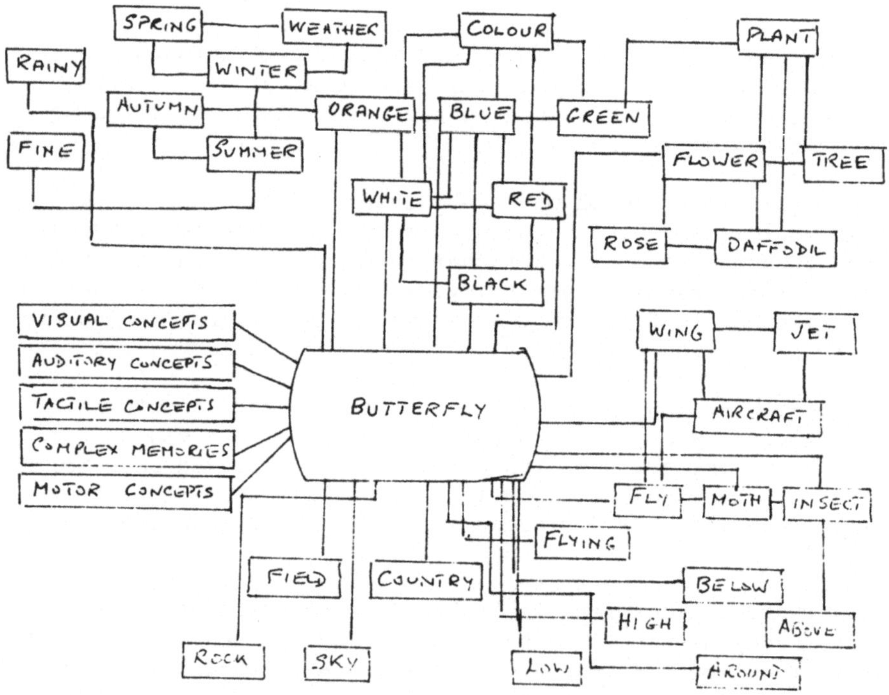

Fig. 2. Some word and concept links with "butterfly".

Fig. 3 is a lay out of some very simple neural logic which in some degree would be capable of answering the question. In considering this circuitry, try to remember that the spaces between the neurons are often too great in proportion to the size of cell shown and also that the diagram is of a very artificial representation, since any neural network would be in three dimensions. (Glial cells, blood supplies and etc. have of course been omitted for simplicity).

Let us suppose then that a person is asked, "What is a butterfly" and that his auditory system has transmitted the signals to the language area. (I suggested further back that the same processing system which handles the reception of speech signals for the auditory system could also handle the initiating of speech and word selection, but that facility is not included in the present circuitry). Also assume that a choice-of-words system, finally triggered perhaps by a combination of "preference loads" and of the random firing of one particular selection-association cell, rather than any other, selects the words chosen, which in this example are "a butterfly is an insect", being the first statement of a possible description of a butterfly.

The function of the network may be described as follows. The cell A is fired by the word-choice and speech-initiation systems. A signal is sent to the motor speech area, via (1), where the correct phoneme sequence is initiated so that the hundred and more muscle systems that produce the spoken sounds can enable the word "a" to be said. At the same time that the signal goes to the phoneme encoder, a signal also goes to an interneuron, B, which would be the neuron between A and any other word cell that could permissibly precede "butterfly", to await a feed-back message from the auditory and motor speech area, along (2) (3) respectively, that the word has been said and completed. The interneuron B is fired by this feed-back signal in combination with a continued signal from A, which stimulus is then inhibited by a feed-back, (4) from B. Cell C, which initiates the spoken word "butterfly", is then fired in combination with the word-choice and speech-initiation system. Notice that there is a connection (5) to cell C from the tactile concept area (in relation to the hands) which would facilitate the initiation of thinking or saying something about a butterfly, especially where one suddenly landed on your hand. There are also connections (6) with the visual system and the visual concept area, such that either the use of the word "butterfly" (in thought) can bring to mind some of the visual concepts one might have of a butterfly, or that the sight of a butterfly can achieve, given appropriate circumstances, the initiation of the thought or spoken word "butterfly". There are connections too, both ways, (7) with the complex events memory, so as to

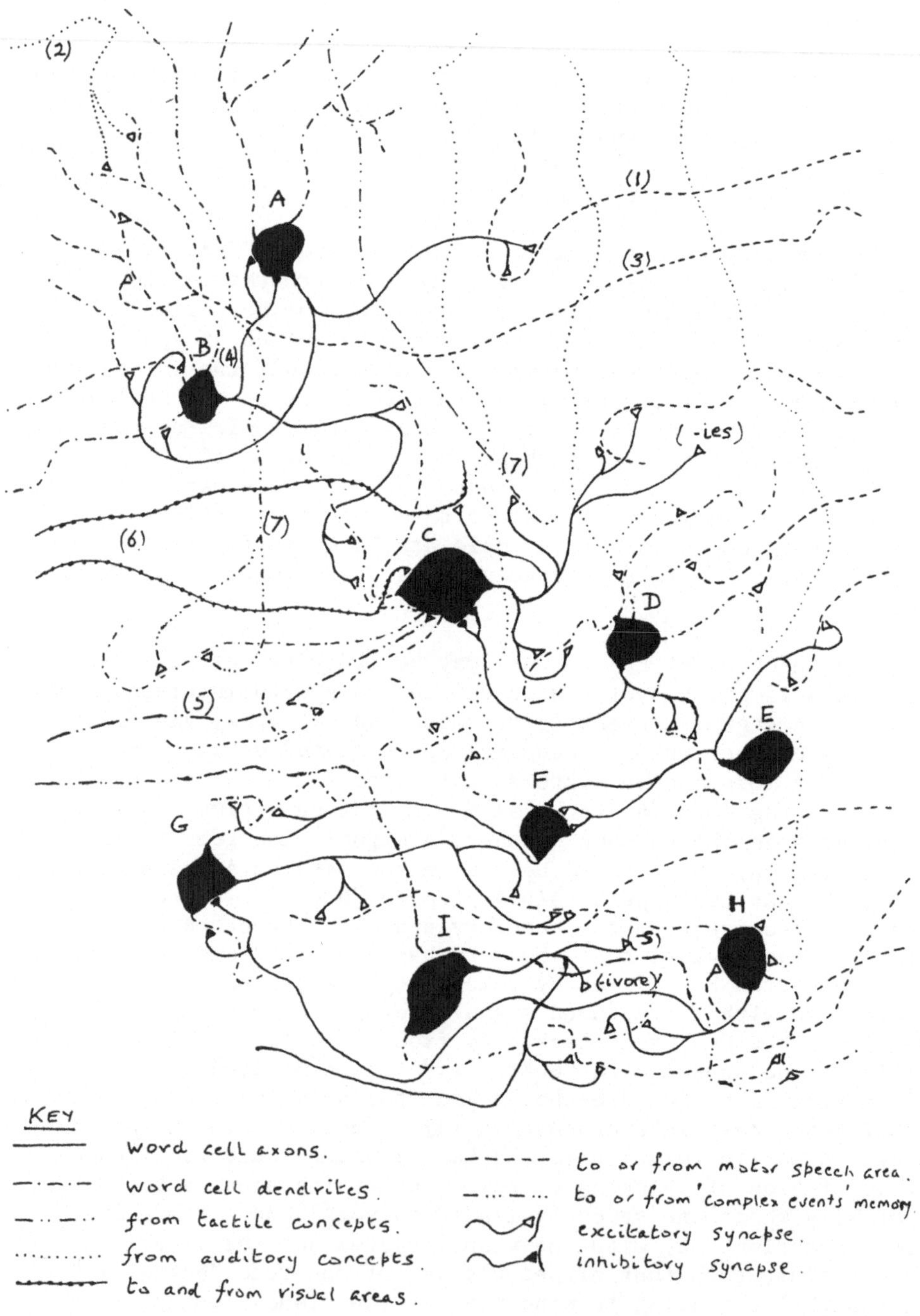

Fig. 3. Some neural logic for word initiating and sequencing.

facilitate either the recall of a complex event in which a butterfly occurred or the storage of the current discussion experience in which the word butterfly occurs.

A facility has been included in the output connections of the neuron C to enable the appropriate choice to be made between the singular and plural endings. (More sophisticated word-ending systems could be provided, to account for the more complex word-ending facilities that we possess in respect of some words, e.g. rough, -age, -est, -ness, -s, -er, -ed, -ly, -en, -ened, -ing). Word-ending connections have also been provided at cell I to facilitate a choice of endings for singular, plural and "-ivore".

The process of the nerve cell C triggering the speech system is followed, as for neuron A, by a feed-back to the next interneuron, D, from the auditory and motor speech areas, and the cell E ("is") is duly fired. With the subsequent firings of neurons G and I the sentence is completed. The logic for the initiation of end-of-sentence processes is not shown, but one of a variety of systems could be used to explain this aspect of our language processing ability.

The rudimentary system shown in Fig. 3, while clearly highly oversimplified, depicts the use of principles of connectivity which can, with extensive parallel processing, be made increasingly more detailed and sophisticated in order to cope with the extremely wide range of the ways in which several questions could be answered and statements commented upon. The same principles could be used in the setting up of logic to represent the neural activities that go toward initiating a question or statement. The system also incorporates the basis of a technique to process concepts as well as words.

If we can even begin to answer the question of how neurophysiologically we process language, we will be very much further on toward providing new ways of handling language (by special parallel processors) and also in understanding what is happening, neurophysiologically, when we think. If the mechanisms of language, which is the media in which most of our knowledge is stored (in literature), are better understood, perhaps we can also hope to cut out a great deal of wasted space in our billions of bookshelves.

This work is being supported by operating grant A2040 from the National Research Council of Canada.

REFERENCES

1. Penfield, W. and L. Roberts., Speech and Brain Mechanisms, Princeton University Press, New Jersey, 1959, (chap. V et al).

2. Luria, A.R., Higher Cortical Functions in Man,Tavistock Publications, London, 1966, parts II (1-5) and III (7-12).

3. Deese, J., The Structure of Associations in Language and Thought, The John Hopkins Press, Baltimore, 1965.

NONLINEAR WAVE PROBLEMS IN NEUROPHYSIOLOGY

Alwyn C. Scott

Department of Electrical and Computer Engineering
The University of Wisconsin
Madison, Wisconsin

PROPAGATION VELOCITY ON A SMOOTH AXON

A characteristic property of the nerve axon is the ability to propagate "energy exchanging" solitary waves (called "spikes"). A set of p.d.e.'s which describe axon propagation are[1]

$$\frac{\partial v}{\partial x} = -ri$$
$$\frac{\partial i}{\partial x} = -c\frac{\partial v}{\partial t} - j_i \qquad (1a,b)$$

where v is the voltage across the membrane, i is the total current through the axon, r is the resistance/length of the axoplasm, c is the capacitance/length of the membrane and j_i is the ion current through the membrane. In 1952 Hodgkin and Huxley developed a phenomenological procedure for computing j_i and demonstrated that (1) could be integrated to yield both the shape and speed of the spike.[2] Since their work and subsequent developments of the Hodgkin-Huxley equations were primarily numerical, it is important to find formulas which relate properties of the spike to parameters of the axon. Here we indicate how the spike velocity may be calculated from an investigation of its leading edge.[3] The leading edge current is primarily due to the flow of sodium ions which can be approximated as in Fig. 1.[4] Then (1) becomes

$$\frac{\partial^2 v}{\partial x^2} - rc\frac{\partial v}{\partial t} - rj_{Na}(v) = 0 \qquad (2)$$

which is a linear p.d.e. both above and below the threshold value $v=V_2$. Thus linear exponential solutions can readily be written above and below threshold. Then application of the power balance condition (Eq. (2) of "Transmission of Information by Solitary Waves") yields the spike velocity as

$$u = \left[\frac{g}{rc^2} \frac{(V_3-V_2)^2}{(V_3-V_1)(V_2-V_1)} \right]^{1/2} \tag{3}$$

This equation should be useful because every parameter can be directly measured for the axon. A corresponding expression has been developed by Richer for the myelinated axon.[5]

STRENGTH-DURATION CURVE FOR THRESHOLD EXCITATION

A classic experiment in electrophysiology is indicated in Fig. 2. Current is introduced into the axon via a glass microelectrode and "firing" is observed on a cathode ray oscilloscope (CRO). If the current is maintained at a constant level, I, for a short time, τ, the condition for excitation of a spike is found to be

$$I\tau = Q_\theta \tag{4}$$

where Q_θ is an experimentally determined constant with the units of electrical charge. Calculations based on the requirement that the membrane voltage change by some fixed amount lead to the condition $I\sqrt{\tau}=\text{const}$, thus this concept is not correct. To obtain (4) we note that if the displacement current through the membrane (cv_t) dominates the ion current (j_i), then (1b) can be written as

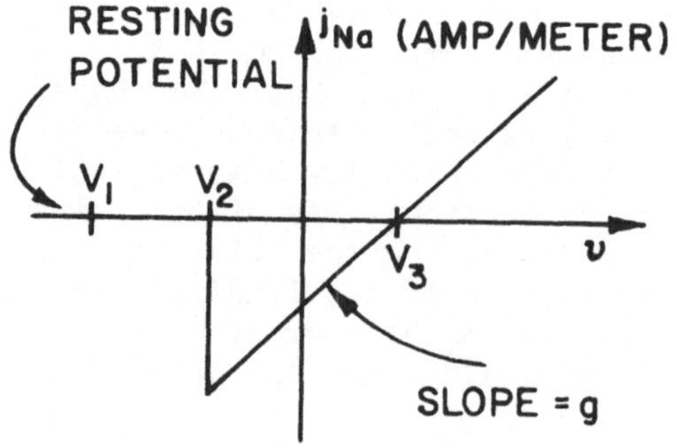

Fig. 1. Sodium ion current vs. membrane voltage

the approximate conservation law[6]

$$\frac{\partial i}{\partial x} + \frac{\partial (cv)}{\partial t} \approx 0 \tag{5}$$

where i is the flow of the conserved quantity. The ion current is small during the leading edge of the pulse,[2] thus we can integrate the flow across the leading edge to get the charge stored

$$Q_o = \int_{\text{leading edge}} i\,dt \tag{6}$$

This is an "approximately" conserved quantity for the approximate conservation law (5). For a spike $v=v(x-ut)$ and using (1a)

$$Q_o = \frac{(V_3-V_1)}{ru} \tag{7}$$

This charge stored in the leading edge of a fully developed action potential Q_o is about twice as large as the threshold charge Q_θ for the squid giant axon considered by Hodgkin and Huxley.[6] This seems to be a reasonable "margin of safety" against accidental extinction of the spike.

INFORMATION PROCESSING IN DENDRITIC TREES

The implication of the preceding section can be summarized by the simple equation

$$Q_\theta = \alpha Q_o \tag{8}$$

where α is a constant of the order of 1/2. This equation may in turn be used to calculate the "logical" character of a branch in

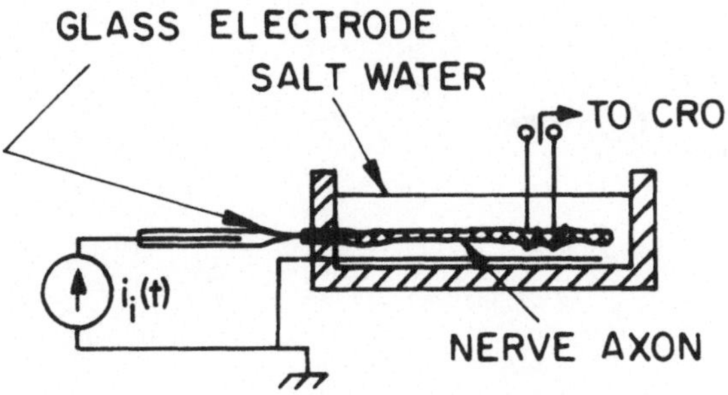

Fig. 2. Stimulation of a nerve axon

a dendritic tree under the assumption that dendritic propagation is active. To carry this through note from (7) that $Q_o \propto d^{3/2}$ where d is the fiber diameter. Coincidentally the characteristic admittance of a fiber is also proportional to $d^{3/2}$.[8] Equating the fraction of the input charge on branch A (see Fig. 3) which enters branch C to the charge necessary to fire branch C, leads to the condition

$$\frac{d_1}{d_2} = \left(\frac{\alpha}{1-\alpha}\right)^{2/3} \tag{9}$$

If the daughter-parent diameter ratio (d_1/d_2) is larger than that given by (9), the branch will act as an "OR" element. Either A "OR" B can fire branch C. If (d_1/d_2) is less, the branch will act as an "AND" element. Such effects have been observed by Tauc and Hughes in axons of the mollusk Aplysia.[9]

These ideas may help to clarify the role which the Purkinje cell plays in the functioning of the cerebellum. There is a considerable body of evidence to indicate that Purkinje cell dendrites do indeed propagate spikes.[10] The question is whether these cells act as simple "Perceptrons" as Albus has suggested,[11] or are much more complex.

CORTICAL WAVES OF INFORMATION

Beurle has suggested some time ago that nonlinear wave effects may play a role in the operation of the neocortex.[12] The basic elements of this theory are quite simple, and it is an interesting extension of the solitary wave concept. Beurle's. approach is essentially statistical since he considers a "neural medium" with a large number of threshold elements per unit volume, but he attempts to keep track of the information carried by a

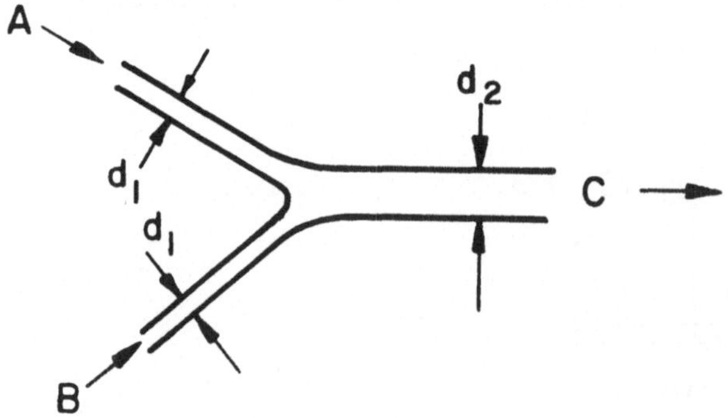

Fig. 3. A simple bifurcation

particular wave. The basic variables, shown in Fig. 4 are: R – the fraction of cells which are sensitive (i.e., have not fired within a refractory period), F – the activity or fraction of cells firing per unit time, Φ – the probability of a sensitive cell being excited above threshold per unit time, and τ – a time delay for firing. Evidently

$$\frac{\partial R}{\partial T} = -F \tag{10}$$

and

$$F(t+\tau) = R(t)\Phi(t) \tag{11}$$

It turns out for various assumptions concerning the probability of interconnection between cells that[13]

$$\Phi = mF \tag{12}$$

where the constant m is essentially the product of the density and the spatial extent of the interconnections. Equation (11) can be written in the form

$$R\Phi - F = \tau \frac{F(t+\tau) - F(t)}{\tau} \approx \tau \frac{\partial F}{\partial t} \tag{13}$$

which, using (12), can be combined with (10) to obtain $\tau dF/dR = (1-mR)$.

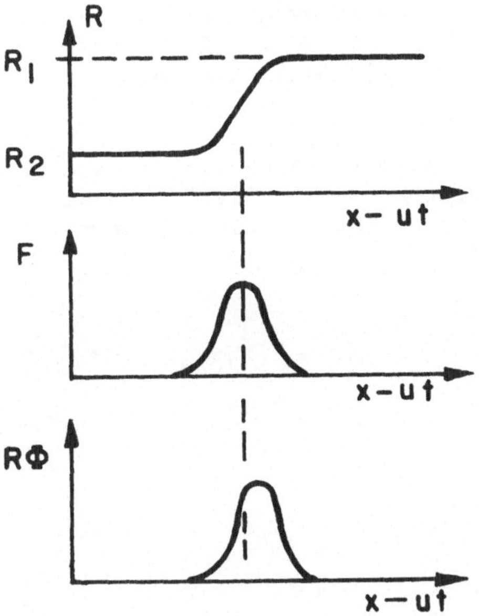

Fig. 4. A wave of information

This integrates to

$$K + R - \frac{m}{2} R^2 = \tau F \tag{14}$$

where K is an arbitrary constant of integration. The two roots of (14) when F=0 are the values R_1 and R_2 in Fig. 4.

It is important to note that this analysis does not determine the velocity of propagation. This is related to the fact, which was discussed by Beurle, that without the assumption of inhibitory interconnections these waves appear to be unstable. Wilson and Cowan have recently extended their "two population" (i.e., excitory and inhibitory neurons) to this case.[14]

One must be careful in assigning a role to such waves of information in the operation of the neocortex. Beurle has suggested a scheme for sequential memory which is related to van Heerden's proposal for three dimensional storage of optical information,[15] and not unlike holographic storage. On the other hand, information waves may be primarily useful as "communication links" between Hebb's "cell assemblies"[16] or Caianiello's "reverberations".[17,18]

In any case Beurle's spatially oriented analysis complements that of Caianiello, de Luca and Ricciardi in which attention is focused upon the interconnection matrix of the neural network.[19] Their result was essentially that the period of reverberation is bounded by $2^K\tau$ where K is the rank of the interconnection matrix. In the simple neuron ring of Fig. 5, K=N. But, from the geometrical character of the net, it is clear that the period is only $N\tau$.

Since it seems appropriate these days to close with a note of ecological awareness, we observe that the "fairy rings" of mushrooms, which excited the superstitions of the Druids, appear to be waves of botanical information.[20,21]

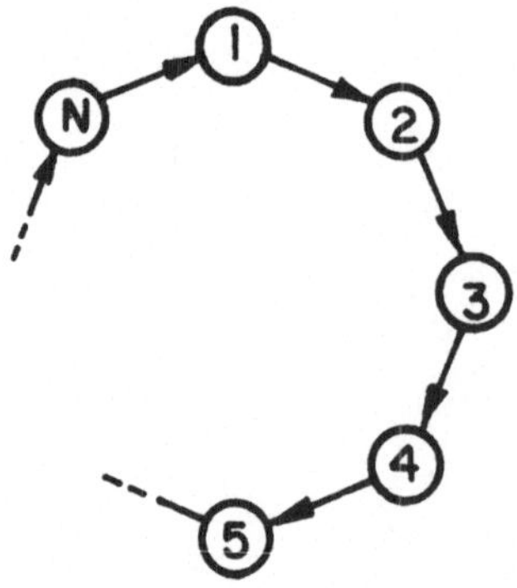

Fig. 5. A neuron ring.

REFERENCES

1. Scott, A.C., Transmission line equivalent for an unmyelinated nerve axon, Math. Biosci., 13, 47, 1972.

2. Hodgkin, A.L. and Huxley, A.F., A quantitative description of membrane current and its application to conduction and excitation in nerve, J. Physiol. (London) 117, 500, 1952.

3. Scott, A.C., Analysis of nonlinear distributed systems, Trans. IRE on Circuit Theory, CT-9, 284, 1962.

4. Moore, J.W., Electronic control of some active bioelectric membranes, Proc. IRE. 47, 1869, 1959.

5. Richer, R., The switch line: A simple lumped transmission line that can support unattenuated propagation, Trans. IEEE on Circuit Theory, CT-13, 388, 1966.

6. Scott, A.C., Strength duration curves for threshold excitation of nerves, Math. Biosci. (to be published).

7. Scott, A.C., Information processing in dendritic trees, Math. Biosci. (to be published).

8. Rall, W., Theory of physiological properties of dendrites, Ann. N.Y. Acad. Sci., 96, 1071, 1962.

9. Tauc, L. and Hughes, G.M., Modes of initiation and propagation of spikes in the branching axons of molluscan central neurons, J. Gen. Physiol., 46, 533, 1963.

10. Llinás and Nicholson, C., Electrophysiological properties of dendrites and somata in Alligator Purkinje cells, J. Neurophysiol., 34, 532, 1971.

11. Albus, J.S., A theory of cerebellar function, Math. Biosci., 10, 25, 1971.

12. Beurle, R.L., Properties of a mass of cells capable of regenerating pulses, Phil. Trans. Roy. Soc. (London) B, 240, 55, 1956.

13. Scott, A.C., (unpublished).

14. Wilson, H.R. and Cowan, J.D., Kybernetic (to be published).

15. van Heerden, P.J., Theory of optical information storage in solids, Applied Optics, 2, 393, 1963.

16. Hebb, D.O., Organization of Behavior, Wiley, New York, 1949.

17. Caianiello, E.R., Outline of a theory of thought processes and thinking machines, J. Theoret. Biol., 2, 204, 1961.

18. Caianiello, E.R., Decision equations and reverberations, Kybernetic, 3, 98, 1966.

19. Caianiello, E.R., de Luca, A. and Ricciardi, L.M., Reverberations and control of neural networks, Kybernetic, 4, 10, 1967.

20. Krieger, L.C.C., The Mushroom Handbook, Dover, New York, 1967.

21. Shantz, H.L. and Piemeisel, R.L., Fungus fairy rings in eastern Colorado and their effect on vegetation, U.S.D.A. J. Agr. Res., 11, 191, 1917.

NEURAL NETS: A BRIEF SURVEY

Eduardo R. Caianiello

Laboratorio di Cibernetica
del C.N.R.
Arco Felice, Napoli.

1. - In a study of parallel as compared to serial computation the mathematical model of nervous activity, subsumed under the name "Neural Net", definitely deserves a mention. First of all historically: the pioneering attempts of people like Wiener, von Neumann, McCulloch and Pitts to bring in this way under the aegis of mathematical and physical reasoning the world of thought did have a tremendous impact on science: the theorems of Kleene, automata theory, whole parts of computing sciences and language theory were deeply affected, when not originated, by them. Then also, when later developments showed, rather unexpectedly, that a number of significant mathematical statements can be made on an appropriately generalized model of the kind, the hope was revived in several scientists - this writer among them - that such model could usefully describe, in a mathematically controllable way, phenomena involving the parallel interaction of very large numbers of decision elements (the "yes"-or-"no" mathematical "neurons") and their serial evolution in time.

Besides the obvious applications to the study of thought processes, the formal equations of Neural Nets and the concepts attached to them are applicable to many classes of phenomena, in economy, genetics, chemical kinetics, cell-space approaches to developmental biology, physics itself (where they seem to provide a description of, say, magnetic facts intermediate between that offered by the Ising model and that of percolation theory).

Further investigations make it now also apparent that the situations described by this model go beyond the scope of thermodynamics, which is only concerned with equilibrium states and fluctuations about them; cyclic, as well as other phenomena, such as for instance adaptation and stability of dynamical behavior, should soon find a unitary and fitting description in terms of Neural Net theory.

It is known since long time that Neural Nets, Automata, Computing Languages are not really different theories. It might well be therefore that mathematical information coming from new developments of Neural Net theory could lead to results relevant also for the extant computing sciences.

This lecture will contain only a bare resumé of essential ideas, results and perspectives. The interested reader will find detailed accounts of past work in the references, and an indication of some forthcoming mathematical developments in the accompanying paper by E. Grimson and this author.

2. - Neural Nets as a Model of Neural Activity.

Our model of a neural network consists mathematically of three parts:

I - Neuronic (or decision) equations (NE)

II - Mnemonic (or evolution) equations (ME)

III - Adiabatic Learning Hypothesis (or Rule)(ALH)

It is to be strongly emphasized, if we wish to refer to the explicative value of the model for intelligent behavior, that,to within wide limits, the specific form of NE or ME is not relevant; many sets of non-linear equations might do as NE, and there is a similar freedom with ME. This will become apparent in the sequel. We begin by describing our model (called, henceforth, "neural network", NN for short) in some more detail.

I - A NN is a system composed of non-linear elements (nodes of the NN) which interact through couplings (characterizing the meshes of the NN) in a manner described by NE; these give its instantaneous behavior: conventionally, a solution of the NE may be termed a "thought" of the NN.

II - The strengths of the couplings may vary themselves as a consequence of the activity of the NN: such changes, when they occur, are described by ME.

III - The rate of change typical of II is "secular" with respect to the times typical of I.

III is required in order to decouple the mathematics of NE and ME; it is suggested by biological evidence [(1), (2)], the denomination "adiabatic" being well familiar from physics; in an artificial net, it should be taken as a prescription or rule when learning against noise is required. Ultimately, it will be imbedded in the form of the ME.

A NN appears thus as a diffuse structure, which contains memory in its meshes, to and from which any number of nodes (up to all) can act as inputs and outputs.

It was shown long ago (2) that, on such premises, qualitative descriptions can be obtained to fit all processes that we usually associate with intelligence: conditioning of various types, reintegration, form analysis, learning, concept formation by abstraction, etc. It was even found, quite unsuspectingly, that a NN has a need for sleep, and that a mathematical kind of "psycho-analysis" can be defined for it!

These results should not change much with models obeying NE different from those used by us (3), (4) : since any logical element can be described in terms of others, there is clearly a high degree of equivalence among possible neural models; our choice was dictated mostly by mathematical convenience.

The irrelevance of the specific NE and ME of a neural model (we only require it, in essence, to satisfy our axioms I, II and III) is very similar to the familiar situation in statistical physics. Intelligence, like, but much more subtly than thermodynamics, is a collective phenomenon with respect to which whole classes of distinct microscopic descriptions will prove equivalent.

This situation is both satisfactory and deluding. On the one hand, it shows that a physical explanation of intelligent animal behavior can be soundly attempted in a variety of ways: this is per se a tremendous progress, although we may be so inured to this idea by now as not to react emotionally to it as we would undoubtedly have done only a few decades ago. On the other hand, it does very little beyond this, with one exception: the behavior of solutions of non-linear equations such as NE can be much more diversified and subtle than may be guessed a priori, so that observing the dynamics of simulated NN's can be highly instructive and lead to the conception of new modes of activity. The structure of the NN and its evolution in time become the next subject of study: these aspects belong, though, rather to the fields of cybernetics or "artificial intelligence", and shall not be discussed here; our main interest in the present context is with NE, whose understanding is, in our view, a necessary preliminary for further progress.

3. - The Formal Neuron.

The typical element of our model is a discriminator with delayed response. If

$$(1) \qquad 1\,[x] = \begin{cases} 1 , & x > 0 \\ 0 , & x \leq 0 \end{cases}$$

denotes the Heaviside step function, s a threshold value, τ a delay, A (t) the (total) applied stimulus, then:

$$(2) \qquad u(t+\tau) = 1 \left[A(t) - \int_{-\infty}^{t} K(t,t')\, u\,(t')\, dt' - s \right]$$

is the neuronic or decision equation describing the element, which henceforth, more or less arbitrarily, we convene to call (formal) neuron (5). K(t,t'), which unless controlled from the outside can be taken simply as K(t-t'), is a kernel describing the refractoriness of the neuron after it has responded. The integral at r.h.s. of (2) may be replaced, as a further approximation, by a sum over past times.

As a model for a single biological neuron, (2) should not certainly be taken too seriously. Objections can of course be met, if one wants to describe a real neuron of known behavior, by using for it a suitable assembly of such formal neurons; a little juggling with connections, and the addition of trivial elements, can expand (2) into anything wanted, still remaining within the frame work of our NE. The real neuron would appear then as a superintegrated circuit, of which elements (2) are the dominant components. Such considerations are not relevant, of course, when studying the behavior of generic large assemblies of neurons; the last remark is intended merely to avoid trivial misunderstandings.

An important question arises, concerning the performance of the formal neuron (2): how does it compare, after all, with that of a real neuron? The answer is much more satisfactory than might have been hoped for, as is shown in ref.(5); there the (periodic) responses under constant stimuli A are computed in closed form for exponentially decaying refractoriness, and found to have frequencies given by:

$$(3) \qquad n(\mu) = \frac{\alpha}{\log\left[1 + \dfrac{\beta(\beta-1)}{\mu(1-\mu)}\right]} \simeq \frac{\alpha}{\beta(\beta-1)}\, \mu(1-\mu) \text{ if } \mu \simeq \frac{1}{2}$$

where

$$K(t;t') = \lambda\, \ell^{-\alpha(t-t')}$$

$$\beta = \ell^{\alpha\tau} \qquad\qquad \lambda > 0$$

$$\mu = \frac{\alpha}{\lambda}\,(A-s)$$

(consistency requirements impose that $0 \leq \mu \leq 1$)

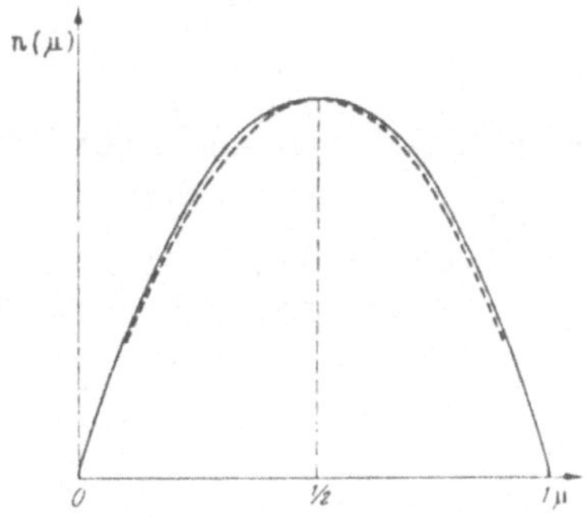

Frequency response as function of μ. The dotted curve is a parabolic approximation which holds for

$$\frac{\beta\,(\beta-1)}{(1-\mu)} << 1$$

Fig. 1

Figure 1 shows that the <u>same</u> frequency may occur for <u>two</u> different values of the stimulus: a feature which caused us much perplexity at the time (1965), but was fully confirmed experimentally, in a qualitative manner of course, in the following years.

As a last remark, it is perhaps worth noting that, as is shown in ref. 5, a thorough study of (2) is possible; in particular, one could obtain in this way a knowledge of the applied stimulus $A(t)$ from the response function $u(t)$, which should be a better way of studying neuronal behavior than with the current approaches, which mix the deterministic operation of an element like (2) with the statistics of its <u>output</u>, instead of <u>input</u>, as should be more natural. Such a study would also clarify the properties of decision elements (2) in general, e.g., as filters or oscillators, and might improve our qualitative and quantitative understanding of circuitry in which linear oscillators and couplings are replaced by non-linear coupled elements.

4. - Neural Nets

We take as NE the following[2]:

$$(4)\quad u_h(t+\tau) = 1\left[\sum_{k,r} a_{hk}^{(r)}\, u_k(t-r\tau) - s_h\right] \qquad (h,k = 1 \ldots N)$$

which generalize the classic McCulloch and Pitts descriptions[3]; the coupling coefficients $a_{hk}^{(r)}$ are taken to be any real numbers (>0 if at time $r\tau$ the connection $k \rightarrow h$ is excitatory, <0 if inhibitory; a_{hh} expresses the action of refractoriness, if any; $\sum_r$ replaces in (4) the $\int_{-\infty}^{t} dt'$ of Eq. (2), as a further simplifying assumption). Time is quantified in intervals τ .

Eq.'s (4) describe discriminators, i.e., threshold elements, whose inputs are linear combinations of the outputs of the NN. Whatever logic elements one wishes to select for the construction of a NN, they can always be described, as was said before, by a NN of type (4) (and conversely): this suffices to explain the equivalence among models mentioned in Sect. 1, and increases the generality of (4).

We add a further formal simplification, which might be removed without affecting the validity of the results cited in the sequel, by assuming that each state $u_1(t)$, $u_2(t), \ldots, u_N(t)$ in (4) depends only on the immediately preceding one, so that (4) reduces, in vector and matrix notation, to:

$$(5)\quad \vec{u}_{m+1} = 1\left[A\,\vec{u}_m - \vec{s}\right]$$

Of course, the N neurons can be split into classes labelled <u>input</u>-, <u>output</u>- and <u>inter-neurons</u>, and A partitioned accordingly, etc. For the present purposes it will suffice to consider <u>autonomous nets</u>, i.e. without inputs and outputs, so that only the setting of the initial state in (5) affects the subsequent operation of the NN.

Enough was written on this subject to make the repetition of motivations or proofs unnecessary (see references).

Note that, by setting

$$\vec{V}_m = A\ \vec{u}_m - \vec{s}$$

($V_{h,m}$ is thus the effective excitation impinging upon neuron h at time τm), eq.'s (5) become

$$(6) \qquad \vec{V}_{m+1} = A\ 1\left[\vec{V}_m\right] - \vec{s}$$

This is an alternative way of looking at the operation of a NN: by recording the excitations $\vec{v}$ instead of the state $\vec{u}$. It is worth noticing that (b) can also be taken as describing elements which assume a discrete and finite number of values, instead of only 0 and 1; in such meaning eq.'s (6) become the equations of a multivalued logic and have been investigated and used for the design of computers at the E.T.L. in Tokyo, under the name of AETL (An Extended Threshold Logic).

The main class of problems in this context concerns the synthesis of NN's of arbitrary size having prescribed dynamical behaviors. Outstanding among them is that of designing an autonomous NN which, no matter how excited at the initial time, will only undergo periodic sequences at states (termed reverberations) of period equal to, or not greater than, a prefixed duration $R\tau$ (after, possibly, a transient phase likewise limited). The connections with mechanisms of biological memory, or with the design of new type of memory for computers, are evident.

Boolean algebra is not of much use; furthermore, a theoretical solution, however complete, is not interesting if it does not afford also a constructive technique well within the limits of practical feasibility (how stringent this requirement can be is immediately realized on remarking that a NN of N neurons has an attainable [6] $R = 2^N$). The form (4) or (5) and (6) of our NE, together with the use of matrix algebra, proved very helpful in this respect [7] - [10]. Although a general answer, of the kind one expects e.g. in the control theory of linear systems, is clearly out of question, sufficient conditions (which amply suffice for all our needs) come out quite simply from elementary considerations of matricial calculus, in a form that might be handled directly to a design engineer. Looking, for simplicity sake, at NE (5), one finds that if K is the rank of the NxN matrix A, then:

1) Eq.'s (5) admit of N-K constants of motion; the sum of all possible periods cannot exceed $2^N - 2^{N-K+1} + 2.$ (10)

2) Reverberations can be limited to periods equal to, or not exceeding, 2^K or a number of the same order, with the addition of simple control devices, or with suitable further specifications on the permissible magnitudes of the couplings a_{hk}.

3) It is possible to obtain, at will, reverberations which are independent of the initial stimulation (after the transient phase), or instead, critically dependent upon it.

Furthermore, regardless of rank:

4) If A is a direct product:

$$A = A_1 \times A_2 \times \ldots \times A_s$$

solutions of (5) can be easily obtained from those of systems (5) associated respectively to $A_1, A_2, \ldots A_s$ (this requires $\sum_k a_{hk} = 2s_h$, which is the condition for (5) to reduce to normal form, as defined in ref.(8), i.e. for a NN to be selfdual).

5) The general synthesis problem can be split into two distinct parts:

a) to decide whether a given sequence of states is realizable by a NN of type (5);

b) to determine, then, the matrix A of the NN which realizes the sequence.

The explicit solution is trivial, but cumbersome for large nets; allowed sequency of no more than N states can of course be realized on inspection. More work seems to be necessary in this direction.

In conclusion, we list some problems, which partly have been, partly are being studied by us:

1) for given NE, to analyze their reverberations in order to learn to associate the response of the NN not with an individual reverberation, but with a whole class of them. This is the first use of statistics on our side, and must of course be pursued further, so as to learn how to elicit a correct response from the NN even if it has some malfunctions;

2) To design a NN, and a way of reading out its dynamical responses, so that failure of neurons (or unwanted changes of connections a_{hk}) does not alter the reliability of the reading (this overlaps with 1) only in part);

3) next in difficulty, to design a NN which, in case of failure of an element, re-adjusts automatically its structure so as to compensate for the failure; i.e. a self-repairing NN;

4) a fascinating gamut of problems appears in relation to the study of communication among neurons of a NN; the exploration, although just begun, shows good promise;

5) can we, in analogy with the procedures of quantum field theory, linearize our N.E.? The answer is positive: we refer to the cited accompanying paper by E. Grimson and the author for some preliminary information.

As a final remark, it is appropriate to mention that a curious phenomenon was exhibited by computer-simulated NN's, designed so as to have very short reverberation periods [11]. It was found that, if $R\tau$ is the period, forced inhibition of some n neurons for $R\tau$ sec changed a reverberation into another one; the same n neurons were inhibited later, again for $R\tau$ sec, and again the reverberation would change into a new one; proceeding in the same way, it soon happened that the final reverberation was such as to remain unaffected by any further inhibitions of those same neurons for $R\tau$ sec. (cyclic behaviours were also, alternatively, observed to occur sometime). This kind of functional adaptation, arising out of a complete deterministic system (no ME at work!), may find an explanation after point 5) in the list above is sufficiently explored. About Mnemonic Equations we shall say nothing here; interesting studies along this line are due to T. Ishihara[12], to whom we refer the reader.

5. - The Role of Probability.

The use of statistics cannot certainly be avoided in any realistic study of NN. Our own use of deterministic models of NN was at all times qualified as being merely a first step on a road known to be fraught with non-deterministic aspects; any study of adaptivity through ME, for instance, would otherwise be inconceivable. Even with a deterministic NN, ruled only by NE, statistics enters in by necessity, as soon as the NN is acted upon by a stochastic environment. Rather than engaging in debates on so vexed a question, we only outline our own approach.

The first question is whether at the time scale at which only NE are relevant, in micro-time we may say, there are things that one can obtain from probabilistic but not from deterministic NN's; the stated restriction means that only wealth of distinct states and sequences thereof in a NN is considered to be relevant to the issue. Our answer is no; a probabilistic description at this level leads to consider as equivalent otherwise distinct states of the NN, and we presume (the proof or disproof of this statement is an open problem left to the interested reader) that, for every probabilistic NN, a smaller equivalent deterministic NN can always be designed which performs without error a same assigned task. The next questions, at the same level, is whether a probabilistic NN can prove more economical than a deterministic one (in realistic cases performance and economy, not respective size, are now relevant). This is very important, because the technological advance most needed in this field might be one that will enable the connections of a NN to grow spontaneously when neurons are suitably stimulated. One can start of course outrightly with statistical treatments of NN's; this seems actually to be the prevalent fashion; our choice is, rather, to consider, perhaps more naively, ways of using NN's and of transmitting signals within, to and from them, which will not be affected by malfunctions (within limits, of course).

Probability certainly plays an important role when it characterizes the external world in which a "robot" (let us thus call loosely a machine whose "nervous system" is a suitable NN) operates. This subject is very relevant and must yet be studied; little is known at present, for instance no valid definition of entropy is known for NN's.

Our approach is to ignore as much as possible, for the time being, all such questions; we intend to take care of them indirectly by studying ways of utilizing NN's which may not be affected by malfunctions; this we try to achieve by introducing tolerances, redundancies, error-correcting devices, by defining as responses of a NN e.g. whole sets of reverberations etc. Roughly speaking, we feel that the most urgent issue of robotics (like with early telephone systems) is to devise first of all something that works, no matter at what cost. Statistics we consider as necessary at later stages, for designing things that work more reliably and economically; as misleading, otherwise.

REFERENCES

1) D. O. Hebb - "The Organization of Behaviour" - (1949). J. Wiley - New York.

2) E. R. Caianiello - "Outline of a Theory of thought processes and thinking machines" - J. Theor. Biol. 1, 204 (1961).

3) W. S. McCulloch and W. Pitts - Bull. Math. Biophys. 5 (1943).

4) J. Simoes da Fonseca - "Bases Neuronais da Vida Psiquica"- Lisbon, 1969.

5) E. R. Caianiello and A. De Luca - Kybernetik 3, 33 (1966).

6) S. Arimoto - "Periodic sequences of states of an autonomous circuit consisting of threshold elements" - Trans. Inst. Elec. Commun. Eng. Japan 1963 (2), p. 17 (in Japanese).

7) E. R. Caianiello, A. De Luca, L.M. Ricciardi - Kybernetik 4, 10-18 (1967).

8) E. R. Caianiello - Kybernetik 3; 98 (1966).

9) A. Aiello, E. Burattini, E. R. Caianiello - Kybernetik 7, 191 (1970).

10) L. Accardi - Kybernetik 8, 163 (1971).

11) E. Burattini and V. Liesis - "Analysys of the Models of Neural Nets" - Kybernetik 10, 38 (1972).

12) T. Ishihara - Math. Jap. 15, 119 (1970).

CELL SPACE APPROACHES IN BIOMATHEMATICS

Tosio KITAGAWA

Honorary Professor of Kyushu University and
Director of International Institute for Advanced Study
of Social Information Science, FUJITSU, Tokyo

1. INTRODUCTION AND SUMMARY

The primary purpose of this paper is to explain some principal aspects of various approaches appealing to certain mathematical formulations based upon cell spaces and to illustrate their implications to biomathematics. In Sects. 2 and 3 we explain our specific investigations on cell space in connection with a family of local transformations satisfying the principle of local majority. In Sect. 4 general considerations and a set of five interpretations of basic notions of cell space approaches are given to explain how and why an abstract and purely mathematical formulation of cell space can have any connection with biological problems and hence can deserve to be a theoretical framework in biological sciences. The general attitude given in Sect. 4 is illustrated in Sect. 5 in which various biological problems, including birth, growth, formation of forms, pattern disintegration, and death, as well as some ecosystem problems are discussed with reference to various works done by many authors. The content of Sections 2 and is a brief summary of a series of the papers by the author[1~10] and his cooperator[11~15], while we refer to the works of other authors Ulam[16], Read[17], Turing[18] and N. Goel et al.[19] in Section 5. We do not intend to give an extensive survey of all the existing results belonging to the realm of what one may call space approaches. However, we shall explain our ideas on how to build up a certain type of cell space approach, on the basis of our own work on the specific cell space approaches that require local majority transformation. The more detailed description can be found in our recent paper[20].

2. FORMULATION OF THE CELL SPACE APPROACH REQUIRING LOCAL MAJORITY TRANSFORMATIONS

[1] Cell Space. We consider a finite, two-dimensional iterative array of finite state cell automata, where each unit cell may be a square, an equilateral triangle or a regular hexagon. Lattice representations may be used for the representation of cell spaces as shown in Fig. 1(a, b, c). Some adequate combination of different unit cells, such as bathroom-tile lattice, is also worthwhile to our considerations in specific problems.

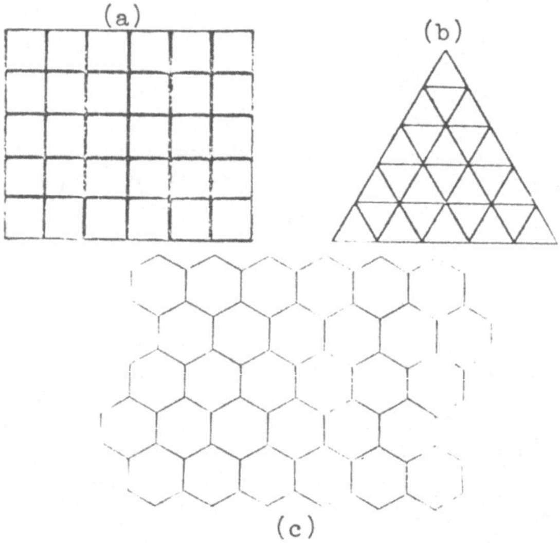

Fig. 1. (a) A 5×6 cell space with square unit cells, (b) [5] cell space with triangular unit cells, and (c) cell space with hexagonal unit cells.

[2] Basic Cell Spaces and Local Transformations. To each cell, located in a cell space defined in the previous paragraph (A), there corresponds one and only one state which belongs to a certain fixed set of states. The whole set of such states, that is, an allocation of each one state to each one cell in the whole cell space, is called a configuration of the cell space. A correspondence T of a configuration C in a cell space to another configuration C' in a cell space is called a mapping transformation in a cell space and is denoted by $T : C \rightarrow C'$. A mapping transformation is called a local mapping transformation (LT) if it is defined with restriction to a subset of the whole cell space, which is called a basic cell space. The examples of basic cell spaces consisting of four unit cells are given in Figure 2. In our cell space approach, our choice of basic cell spaces is based upon the following two considerations:

(i) To choose a cell set with the simplest form and with the minimum number of cells so as to make possible the deepest observations on the application of LT to our cell space.

(ii) To choose a cell set without adhering to any traditional pattern so as to secure all the possibilities of discussing any uncultivated area of investigation on cell space configurations.

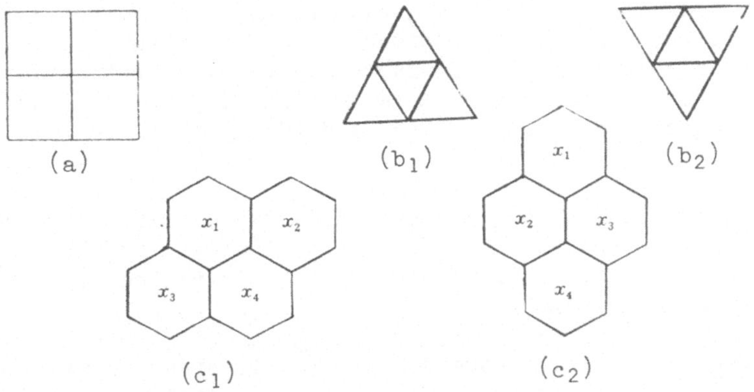

Fig. 2. Basic cell spaces consisting of four unit cells. (a) 2×2 basic cell space with square unit cells, (b_1) basic cell space with triangular unit cells, (b_2) basic cell space with triangular unit cells, (c_1) basic cell space with hexagonal unit cells, and (c_2) basic cell space with hexagonal unit cells.

Now let us define a specific local transformation, which we shall call a local transformation satisfying the principle of local majority. Let us explain an LMT with reference to a basic cell space consisting of four unit cells. Let two configurations in the same basic cell space be denoted as

$$C(X) = \begin{pmatrix} x_1 & x_2 \\ x_3 & x_4 \end{pmatrix}, \qquad C(Y) = \begin{pmatrix} y_1 & y_2 \\ y_3 & y_4 \end{pmatrix}, \tag{2.1}$$

where each x_i and each y_i may be either 1 or 0. A local mapping transformation $C(X) \xrightarrow{1} C(Y)$ is said to satisfy the principle of local majority, denoted by LMT, when

$$y_j = \begin{cases} 1 & \text{if } S(x) = 3,\ 4, \\ x_j & \text{if } S(x) = 2, \\ 0 & \text{if } S(x) = 0,\ 1, \end{cases} \tag{2.2}$$

where

$$S(x) \equiv \sum_{i=1}^{4} x_i \,. \tag{2.3}$$

LMT can be defined in terms of threshold logic. In fact (2.2) with (2.3) is equivalent to

$$y_j = 1\left(x_j + \sum_{i=1}^{4} x_i - 2\right), \tag{2.4}$$

where $1(u) = 1$ for $u > 0$ and $1(u) = 0$ for $u \leqslant 0$.

LMT can be defined also for triangular and hexagonal basic spaces as shown in Fig. 2(b_1, b_2, c_1, and c_2). It is noted that there may be several generalizations of LMT for various directions. The size of basic cell spaces may be greater than 4. There may be also two switching constants instead of just one, namely one switching constant, 2, in (2.2) and (2.4).

[3] Firing Systems. There are basically two different procedures of firing, namely sequential and parallel firing. For the sake of brevity let us explain what we mean by sequential and parallel firing in following Examples.

Example 2.1. Let us consider a 3×3 cell space consisting of nine sequare cells as shown in Fig. 3(a). A configuration C and a set of four central points each of which is associated to each 2×2 basic cell space, denoted by q_i (i = 1, 2, 3, 4), as in Fig. 3(a), and can be abbreviated by Fig. 3(b). Now any firing procedure in the 3×3 cell space can be enunciated by a choice procedure among these four points, q_1, q_2, q_3, and q_4.

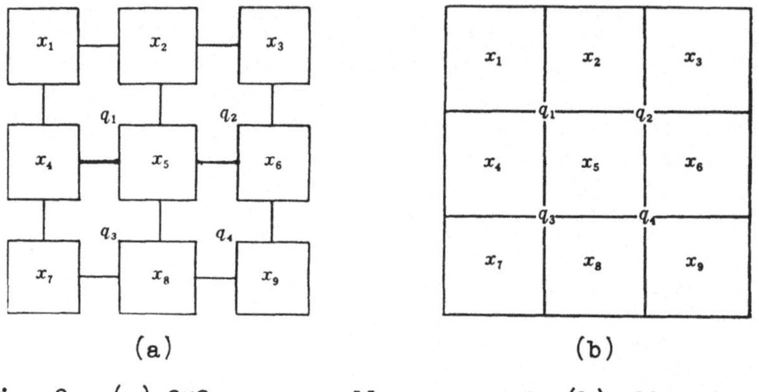

Fig. 3. (a) 3×3 square cell space, and (b) abbreviated notion of 3×3 square cell space.

The main difficulty for simultaneous firing of at least two central points among these four points, q_i (i = 1, 2, 3, 4), comes from the fact that we need some coordination scheme for two sometimes mutually contradictory outcomes. For instance

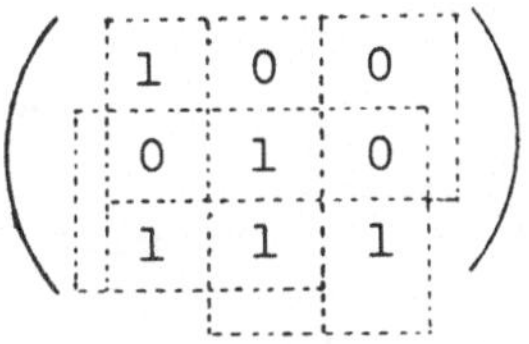

$$\begin{pmatrix} 1 & 0 & 0 \\ 0 & 1 & 0 \\ 1 & 1 & 1 \end{pmatrix} \qquad LT_{q_1}: \begin{pmatrix} 1 & 0 \\ 0 & 1 \end{pmatrix} \rightarrow \begin{pmatrix} 1 & 0 \\ 0 & 1 \end{pmatrix}$$

$$LT_{q_2}: \begin{pmatrix} 0 & 0 \\ 1 & 0 \end{pmatrix} \rightarrow \begin{pmatrix} 0 & 0 \\ 0 & 0 \end{pmatrix}$$

$$LT_{q_3}: \begin{pmatrix} 1 & 0 \\ 1 & 1 \end{pmatrix} \rightarrow \begin{pmatrix} 1 & 1 \\ 1 & 1 \end{pmatrix} \qquad (2.5)$$

$$LT_{q_4}: \begin{pmatrix} 0 & 1 \\ 1 & 1 \end{pmatrix} \rightarrow \begin{pmatrix} 1 & 1 \\ 1 & 1 \end{pmatrix}$$

The simultaneous application of LT_{q_1} and LT_{q_2} is contradictory to each other: $LT_{q_1} \oplus LT_{q_2} = \Lambda$, while that of LT_{q_1} and LT_{q_3} is compatible to each other:

$$LT_{q_1} \oplus LT_{q_3} = \begin{pmatrix} 1 & 0 & 0 \\ 0 & 1 & 1 \\ 1 & 1 & 1 \end{pmatrix} . \qquad (2.6)$$

Similarly we have

$$LT_{q_1} \oplus LT_{q_4} = \Lambda$$
$$LT_{q_2} \oplus LT_{q_3} = \Lambda$$
$$LT_{q_2} \oplus LT_{q_4} = \Lambda \qquad (2.7)$$
$$LT_{q_3} \oplus LT_{q_4} = \begin{pmatrix} 1 & 0 & 1 \\ 1 & 1 & 1 \\ 1 & 1 & 1 \end{pmatrix}$$

(b) A configuration C is said to be stable if it does not change under any application of LMT to any basic cell space belonging to the cell space. It is shown that stable configurations are nothing but absorbing states in the sense of Markovian processes.

(c) In an $m \times n$ cell space the number of all possible stable configurations is 2^{m+n-1} (Theorem 1 in Ref. 9). The number of all possible configurations is equal to 2^{mn}.

(d) Theorem 3 in Kitagawa and Yamaguchi[9] show that a stable configuration in an $m \times n$ cell space has the following pattern.

(1) There is a pair of partitions such that

$$m = m_1 + m_2 + \cdots + m_k,$$
$$n = n_1 + n_2 + \cdots + n_l, \qquad (2.11)$$

with positive integers m_i and n_j, $1 \leqslant i \leqslant k$, $1 \leqslant j \leqslant l$ where k and l are subject to

$$1 \leqslant k \leqslant m, \qquad 1 \leqslant l \leqslant n. \tag{2.12}$$

(2) In correspondence with the pair of partitions given in (1), the whole m×n cell space is divided in kl subspaces, each of which will be denoted by $S(m_i, n_j)(i = 1, 2,\ldots, k;\ j = 1, 2,\ldots, l)$.

(3) (a) The elements of each subspace $S(m_i, n_j)$ are entirely either 1 or zero. A subspace $S(m_i, n_j)$, all of whose elements are equal to 1 will be called Type I, while a subspace, all of whose elements are equal to zero will be called Type 0.

(b) Subspaces of these two types occur alternatively in an m×n cell space (see Fig. 4).

	n_1	n_2	n_3	n_4	n_5	n_6
m_1	1	0	1	0	1	0
m_2	0	1	0	1	0	1
m_3	1	0	1	0	1	0
m_4	0	1	0	1	0	1

Fig. 4. Pattern of stable configuration.

3. CELL SPACE UNDER THE APPLICATIONS OF LOCAL MAJORITY TRANSFORMATION

Here we shall ennunciate some characteristic aspects of our topics in cell space approach under the applications of local majority transformation.

3.1. Configuration Lineage. A configuration C_2 is said to be a direct descendant of a configuration C_1 if there is an LMT by which C_1 is transformed to C_2 which is different form C_1. In this case C_1 is said to be a direct ancestor of C_2. We may seek for direct descendants of direct descendants and so on. This procedure gives us the descendant tree of a configuration. Similarly, we can have an ancestor tree of a configuration. Since the number of all the configurations in an m×n cell space for $(1 \leqslant m, n \leqslant \infty)$ is finite, we shall attain a finite set of configurations each of which has no direct ancestor. In particular there is a set of configurations, called an Eden garden, each of which has no direct ancestor, and no LMT exists which can transform the configuration into itself.

Stable configurations can be characterized as configurations without direct descendants. Isolated stable configurations are stable configurations without direct ancestors. Thus, stable configurations and isolated stable configurations correspond to configurations with no direct ancestors, and to Eden garden, respectively. Kitagawa[6] gave a detailed obsevation of ancestor trees in the 2×2, 2×3 and 3×3 cell spaces. In view of these observations, Kitagawa[6] proceeded to a general discussion of ancestor-descendant relation, that is, lineage of all the configurations in a general $m\times n$ cell space. The result is extremely simple in the sense that all the configurations belong to the same lineage, except the set of isolated stable configurations each constitutes one lineage respectively, as shown in Theorem II in Ref. 6.

3.2. Determinative Subspace for Stable Configuration. Let us start with an $m\times n$ cell space with square unit cells. The set of all possible configurations, whose total number is 2^{mn}, is chaotic and irregular. However, applications of LMT give us a picture of significance leading finally to the set of all stable configurations, whose total number is 2^{m+n-1}. This constitutes a pattern of regularities within chaos in a certain sense. In view of the significance of these pattern formations, it will be important to find a subspace, called a determinative subspace, in an $m\times n$ cell space whose restriction of any configuration determines one and only one configuration in the whole $m\times n$ cell space.

The simplest example of a determinative subspace in an $m\times n$ cell space is given by the set of all the cells belonging either to the first column and/or to the first row. In fact the states of all the other cells $(m - 1)(n - 1)$ are uniquely determined because our configuration is stable under any application of LMT.

Now there follows a sequence of the problems regarding determinative subspaces:

(i) How can we construct a set of determinative subspace?

(ii) What is the process of determination of the states of all the other cells starting with any assigned configuration restricted in a given determinative subspace?

The process of determinations mentioned in (ii) can better be called the process of generation of stable configuration. A determinative subspace is called generative if there exists an effective process of generation for a stable configuration that is uniquely determined. A determinative subspace which is not generative in this sense is called nongenerative.

Our results obtained in this connection are summarized as follows:

(a) Kitagawa[2] discussed stable configurations in cell spaces whose unit cells are either triangular or hexagonal and introduced the notion of generations in connection with animal growth.

(b) Yamaguchi[12] introduced the notion of determinative

subspaces and gave a structural representation of all the stable configurations with reference to a specific determinative subspace $D_0^{(n)}$ in a triangular cell space $\Delta^{(n)}$.

(c) Kitagawa and Yamaguchi[10] gave various examples of determinative subspaces in a triangular cell space $\Delta^{(n)}$, as shown in Figs. 1, 2, 3, in this paper. These examples lead the authors to the process of generation already mentioned, and also to the construction of various determinative subspaces for an m×n cell space, and for a multiple connected cell space whose unit cells are squares (see Ref. 10, Sec. 3).

(d) In the latter part of Ref. 3, the author discussed the problem of determining the whole of cells whose configuration is uniquely determined from an assigned configuration restricted in a given set of cells. In this way, the author introduced the notion of naturally bounded cell spaces. It was in this connection that Kitagawa[3] introduced the definitions of generativeness and nongenerativeness of determinative subspaces.

(e) Yamaguchi[13] discussed the structural aspects of determinative subspaces in a $\Delta^{(n)}$ cell space and gave construction procedures for any generative subspace and also for any nongenerative one. For this purpose Yamaguchi[13] introduced the notion of spiny subspace and elementary subsets.

(f) Kitagawa[4] gave several examples for construction of generative determinative subspaces in a $\Delta^{(n)}$ cell space, with an emphasis on an answer to the question: what is a naturally bounded cell space in which stable configurations are generated from a given determinative subspace? The examples given here give a set of generative linear sets of stable configurations. These yield us a contrast between the construction procedure given by Yamaguchi[13] which is more or less concerned with the generative kern, that is with a start of a compact set of unit cells and additional unit cells suitably chosen at each stage of the generation process.

In view of naturally bounded spaces which were encountered in these examples, Kitagawa[4] stated a theorem proving that in a cell space whose unit cells are triangular, the size of determinative subspace of any convex polygon is equal to the number of unit cells located on the pheriphery of the convex polygon. It is also noted that Yamaguchi[14] showed that the condition of convexity in this theorem is not required in order to attain the same result.

(g) Kitagawa[8] showed that any stable configuration in an m×n parallel quadrangle cell space with triangular unit cells has a pattern similar to that given in Theorem 3 in Ref. 9, which we have already explained. This fact shows that, in spite of several different features of stable configurations in two cell spaces with their respective square and triangle unit cells, there does exist a transparent universality which is valid to both these cell spaces. Also we are reminded of the importance of the notion of naturally bounded cell spaces.

3.3. Structural Aspect.

[1] Introduction of Inhibition State ϕ. Besides two states, 0 and 1, which were already used in our considerations, we introduce a new state ϕ which is called an inhibition state such that any configurations in a 2×2 basic cell space having at least one cell whose state is ϕ, and is invariant under any application of LMT (Definition 4, in Ref. 9).

An introduction of inhibition states in an m×n cell space induces a certain characteristic oscillatory phenomenon which has not yet observed. (See Ref. 9, Examples 4 and 5). Kitagawa[5] discussed, in detail, oscillatory phenomena to be observed in a 4×4 cell space with two inhibition states ϕ as follows:

$$C = \begin{pmatrix} \phi & 0 & 0 & 0 \\ 1 & a & b & 0 \\ 1 & c & d & 0 \\ 1 & 1 & 1 & \phi \end{pmatrix} \equiv C(a, b, c, d), \qquad (3.4)$$

where a, b, c, and d may take 0 and 1, respectively, and hence at the begnining there are 2^4 configurations. Under a certain probabilistic scheme the following assertions are shown to be valid:

(i) From any starting configuration C(a, b, c, d), we reach, with certain positive probability, one of the six configurations C(0, 0, 0, 0), C(0, 0, 1, 0), C(1, 0, 1, 0), C(0, 0, 1,.1), C(1, 0, 1, 1), and C(1, 1, 1, 1).

(ii) Once we reach one of the six configurations mentioned in (i), we shall never get out of these six configurations.

(iii) We shall not be absorbed in any of these six configurations, but, with positive probability, oscillatory phenomena will occur since transitions among these six configurations continue.

Based on this specific observations, Kitagawa[6] showed that oscillatory phenomena occur in a more generalized formulation of an (m+2) × (n+2) cell space with two inhibition states ϕ.

The rôles of inhibition states ϕ are not however limited to involving oscillatory phenomena. In fact several interesting phenomena are introduced by means of inhibition states ϕ:

(a) Partition of cell space into a set of mutually independent cell subspaces

(b) Road construction for information transmission

(c) Rôles of ϕ in a multiple connected cell space

(d) Time changes in inhibition states ϕ and a hierachical system of inhibitions

The details are given in Kitagawa[2].

[2] Influences from the outer world to configurations of cell space. We are interested in a certain set of features of living organisms when they are subject to influences from the outer world,

that is, envirouments to which they have adapted. Some of these features can be observed in our m×n cell space when the separation principle is introduced so as to yield a structure with a distinction of the interior, the boundary, and the environment. See Kitagawa[2].

[3] Structures and firing systems in the case of an m×n cell space. A choice of the firing systems can also be performed with reference to various sampling schemes such as (a) multistage sampling scheme and (b) a cluster sampling scheme, besides the purely independent random sampling scheme. See Kitagawa[2].

4. CELL SPACE APPROACH IN BIOMATHEMATICS

It is our principal research strategy in using the cell space approach that we must concentrate on dealing with mathematical models without any specific connection with real phenomena and without any reliance with direct experimental observations, until we reach a certain stage of development when we can and must consider an interpretation of the mathematics.

Interpretation will then lead us to establishing a correspondence between our mathematical model and natural phenomena, particularly biological phenomena. These connections will give us the possibilities of finding new ways to development of biomathematics, and hence contribute to making clear the road that connects the two areas, biology and mathematics.

In order that any branch of biomathematics can serve as theoretical vehicle to biological science, it should have some connection with observational specifications through which data can be accumulated. Biomathematics will prepare to present a general scheme of hypotheses which can be tested by biologists in the laboratory and in field work.

So far as cell space approaches are concerned, there is a system of five interpretations which will assist us in making clearer and more definite the general views illustrated above.

INTERPRETATION 1. Because of the very fact that our cell space approach is free from any physical aspects, including matter and energy, and is exclusively concerned with functional aspects of abstract automata, we can and must reserve all the possibilities for interpretating fundamental notions about cells, as well as other fundamental notions about other biological phenomema.

In fact it is possible for us to interpret the meaning of a cell according to almost every level of biological existence, such as genes, cells, tissues, organs, individuall units, groups of individuals, communities, and ecosystems, which constitute a hierarchical classification of biological existence as a whole.

INTERPRETATION 2. Two possible states, 1 and 0, of each cell in our cell space approach admit various different interpretations. For instance either of the following two interpretations are of particular importance.

(i) At some biological level of existence, the state 1 represents an existence of an individual unit, while the state 0 represents the absence of it.

(ii) At other biological levels of existence, the states 1 and 0 refer to one of two mutually different kinds of biological existences.

INTERPRETATION 3. In connection with Interpretation 2(ii), cell space should be considered as a field which consists of an iterative array of elementary domains each of which may or may not contain at most one cell (individual unit).

INTERPRETATION 4. A basic cell space may be considered as a set of finite-number elementary domain where, under the condition of firing, each state of cells belonging to the elementary domain of the basic cell space, is subject to an application of local mapping transformation defined over the basic cell space. A firing in a basic cell space is the start of mutual interference among the cells belonging to the basic cell space, and it can be interpretated according to the biological level of the cells themselves. For instance when a cell represents an individual unit having a certain mobility and a certain recognition ability, the phenomena of firing should be involved just by recognition of other unit individuals when they are in the same basic cell space.

INTERPRETATION 5. Changes of configurations in cell space under applications of local mapping transformations may induce a quantitative change regarding the number of unit individuals and/or the number of those having a certain specific state. In these cases the possibility occurs of interpreting changes of configurations in a cell space as biological phenomena such as genesis, growth, breakdown, disintegration, and death in biological population. In Interpretation 2(ii), it is possible to give a certain picture of coexistence and struggle for existence. Furthermore, a structural specification of cell space makes it possible for the cell space approach to introduce an evolution of local mapping transformations with time and hence to be subject to a self-organizing process.

5. BIOLOGICAL PROBLEMS TO BE TREATED IN CELL SPACE APPROACHES

The object of this section is to give several examples of biological problems which can be and sometimes should be most adequately treated within the framework of a cell space approach. According to our interpretation system given in the previous section, it turns out that not a few biological problems discussed by several authors can be recognized as belonging to the realm of cell space approach, as we shall explain.

Here we shall be content with giving a list of topics which we think to belong to the realm of cell space approaches.

[1] Growth problems. Ulam[16] and Read[17]

[2] Morphogenesis. Turing[18] Goel et al.[19] and snow crystallization phenomena

[3] Population of individuals

(a) Deep structure of distribution

(b) Territory in ecology

[4] Local law vs global law regarding population growth and death

[5] Coexistence, mixing, and mutual inbedding of individual units of the different kinds Goel et al.[19]

[6] Mutual imbedding Read[17]

Besides these well known approaches, we may give a game theoretical formulation to be adopted in the realm of the cell space approach. We proceed in the following four steps.

(i) We introduce a new game called YUGO which is defined in connection with LMT.

(ii) We make an analysis of human ability in playing any game, which will lead us to an analysis of the recognition ability and mobility of animals which join the game called struggle for existence.

(iii) In order to give an adequate mathematical formulation of game theoretical approach in cell space, we introduce various sorts of biorobots each of which is an individual unit which has a certain set of recognition abilities and mobilities.

(iv) It is suggested to introduce a game of new types which will serve as a mathematical model for dealing with the problems of ecological aspects.

The details are given in Kitagawa[20]. It is noted that we are now getting out of the realm of classical game theory due to von Neumann and Morgenstern[21].

REFERENCES

1. Kitagawa, T., A contribution to the methodology of biomathematics, Math. Biosci., 12, 329, 1971.
2. Kitagawa, T., Prolegomena to cell space approaches, Mem. Fac. Sci. Kyushu Univ., Ser. A, 26, 1, 1972.
3. Kitagawa, T., The second prolegomena to cell space approaches, Mem. Fac. Sci. Kyushu Univ., Ser. A, 26, 74, 1972.
4. Kitagawa, T., The size of generative determinative subspace of convex polygon in a $\Delta^{(n)}$ cell space, Research Report of Institute of Fundamental Information Science Kyushu Univ., No. 18, 1972.
5. Kitagawa, T., Oscillatory phenomena in the cell space with inhibition states ϕ, Bull. Math. Stat., 15, 57, 1972.
6. Kitagawa, T., An m×n cell space surrounded by time invariant

boundaries and two inhibition states, Bull. Math. Stat., 15, 67, 1972.

7. Kitagawa, T., Generative and generalogical classifications of all the configurations in an m×n cell space under applications of local majority transformation, Bull. Math. Stat., 15, 85, 1972.
8. Kitagawa, T., Stable configurations under local majority transformations in an m×n parallel quadrangle with triangle unit cells, Research Report of Institute of Fundamental Information Science Kyushu Univ., No. 24 (to appear).
9. Kitagawa, T., and Yamaguchi, M., Local majority transformations in cell space, Bull. Math. Stat., 14, 61, 1971.
10. Kitagawa, T., and Yamaguchi, M., Determinative subspace for stable configurations under local majority transformations on cell space, Research Report of Institute of Fundamental Information Science Kyushu Univ., No. 15, 1970.
11. Yamaguchi, M., The stability index of stable configurations under local majority transformation on cell space, Bull. Math. Stat., 14, 83, 1971.
12. Yamaguchi, M., Stable configuration under local majority transformations on cell space, Bull. Math. Stat., 14, 93, 1971.
13. Yamaguchi, M., Structure of determinative subspace in triangular cell space, Bull. Math. Stat., 14, 107, 1971.
14. Yamaguchi, M., The size of determinative subspace of triangular cell space, Research Report of Institute of Fundamental Information Science Kyushu Univ., No. 21, 1972.
15. Yamaguchi, M., Some properties of determinative subspaces in cell space, Research Report of Institute of Fundamental Information Science Kyushu Univ., No. 25, 1972.
16. Ulam, M.S., On some mathematical problems connected with patterns of growth of figures, Proc. Symposia in Applied Math., 14, 215, 1962.
17. Read, R.C., Contributions to the cell growth problem, Canadian Journ. Math., 14, 1, 1962.
18. Turing, A.M., The chemical basis of morphogenesis, Phil. Trans. 237, B, 641, 37, 1952.
19. Goel, N., Campbell, R.D., Gordon, R., Rosen, R., Martinez, H., and Y'cas, M., Self-sorting isotropic cells, Journ. Theor. Biol., 28, 423, 1970.
20. Kitagawa, T., Cell space approaches in biomathematics, Math. Biosci., 17, 1973 (in press).
21. von Neumann, J., and Morgenstern, O., Theory of games and economic behaviors, 3rd Ed., John Wiley & Sons, New York, London, Sydney, 1953, 641.

SOME CONSIDERATIONS ON DYNAMICAL SYSTEMS AND OPERATORS ASSOCIATED WITH A SINGLE NEURONIC EQUATION

Tosio KITAGAWA

Honorary Professor of Kyushu University and Director of International Institute for Advanced Study of Social Information Service, FUJITSU, Tokyo

1. INTRODUCTION AND SUMMARY

In this paper we are concerned with a single neuronic equation

$$x(t+1) = 1\left[\sum_{k=0}^{n-1} a_k x(t-k) - \theta\right], \tag{1.1}$$

where $\mathbf{a} = (a_0, a_1, .., a_{n-1})$ and θ are assigned and

$$1(u) = \begin{cases} 1 & \text{for } u > 0 \\ 0 & \text{for } u \leqq 0. \end{cases} \tag{1.2}$$

Let us introduce an n-dimensional vector, which is called a state configuration,

$$\boldsymbol{\delta}^{(t)} = (\delta_t, \delta_{t-1}, \delta_{t-2}, \ldots, \delta_{t-(n-2)}, \delta_{t-(n-1)}), \tag{1.3}$$

where each component δ takes either one value of 0 and 1, and we define the inner product

$$(\mathbf{a}, \boldsymbol{\delta}^{(t)}) \equiv \sum_{k=0}^{n-1} a_k \delta_{t-k}.$$

By introducing the real-valued concurrence function

$$z(\boldsymbol{\delta}^{(t)}) \equiv (\mathbf{a}, \boldsymbol{\delta}^{(t)}) - \theta \tag{1.4}$$

the functional equation (1) is now reduced to

$$\delta_{t+1} = 1\left[z(\delta^{(t)})\right] \equiv f(\delta^{(t)}). \qquad (1.5)$$

We may and we shall introduce a transformation $\mathcal{L}$ of n-dimensional vectors defined by

$$\mathcal{L}\begin{bmatrix}\delta_t \\ \delta t-1 \\ \vdots \\ \delta t-(n-1)\end{bmatrix} = \begin{bmatrix} f(\delta^{(t)}) \\ \delta_t \\ \vdots \\ \delta\ t-(n-2)\end{bmatrix}. \qquad (1.6)$$

In view of the fact that the operator $\mathcal{L}$ is commutative with translations of t, we may and we shall confine our discussion to the case when t = n-1, that is, essentially to a digraph (X,Γ) where X is the set of all 2^n state configurations, and Γ is the directed connection, that is, the set of arcs (edges), defined by the map

$$\Gamma:\ \delta^{(n)} \to \mathcal{L}(\delta^{(n)}). \qquad (1.7)$$

Since the introduction of neuronic equations by Caianiello[1], a lot of investigations have been by him[2,3] and his colleagues working at the Laboratorio di Cibernetica del C.N.R., Napoli, including de Luca[4,5], Aiello, Burattini, and Caianiello[6], Caianiello, de Luca and Ricciardi[7], as well as many Japanese engineers and mathematicians, such as Nagumo and Sato[8], Amari[9,10], Suzuki, Katsuno and Matano[11], Ishihara[12], and Yamaguchi[13, 14]. In contract to most of these considerations, our attitude for discussing the functional equation (1.1) is to take into consideration all the possible behaviors of dynamical systems defined by (1.1). We discuss not only each individual digraph but also a family of digraphs, and transitions among digraphs are one of our main concerns. In fact we are concerned with a certain family of transition phenomena:

(i) convergences to stable configurations

(ii) convergences to reverberation cycles, and

(iii) disintegration of reverberation cycles into other types of configurations.

For this purpose we introduce a family of operators:

Definition 1.1. The following set of operators $\mathcal{L}_\omega$, $\mathcal{L}_{\bar{\omega}}$,

$\{\mathcal{L}_{\alpha\ell}\}$, $\{\mathcal{L}_{\bar{\alpha}\ell}\}$, $\{\mathcal{L}_{\beta\ell}\}$, $\{\mathcal{L}_{\bar{\beta}\ell}\}$ $(\ell = 0, 1, 2, .., n-1)$ is introduced as a set of special cases of the operator $\mathcal{L}$ defined by (1.6) each of which is specified by their respective $f(\delta^{(n)})$ in the following way:

(1) $\mathcal{L}_{\omega} : f(\delta^{(n)}) = 1$ for all $\delta^{(n)}$ in X_n.

(2) $\mathcal{L}_{\bar{\omega}} : f(\delta^{(n)}) = 0$ for all $\delta^{(n)}$ in X_n.

(3) $\mathcal{L}_{\alpha\ell} : f(\delta^{(n)}) = \delta_\ell$ for all $\delta^{(n)}$ in X_n.

(4) $\mathcal{L}_{\bar{\alpha}\ell} : f(\delta^{(n)}) = \bar{\delta}_\ell$ for all $\delta^{(n)}$ in X_n.

(5) $\mathcal{L}_{\beta\ell} : f(\delta^{(n)}) = \delta_\ell \ \delta_{\ell-1} \cdots \delta_1 \delta_0$ for all $\delta^{(n)}$ in X_n.

(6) $\mathcal{L}_{\bar{\beta}\ell} : f(\delta^{(n)}) = \overline{\delta_\ell \ \delta_{\ell-1} \cdots \delta_1 \delta_0}$ for all $\delta^{(n)}$ in X_n.

Section 2 is devoted to the discussions of reverberation cycles for $\mathcal{L}_{\bar{\alpha}0}$. In Section 3 we are concerned with the dynamical behaviours induced by applications of $\mathcal{L}_{\bar{\alpha}\, n-\ell}$, where, besides reverberation cycles, we have to discuss some sort of transitory phenomena as well as. In Section 4 we turn to the discussion of transition phenomena of the n-dimensional state configurations under applications of $\mathcal{L}_{\bar{\beta}\ell}$. We are concerned not only with each individual digraph $(X_n, \Gamma_{\bar{\beta}\ell})$ but also with a family of digraphs $(X_n, \Gamma_{\bar{\beta}\ell})$ when ℓ runs through the set $\{0, 1, .., n-1\}$.

The last Section 5 is devoted to the discussion to make clear the relationship between the family of operators and the linear threshold functions associated with the neuronic equation of a single neuron.

This paper is a partial summary of the recent papers[15,16], where the reader can find a systematic and detailed account of the whole content which we are now explaining.

2. REVERBERATION CYCLES FOR $\mathcal{L}_{\bar{\alpha}0}$

In this Section we are entirely concerned with the operator $\mathcal{L}_{\bar{\alpha}0}$ such that, for any $\delta^{(n)} = (\delta_{n-1}, \delta_{n-2}, .., \delta_1, \delta_0)$,

$$\mathcal{L}\bar{a}_0(\delta_{n-1}, \delta_{n-2}, .., \delta_1, \delta_0)$$

$$= (\bar{\delta}_0, \delta_{n-1}, \delta_{n-2}, .., \delta_2, \delta_1) \qquad (2.1)$$

We introduce the following definitions:

Definition 2.1. A representation of an n-dimensional state vector $\delta^{(n)} = (\delta_{n-1}, \delta_{n-2}, .., \delta_1, \delta_0)$ by a catenation of a sequence q k-dimensional state vectors

$$\delta_1^{(k)}, \ \delta_2^{(k)}, \ \ldots, \ \delta_q^{(k)} \qquad (2.2)$$

such that, for j = 1, 2, .., q,

$$\delta_j^{(k)} = (\delta_{n-(j-1)k-1}, \delta_{n-(j-1)k-2}, .., \delta_{n-jk}) \qquad (2.3)$$

with

$$\delta^{(n)} = \delta_1^{(k)} \delta_2^{(k)} \ldots \delta_q^{(k)}, \qquad (2.4)$$

and qk = n is called to be a regular (k, m) articular representation of $\delta^{(n)}$ by means of q articulars with length k when the following three conditions are satisfied:

$$\left.\begin{array}{ll} \text{(i)} & n = kq = k(2m+1) \\ \text{(ii)} & \delta_1^{(k)} = \delta_3^{(k)} = .. = \delta_{2j+1}^{(k)} = \ldots.. \\ & = \delta_{2m-1}^{(k)} = \delta_{2m+1}^{(k)} = \delta^{(k)}, \text{ say} \\ \text{(iii)} & \delta_2^{(k)} = \delta_4^{(k)} = \ldots = \delta_{2j}^{(k)} = \ldots \\ & = \delta_{2m-2}^{(k)} = \delta_{2m}^{(k)} = \bar{\delta}^{(k)} \end{array}\right\} \qquad (2.5)$$

It is essential to observe

Proposition 2.1. An n-dimensional state configuration $\delta^{(n)}$ belongs to a reverberation cycle of exactly 2k length if and only if there is a regular (k, m) articular representation of $\delta^{(n)}$ with the constituent articular $\delta^{(k)}$ for which there does not exist any

(k_1, m_1) articular representation $k = (2m_1+1)k_1$ with a non negative integer m_1 and a positive integer k_1.

Example 2.1. The reverberation cycles of exact length 2ℓ ($\ell = 1, 2, 3$)

$$\text{(i)} \quad \ell = 1 \quad \delta^{(2m+1)} = (\delta^{(1)}\,\bar{\delta}^{(1)})^m\,\delta^{(1)} = (10)^m 1,$$

$$\text{(ii)} \quad \ell = 2 \quad \delta^{\{2(2m+1)\}} = (\delta^{(2)}\,\bar{\delta}^{(2)})^m\,\delta^{(2)} = (0^2 1^2)^m 0^2,$$

$$\text{(iii)} \quad \ell = 3 \quad \delta^{\{3(2m+1)\}} = (\delta^{(3)}\,\bar{\delta}^{(3)})^m\,\delta^{(3)} = (0^3 1^3)^m 0^3.$$

Now we proceed to give all the types and the total numbers of reverberation cycles for n-dimensional state configurations under the operator $\mathcal{L}_{\bar{\alpha}_0}$.

For this purpose we introduce

Definition 2.2. For any set of n different prime numbers $\{p_i\}$ ($p_i > 2$, $i = 1, 2, \ldots r$), any set of r nonnegative integers $\{\nu_i\}$ and positive integer G, we define

$$\text{(1)} \quad \Delta^{(1)}\binom{p_i}{\nu_i}2^G \equiv \begin{cases} 2^{p_i^{\nu_i}G} - 2^{p_i^{\nu_i-1}G}, & \text{for } \nu_i \geqq 1, \\ 2^G, & \text{for } \nu_i = 0, \end{cases}$$

$$\text{(2)} \quad \Delta^{(2)}\begin{pmatrix} p_i & p_j \\ \nu_i & \nu_j \end{pmatrix}2^G \equiv \Delta^{(1)}\binom{p_i}{\nu_i}\left\{\Delta^{(1)}\binom{p_j}{\nu_j}2^G\right\},$$

$$\equiv \begin{cases} \Delta^{(1)}\binom{p_i}{\nu_i}\{2^{p_j^{\nu_j}G} - 2^{p_j^{\nu_j-1}G}\}, & \text{for } \nu_j \geqq 1, \\ \Delta^{(1)}\binom{p_i}{\nu_i}2^G, & \text{for } \nu_j = 0, \end{cases}$$

$$\equiv \begin{cases} \Delta^{(1)}\binom{p_i}{\nu_i}2^{p_j^{\nu_j}G} - \Delta^{(1)}\binom{p_i}{\nu_i}2^{p_j^{\nu_j-1}G}, & \text{for } \nu_j \geqq 1, \\ \Delta^{(1)}\binom{p_i}{\nu_i}2^G, & \text{for } \nu_j = 0 \end{cases}$$

$$\text{(3)} \quad \Delta^{(k)}\begin{pmatrix} p_1 & p_2 & \cdots & p_k \\ \nu_1 & \nu_2 & \cdots & \nu_k \end{pmatrix}$$

$$\equiv \Delta^{(k-1)}\begin{pmatrix} p_1 & p_2 & \cdots & p_{k-1} \\ \nu_1 & \nu_2 & \cdots & \nu_{k-1} \end{pmatrix}\left\{\Delta^{(1)}\binom{p_k}{k}2^G\right\},$$

$$\equiv \begin{cases} \Delta^{(k-1)}\begin{pmatrix} p_1 & p_2 & \cdots & p_{k-1} \\ \nu_1 & \nu_2 & \cdots & \nu_{k-1} \end{pmatrix}\{2^{p_k{}^{\nu_k}G} - 2^{p_k{}^{\nu_k - 1}G}\}, & \text{for } \nu_k \geqq 1, \\ \Delta^{(k-1)}\begin{pmatrix} p_1 & p_2 & \cdots & p_{k-1} \\ \nu_1 & \nu_2 & \cdots & \nu_{k-1} \end{pmatrix}2^G, & \text{for } \nu_k = 0, \end{cases}$$

$$\equiv \begin{cases} \Delta^{(k-1)}\begin{pmatrix} p_1 & p_2 & \cdots & p_{k-1} \\ \nu_1 & \nu_2 & \cdots & \nu_{k-1} \end{pmatrix}2^{p_k{}^{\nu_k}G}, & \\ -\Delta^{(k-1)}\begin{pmatrix} p_1 & p_2 & \cdots & p_{k-1} \\ \nu_1 & \nu_2 & \cdots & \nu_{k-1} \end{pmatrix}2^{p_k{}^{\nu_k - 1}G}, & \text{for } \nu_k \geqq 1, \\ \Delta^{(k-1)}\begin{pmatrix} p_1 & p_2 & \cdots & p_{k-1} \\ \nu_1 & \nu_2 & \cdots & \nu_{k-1} \end{pmatrix}2^G, & \text{for } \nu_k = 0. \end{cases}$$

Now we can enunciate the following proposition:

<u>Proposition 2.2.</u> In the case $n = 2^{\alpha} p_1^{\beta_1} p_2^{\beta_2} \ldots p_r^{\beta_r}$ where p_i are prime numbers such that $3 \leqq p_1 < p_2 < \cdots < p_r$ and $\beta_i \geqq 1 (i = 1, 2, \ldots, r)$, we have following assertions.

(1) Each n-dimensional state configuration δ belongs to one and only one reverberation cycle of exact length $2k(\delta)$ where $k(\delta)$ is a divisor of n with the form

$$k(\delta) = 2^{n} p_{i_1}^{r_{i_1}} p_{i_2}^{r_{i_2}} \ldots p_{i_\nu}^{r_{i_\nu}}, \tag{2.6}$$

where

$$\eta = 0, \alpha, \tag{2.7}$$

$$1 \leq r_{i_j} \leq \beta_{i_j} \quad (j = 1, 2, \ldots, \nu), \tag{2.8}$$

and $(i_1, i_2, \ldots, i_\nu)$ is subset of $(1, 2, \ldots, n)$.

(2) To each assigned triplet of a nonnegative integer ν, a subset $(i_1, i_2, \ldots, i_\nu)$ of the set $(1, 2, \ldots, n)$, with the condition $1 \leqq i_1 < i_2 < \ldots < i_\nu \leqq n$, and a set of ν positive integers $(r_{i_1}, r_{i_2}, \ldots, r_{i_\nu})$ such that $1 \leqq r_{i_j} \leqq \beta_{i_j}$ $(j = 1, 2, \ldots, \nu)$ and to each η which may be 0 or α, there exist

$$\Delta^{(\nu)}\begin{pmatrix} p_{i_1} & p_{i_2} & \ldots & p_{i_\nu} \\ r_{i_1} & r_{i_2} & \ldots & r_{i_\nu} \end{pmatrix} 2^{2\eta}, \tag{2.9}$$

different reverberation cycles of the exact length $2^{\eta} p_{i_1}^{r_{i_1}} p_{i_2}^{r_{i_2}} \cdots p_{i_\nu}^{r_{i_\nu}}$. That is to say, denoting by $N(\ell)$ the total number of reverberation cycles having the exact length ℓ we have the result that $N[2k(\delta)]$ is equal to (2.9) for any $k(\delta)$ given by (2.6), (2.7) and (2.8).

Example 2.2. The types and the numbers of reverberation cycles of n-dimensional state configurations under applications of $\mathcal{L}\bar{\alpha}_0$ for n = 2~20. The applications of our Proposition yield a complete enumeration of all the possible types and their numbers of reverberation cycles as shown in the Tables 1-4, where the following notation is introduced:

(1) LRC = ℓ = Length of reverberation cycle.

(2) NSC = $N(\ell)$ = Number of n-dimensional state configurations belonging to a reverberation cycle of the specified exact length ℓ.

(3) NRC = $N(\ell)/\ell$ = Number of reverberation cycles having the specified exact length ℓ.

n	LRV	NSC	NRC
2^{α}	$2^{\alpha+1}$	$2^{2^{\alpha}}$	$2^{2^{\alpha}-(\alpha+1)}$
2	4	4	1
4	8	16	2
8	16	256	16
16	32	65536	248

TABLE 1. Types and Numbers of Reverberation Cycles of n-dimensional State Configurations Under Application of $\mathcal{L}_{\bar{\alpha}_0}$ for $n = 2 \sim 20$ (1) $n = 2^{\alpha}$ ($\alpha = 1,2,3,4$)

n	LRV	NSC	NRC
3	2 6	2 6	1 1
5	2 10	2 30	1 3
7	2 14	2 126	1 9
11	2 22	2 2046	1 93
13	2 26	2 8190	1 315
17	2 34	2 131070	1 3855
19	2 38	2 524286	1 13797

TABLE 2. Types and Numbers of Reverberation Cycles of n-Dimensional State Configurations Under Application of $\mathcal{L}_{\bar{\alpha}_0}$ for $n = 2 \sim 20$ (2) $n = p$(prime number ≥ 3)

n	LRV	NSC	NRC
6 = 2.3	4	4	4
	12	60	5
10 = 2.5	4	4	1
	20	1020	51
12 = $2^2.3$	8	16	2
	24	4080	170
14 = 2.7	4	4	1
	28	16380	585
20 = $2^2.5$	8	16	2
	40	126076	3174

TABLE 3. Types and Numbers of Reverberation Cycles of n-Dimensional State Configurations Under Application of $\mathcal{L}_{\bar{\alpha}_0}$ (3) $2^{\alpha}p$

n	LRV	NSC	NRC
9 = 3^2	2	2	1
	6	6	1
	18	504	28
15 = 3.5	2	2	1
	6	6	1
	10	30	3
	30	32730	1091
18 = 2.3^2	4	4	1
	12	60	5
	36	262080	7280

TABLE 4. Types and Numbers of Reverberation Cycles of n-Dimensional State Configurations Under Application of $\mathcal{L}_{\bar{\alpha}_0}$ (4) 9, 15, 18

Yamaguchi[14] gave an equivalent result under another formula to the effect that

$$N(2k(\delta)) = \frac{1}{2k(\delta)} \sum_{\substack{d/2k(\delta) \\ d:\ odd}} \mu(d) 2^{k(\delta)/d} \tag{2.10}$$

where $\mu(d)$ denotes Möbius function in number theory.

3. $\delta^{(n)*}$-REVERBERATION CYCLES OF n-DIMENSIONAL STATE CONFIGURATIONS UNDER APPLICATIONS OF $\mathcal{L}_{\bar{a}_{n-\ell}}$

In this case there exists a set of state configurations that are not a member of any reverberation cycle but are transitory state configurations in the sense that repeated applications of $\bar{\mathcal{L}}_{\bar{a}_{n-\ell}}$ will carry each of them into some state configuration, which is a member of some reverberation cycle called $\delta^{(n)*}$-reverberation cycle. Here again an articular representation is equally important as in the case of $\mathcal{L}_{\bar{a}_0}$.

However here we shall not enter into any detailed discussion but we shall give the following example as an illustration.

Example 3.1. $\delta^{(n)*}$- reverberation cycles under the application of $\mathcal{L}_{\bar{a}_{n-2}}$.

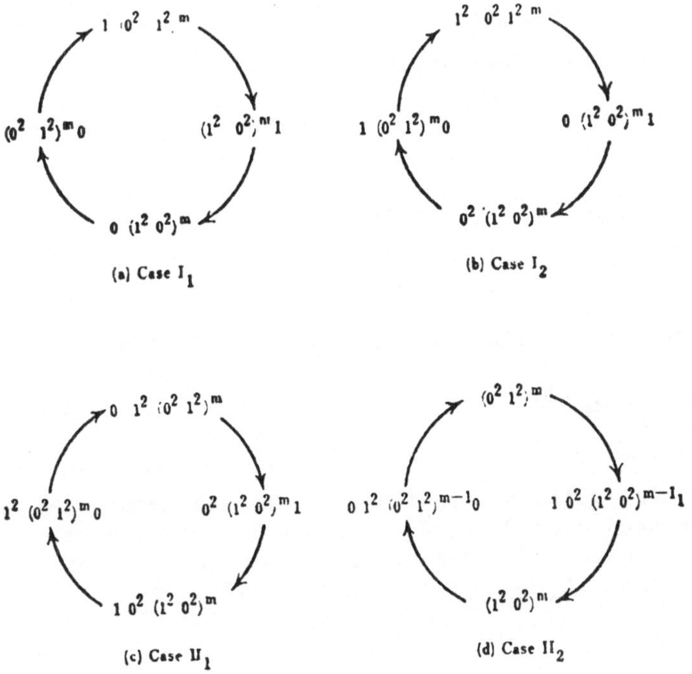

Fig. 1. $\delta^{(n)*}$- reverberation cycles under the applications of $\mathcal{L}\bar{\alpha}_{n-2}$.

4. TRANSITIONS OF n-DIMENSIONAL STATE CONFIGURATIONS UNDER APPLI-CATIONS OF $\mathcal{L}\bar{\beta}_\ell$

In this paragraph we shall introduce a set of operators $\{\mathcal{L}\bar{\beta}_\ell\}$ ($\ell = 0,1,2, \ldots, n-1$) and we shall give some introductory observations on the mutual relationship among these operators by referring to simple examples.

<u>Definition 4.1.</u> An operator $\mathcal{L}\bar{\beta}_\ell$ to each n-dimensional state configuration $\delta = (\delta_{n-1}, \delta_{n-2}, \ldots, \delta_1, \delta_0)$ is defined by

$$\mathcal{L}\bar{\beta}_\ell(\delta_{n-1}, \delta_{n-2}, \ldots, \delta_1, \delta_0) \tag{4.1}$$

$$= (1, \delta_{n-1}, \delta_{n-3}, \ldots, \delta_2, \delta_1), \text{ if } \delta_\ell\,\delta_{\ell-1} \cdots \delta_0 = 0$$

$$= (0, \delta_{n-1}, \delta_{n-2}, \ldots, \delta_2, \delta_1), \text{ if } \delta_\ell\,\delta_{\ell-1} \cdots \delta_0 = 1.$$

Definition 4.2. A digraph $(X_n, \Gamma_{\bar{\beta}_\ell})$ is called a $\mathcal{L}_{\bar{\beta}_\ell}$-digraph in the n-dimensional state configuration when

(1) X_n is the set of all the n-dimensional state configurations

$$\{\delta = (\delta_{n-1}, \delta_{n-2}, \dots, \delta_1, \delta_0)\},$$

and

(2) $\Gamma_{\bar{\beta}_\ell}$ is the set of all the directed arcs defined by

$$\Gamma_{\bar{\beta}_\ell} : \delta \to \mathcal{L}_{\beta_\ell}(\delta). \tag{4.2}$$

We are interested not only in each digraph $(X, \Gamma_{\bar{\beta}_\ell})$ for each assigned ℓ, but also with the transitory phenomena among these n digraphs when ℓ ranges over 0, 1, 2, ..., n - 1. In fact the following simple examples are suggestive by showing a set of bridges from the graph $(X_n, \Gamma_{\bar{\beta}_{n-1}})$ to the graph $(X_n, \Gamma_{\bar{\beta}_0})$.

<u>Example 4.1.</u> The family of digraphs $(X_4, \Gamma_{\bar{\beta}_\ell})$ (ℓ = 0, 1, 2, 3).

Each digraph $(X_4, \Gamma_{\bar{\beta}_\ell})$ (ℓ = 3, 2, 1, 0) is shown in Fig. 2(a), (b), and (c).

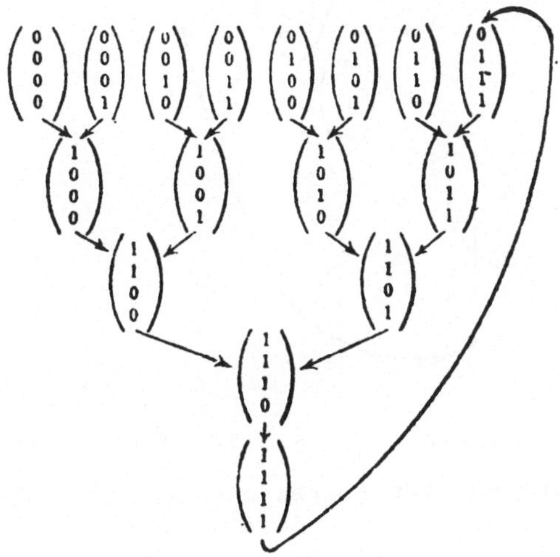

Fig. 2(a). The digraph $(X_4, \Gamma_{\bar{\beta}_3})$ for $\bar{\beta}_3$ applied on 4-dimensional state configuration space.

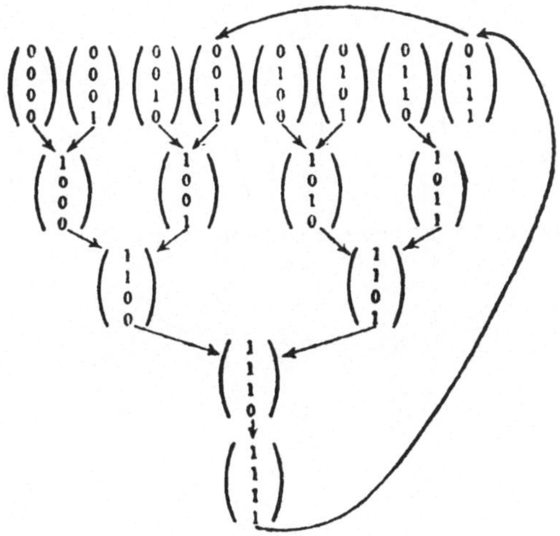

Fig. 2(b). The digraph $(X_4, \Gamma_{\bar{\beta}_2})$ for $\bar{\beta}_2$ applied on 4-dimensional state configuration space.

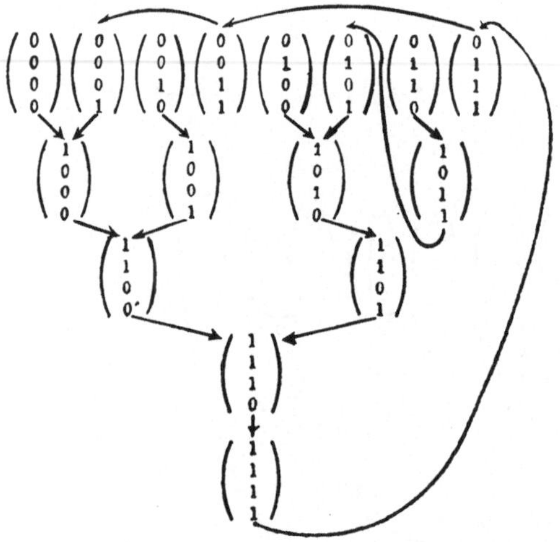

Fig. 2(c). The digraph $(X_4, \Gamma_{\bar{\beta}_1})$ for $\mathcal{L}_{\bar{\beta}_1}$ applied on 4-dimensional state configuration space.

A direct observation on the digraph $(X_4, \Gamma_{\bar{\beta}_1})$ shows that there are a reverberation cycle and a set of state configurations each of which attains at one state configuration belonging to the reverberation cycle. By denoting state configuration in the following way,

$$\begin{pmatrix} 0 \\ 0 \\ 0 \\ 1 \end{pmatrix} = 0^3\ 1, \quad \begin{pmatrix} 0 \\ 1 \\ 0 \\ 0 \end{pmatrix} = 0\ 1\ 0^2, \quad \begin{pmatrix} 1 \\ 0 \\ 0 \\ 1 \end{pmatrix} = 1\ 0^2\ 1, \quad \begin{pmatrix} 1 \\ 0 \\ 1 \\ 0 \end{pmatrix} = (1\,0)^2,$$

Fig. 2(d) yields the digraph representation [Fig. 2(d)].

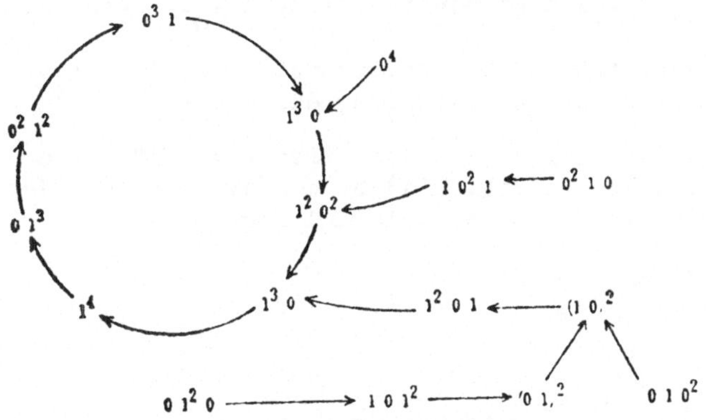

Fig. 2(d). The digraph $(X_4, \Gamma_{\bar{\beta}_1})$ for $\mathcal{L}_{\bar{\beta}_1}$ applied on 4-dimensional state configuration space.

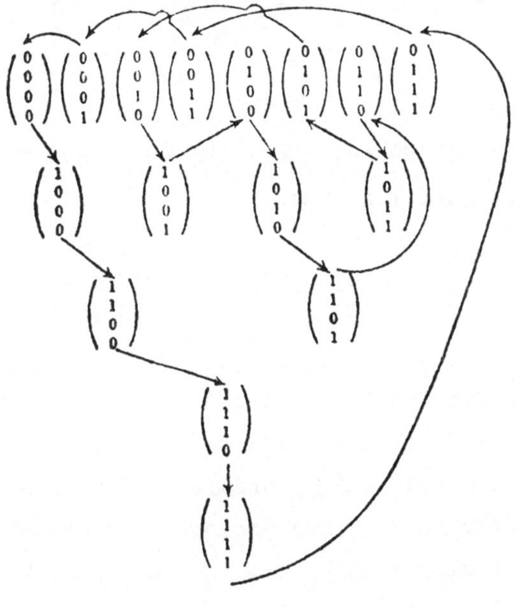

Fig. 2(e). The digraph $(X_4, \Gamma_{\bar{\beta}_0})$ for $\mathcal{L}_{\bar{\beta}_0}$ applied on 4-dimensional state configuration space.

In this case there are two reverberation cycles as we have shown in Sec. 2 for $\mathcal{L}_{\bar{\alpha}_0}$ which is equivalent to $\mathcal{L}_{\bar{\beta}_0}$. Fig. 2(f) shows the clearer description of two reverberation cycles as well as their connections with the corresponding digraph $(X_4, \Gamma_{\bar{\beta}_1})$ given in Fig. 2(d) by denoting an arc lost from Fig. 2(d) by a dotted arc and an added arc to Fig. 2(d) by a heavy arc in order to obtain Fig. 2(e) and (f) from the original Fig. 2(d).

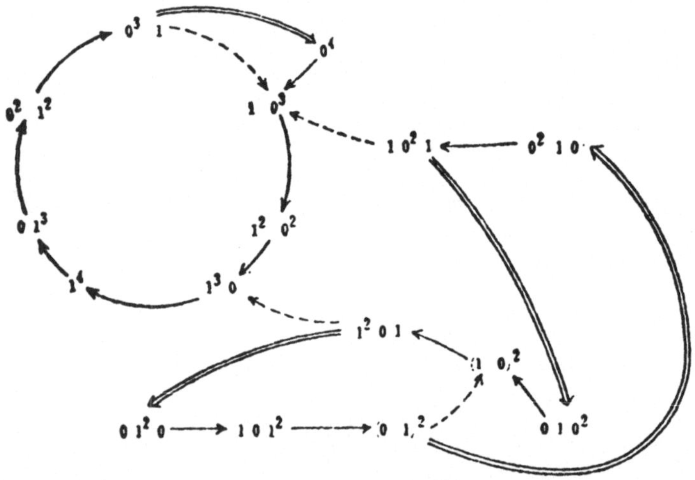

Fig. 2(f). The digraph $(X_4, \Gamma_{\bar{\beta}_0})$ for $\mathcal{L}_{\bar{\beta}_0}$ applied on 4-dimensional state configuration space.

5. THE FAMILY OF OPERATORS AND THE LINEAR THRESHOLD FUNCTIONS ASSOCIATED WITH THE NEURONIC EQUATION OF A SINGLE NEURON

Throughout these previous Sections we have been substantially concerned with the specific family of operators $\mathcal{L}_\omega$, $\mathcal{L}_{\bar{\omega}}$, $\mathcal{L}_{\alpha_0}$, $\mathcal{L}_{\bar{\alpha}_0}$, $\mathcal{L}_{\beta_\ell}$, and $\mathcal{L}_{\bar{\beta}_\ell}$ ($\ell = 1,2, \ldots, n-1$). In fact, although our explanations have been limited to $\mathcal{L}_{\bar{\alpha}_0}$ and $\mathcal{L}_{\bar{\beta}_\ell}$, situations are simpler for the cases of $\mathcal{L}_\omega$ and $\mathcal{L}_{\bar{\omega}}$, and analogous observations are valid between $\mathcal{L}_{\bar{\alpha}_0}$ $(=\mathcal{L}_{\bar{\beta}_0})$ and $\mathcal{L}_{\alpha_0}$ $(=\mathcal{L}_{\beta_0})$, and also between $\mathcal{L}_{\bar{\beta}_\ell}$ and $\mathcal{L}_{\beta_\ell}$ respectively. At this stage we wish to make clear the conditions imposed on the set of coefficients $\{a_{n-1-k}\}$ $(k = 0,1,2, \ldots, n-1)$ and the parameter θ or even the function $\theta(t)$ by which the translatable functional operator defined in (1.1) becomes equivalent to one member of the operator family consisting of $\mathcal{L}_\omega$, $\mathcal{L}_{\bar{\omega}}$, $\mathcal{L}_{\beta_\ell}$, and $\mathcal{L}_{\bar{\beta}_\ell}$ ($\ell = 0,1,2, \ldots, n-1$). This topic will be discussed in a future paper[16]. It will be shown that the uses of the operator family have a certain significance provided that specific considerations are given in composing these operators. In this paper, Sec. 5, an illustration of the systematic uses of the operator family is given with regard to the special case of (1.1) when $n - 1 = 1, 2$.

Example 5.1. We are concerned with the neuronic equation

$$x(t+1) = 1\,[a_0 x(t) + a_1 x(t-1) - \theta], \tag{5.1}$$

which yields

$$z(0,0) = -\theta, \qquad z(1,0) = a_0 - \theta,$$
$$z(0,1) = a_1 - \theta, \qquad z(1,1) = a_0 + a_1 - \theta.$$

We shall take into consideration all possible combinations of a_1, a_0, and θ. Regarding the pair a_1 and a_0, there are eight situations such that

$$\begin{array}{lll}
\text{Case I:} & 0 < a_1 \leqq a_0 < a_0 + a_1 & : (0)\ 1\ 0\ 2 \\
\text{Case II:} & 0 \leqq a_0 < a_1 \leqq a_0 + a_1 & : (0)\ 0\ 1\ 2 \\
\text{Case III:} & a_0 < 0 \leqq a_0 + a_1 < a_1 & : 0\,(0)\ 2\ 1 \\
\text{Case IV:} & a_0 < a_0 + a_1 < 0 \leqq a_1 & : 0\ 2\ (0)\,1 \\
\text{Case V:} & a_0 + a_1 < a_0 \leqq a_1 < 0 & : 2\ 0\ 1\ (0) \\
\text{Case VI:} & a_0 + a_1 < a_1 < a_0 \leqq 0 & : 2\ 1\ 0\ (0) \\
\text{Case VII:} & a_1 < a_0 + a_1 \leqq 0 < a_0 & : 1\ 2\ (0)\ 0 \\
\text{Case VIII:} & a_1 \leqq 0 < a_0 + a_1 < a_0 & : 1\ (0)\ 2\ 0,
\end{array} \tag{5.2}$$

where 0, 1, and 2 denote a_0, a_1, and $a_0 + a_1$ respectively, while (0) denotes the value 0.

In each of the eight cases, there is a division of the whole real axis of θ into five subintervals $\{J_h\}$ ($h = 1,2,3,4,5$) in each of which we have the representation of the linear threshold functions by means of $\mathcal{L}_\omega$, $\mathcal{L}_{\bar\omega}$, $\mathcal{L}_{\beta_\ell}$, $\mathcal{L}_{\bar\beta_\ell}$ ($\ell = 0,1$) as shown in Table 5.

We are not going to give any systematic approach to operator representations of $1[z(\delta)]$ for general situation. Nevertheless we have to add one more example in which simple descriptions given in Example 10.1 do not hold, but there is a need for introducing the notion of conditional operator representations of $1[z(\delta)]$ for certain cases.

	Subinterval				
Case	J_1	J_2	J_3	J_4	J_5
I. (0)102	ω	$\bar{\beta}(\bar{1},\bar{0})$	$\beta(1)$	$\beta(1,0)$	$\bar{\omega}$
II. (0)012	ω	$\bar{\beta}(\bar{1},\bar{0})$	$\beta(0)$	$\beta(1,0)$	$\bar{\omega}$
III. 0(0)21	ω	$\bar{\beta}(1,\bar{0})$	$\beta(1)$	$\beta(\bar{1},0)$	$\bar{\omega}$
IV. 02(0)1	ω	$\bar{\beta}(1,\bar{0})$	$\bar{\beta}(1)$	$\beta(\bar{1},0)$	$\bar{\omega}$
V. 201(0)	ω	$\bar{\beta}(1,0)$	$\bar{\beta}(1)$	$\beta(\bar{1},\bar{0})$	$\bar{\omega}$
VI. 210(0)	ω	$\bar{\beta}(1,0)$	$\bar{\beta}(0)$	$\beta(\bar{1},\bar{0})$	$\bar{\omega}$
VII. 12(0)0	ω	$\bar{\beta}(\bar{1},0)$	$\bar{\beta}(0)$	$\beta(1,\bar{0})$	$\bar{\omega}$
VIII. 1(0)20	ω	$\bar{\beta}(\bar{1},0)$	$\beta(1)$	$\beta(1,0)$	$\bar{\omega}$

TABLE 5. Operator Representations of Linear Threshold Function Given by (1)

<u>Example 5.2.</u> Let us consider the neuronic equation

$$x(t+1) = 1[a_0x(t) + a_1x(t-1) + a_2x(t-2) - \theta].$$

For the present purpose just mentioned, let us consider the case when

$$0 < a_2 < a_1 < a_1 + a_2 < a_0 < a_0 + a_2$$

$$< a_0 + a_1 < a_0 + a_1 + a_2,$$

which defines the subdivision of the whole θ axis into a family of nine subintervals $\{J_h\}$ $(h = 1,2, \ldots, 9)$ where $J_1 = (-\infty, 0)$, $J_2 = [0, a_2)$, $J_3 = [a_2, a_1)$, $J_4 = [a_1, a_1 + a_2)$, $J_5 = [a_1 + a_2, a_0)$, $J_6 = [a_0, a_0 + a_2)$, $J_7 = [a_0 + a_2, a_0 + a_1)$, $J_8 = [a_0 + a_1, a_0 + a_1 + a_2)$, $J_9 = [a_0 + a_1 + a_2, \infty)$. We can discuss the operator representations in each of these subintervals. In the subinterval J_4, for instance, two cases (a) $\delta_2 = 1$ and (b) $\delta_2 = 0$ should be distinguished. In the case (a) we have $z(1, \delta_1, \delta_0) > 0$ for all (δ_1, δ_0) in X_2, which can be denoted by $\mathcal{L}_{\omega:2}$ as conditional representation, where 2 denotes the condition that $\delta_2 = 1$, in the case (b) we have $z(0, 1, 1) > 0$ and $z(0, \delta_1, \delta_0) \leq 0$ for all $(\delta_1, \delta_0) \neq (1, 1)$ in X_2, which can be denoted by $\mathcal{L}_{\beta(1,0):\bar{2}}$, where $\bar{2}$ denotes the condition that $\delta_2 = 0$.

Subinterval	Operator representation	Operator representation	Subinterval
J_1	$\mathcal{L}_{\omega}$	$\mathcal{L}_{\bar{\omega}}$	J_9
J_2	$\mathcal{L}_{\bar{\beta}(\bar{2},\bar{1},\bar{0})}$	$\mathcal{L}_{\beta}(2,1,0)$	J_8
J_3	$\mathcal{L}_{\bar{\beta}(\bar{2},\bar{1})}$	$\mathcal{L}_{\beta}(2,1)$	J_7
	$\mathcal{L}_{\omega:2}$	$\mathcal{L}_{\bar{\omega}:\bar{2}}$	J_6
J_4	$\mathcal{L}_{\beta(1,0):\bar{2}}$	$\mathcal{L}_{\bar{\beta}(\bar{1},\bar{0}):2}$	
J_5		$\mathcal{L}_{\beta(2)}$	J_5

TABLE 6. Operator representations in each subinterval of $1\,[z(\delta_2, \delta_1, \delta_0)]$

REFERENCES

1. Caianiello, E.R., Outline of a thought-processes and thinking machines, J. Theoret. Biol. 2, 204, 1961.

2. Caianiello, E.R., Decision equations and reverberations, Kybernetik 3, 98, 1966.

3. Caianiello, E.R., Brain models, natural languages and robots, Japan Industrial Technological Association Symposium on Information Processing Systems, Tokyo, March, 1972.

4. De Luca, A., On some dynamical properties of linear and affine networks, Kybernetik 8, 123, 1971.

5. De Luca, A., On some representations of Boolean functions. Application to the theory of switching elements nets, Kybernetik 9, 1, 1971.

6. Aiello, A., Burattini, E. and Caianiello, E.R., Synthesis of reverberating neural networks, Kybernetik, 7, 191, 1970.

7. Caianiello, E.R., De Luca, A. and Ricciardi, L.M., Reverberations and control of neural networks, Kybernetik 4, 10, 1967.

8. Nagumo, J. and Sato, S., On a response characteristic of a mathematical neuron model, Kybernetik 10, 155, 1972.

9. Amari, S., Characteristics of randomly connected threshold-element networks and networks and network systems, Proc. IEEE 59, 35, 1971.

10. Amari, S., Learning of patterns by self-organizing nets of threshold elements, (in Japanese) Densi Tsushin Gakukai, Document EC 71, 1971-11.

11. Suzuki, R., Katsuno, I. and Matano, K., Dynamics of "neuron ring", computer simulation of nervous system of starfish, Kybernetik 8, 39, 1971.

12. Ishihara, T., Local reverberations in the nervous system and conditional reflex, Math. Biosci. 12, 23, 1971.

13. Yamaguchi, M., Characterization of Operators and their reverberation cycles associated with a single neural equation, Research Report, No. 31, Research Institute of Fundamental Information Science, June 1972.

14. Yamaguchi, M., Some contribution to dynamical features on biomathematical systems, Thesis, Faculty of Science, Kyushu University, September 1972.

15. Kitagawa, T., Operator representations of linear threshold functions associated with a single neuronic equation, Research Report, Research Institute of Fundamental Information Science, September 1973, (in preperation).

16. Kitagawa, T., Dynamical systems and operators associated with a single neuronic equation, Math. Biosciences, 17, 1973 (in press).

TENSORIAL LINEARIZATION OF THRESHOLD FUNCTIONS

Eduardo R. Caianiello and W.E.L. Grimson

Laboratorio di Cibernetica del CNR
Arco Felice, Napoli

1. - The importance of threshold (or linearly separable, or pseudo-Boolean) functions in the study of parallel processes can hardly be overemphasized. As is the case with particle physics, where one gets quite often more from limiting cases, such as the "non-relativistic" and the "extreme relativistic" approximations than from the exact theory, so with parallel processes one can obtain more from what we may call the "linear" and the "extreme non-linear" approximations. The first is well familiar e.g. from systems theory, by the second we mean here the study of threshold functions, either per se or as expressions for other logical functions.

Nets dynamically described by such functions were in particular taken as the basis for models of nervous activity (hence the name "neural nets") by the first named author and collaborators; the study of periodic sequences of states (reverberations) could thus be put under mathematical control and some relevant questions posed and answered [1),2)]. The possible interest of such approach for new computer technologies is beginning to be explored (cf.e.g. the AETL of Japan's Electrotechnical Laboratory).

A new development began with works in which threshold functions (as well as all boolean functions) were linearized in the tensor space of their binary variables[3),4)]. The present report shows that, at least for small numbers N of binary variables, the computation of the coefficients of the tensor expansion of a threshold function in terms of the Chow parameters, which gives hopelessly non-linear expressions, can in fact be split into two distinct processes; the first is purely combinatoric and quite trivial, the second is computational and entirely linear. Some "polygonal representations" of threshold functions present themselves as a simple way of classifying threshold functions for small N. The extension to values of $N > 5$ of the second result is not certain and by no means obvious and is yet to be investigated (it stays true also for $N = 6$); the first one is however true in general and is of interest in understanding the nature of threshold functions.

Using a set of Boolean variables $\xi = \{\xi_1, \ldots, \xi_N\}$ defined as taking the values ± 1, it is fairly trivial to establish that a Boolean function

$$\sigma\,|F(\xi)| = \begin{cases} +1 & F(\xi^+) > 1 \\ -1 & F(\xi^-) < 0 \end{cases}$$

where $F(\xi)$, subject only to the condition $F(\xi)$ always $\neq 0$, may be expanded into a tensor space so that

$$\sigma\,|F(\xi)| = \sum_{\alpha=0}^{2^N-1} f_\alpha \, \eta_\alpha$$

where

$$\eta_\alpha = \begin{cases} 1 & \alpha = 0 \\ \xi_{i}, \ldots, \xi_{i_h} & \alpha \neq 0, \quad 1 \leq h \leq N \end{cases}$$

$$f_\alpha = \langle \eta_\alpha \sigma |F(\xi)| \rangle \frac{1}{2^N} \tilde{\Sigma}\, \eta_\alpha \sigma |F(\xi)| = \frac{1}{2^N} \tilde{\Sigma}\, \sigma |\eta_\alpha F(\xi)|$$

with $\tilde{\Sigma}$ being the trace operations $\Sigma \,\forall\, \xi_1 = \pm 1, \ldots, \xi_N = \pm 1$. Using this notation, since we are concerned with threshold functions we may write, without loss of generality,

$$\sigma |F(\xi)| = \sigma \Big| \sum_{n=1}^{N} x_n \xi_n \Big| , \quad \sum_{n=1}^{N} x_n \xi_n \neq 0$$

Now there are 2^N coefficients f_α rather than the N coefficients $x_1,\ldots x_N$ which we had before expansion. If we fix $\xi_N = -1$, we have threshold $x_N \neq 0$ and $M = N-1$ variables. It is known that only M+1 parameters (Chow's parameters) are necessary to characterize a threshold function of M variables and a threshold. There will exist therefore relations connecting the f_α 's to $f_1,\ldots f_N$.

Unfortunately, these relations appear to be hopelessly non-linear; e.g., for N = 3 we find [4)]

$$f_{123} = \frac{-3f_1f_2f_3}{f_1^2+f_2^2+f_3^2}$$

or [3)]

$$f_{123} = -4f_1f_2f_3$$

(the two expressions are <u>both</u> correct since f_1,f_2,f_3 can assume only very few discrete values).

It is interesting to point out that $f_1,\ldots f_N$ actually are representations for Chow's parameters given in our (1,-1) configuration rather than the (1,0) configuration used by Chow. Chow's parameters are

$$m \equiv \langle G(\mu) \rangle$$

$$a_i \equiv \langle G(\mu)\,\mu_i \rangle \qquad 1 \leq i \leq M$$

where G is a function of M variables $\mu_i = 1,0$. Clearly, since we are using a (1,-1) configuration, we would have

$$\xi_i = 2\mu_i - 1 \qquad 1 \leq i \leq M$$

and $$\sigma|F(\xi)| = 2G(\mu))-1$$

Remembering that we have no threshold and thus taking a function of N = M+1 variables, we have by substitution

$$m = \left\langle \frac{\sigma\left|\sum_{n=1}^{N} x_n \xi_n\right| + 1}{2} \right\rangle$$

$$a_i = \left\langle \frac{\sigma\left|\sum_{n=1}^{N} x_n \xi_n\right| + 1}{2} \; \frac{\xi_i + 1}{2} \right\rangle$$

where the symbols $\langle\ \rangle$ indicate a summation over all possible values of ξ .

Clearly, then

$$2m - 2^N = \langle \sigma\left|\sum_{n=1}^{N} x_n \xi_n\right| \rangle$$

$$4a_i - 2m = \langle \xi_i \sigma\left|\sum_{n=1}^{N} x_n \xi_n\right| \rangle + \langle \xi_i \rangle$$

Of course, $\langle \xi_i \rangle = 0$ since, by definition, ξ_i takes the value 1 as many times as it takes the value -1.

We have thus

$$f_i = \langle \xi_i \sigma\left|\sum_{n=1}^{N} x_n \xi_n\right| \rangle$$

and $f_o = \langle \sigma\left|\sum_{n=1}^{N} x_n \xi_n\right| \rangle$

so that

$$f_o = 2m - 2^N$$

$$f_i = 4a_i - 2m$$

f_o vanishes ($\alpha = 0$ is even) so that we are left with $N = M+1$ characteristic parameters.

$$f_i = 4a_i - 2^N \qquad i = 1,2,\dots,N' = N+1$$

Thus, without specifically designing our system to include such, we have Chow's parameters falling out for free. Indeed, such a representation for $f_1,\dots,f_N$ is perhaps better defined and more natural than Chow's, i.e.

$$f_i = \frac{1}{2^N} \tilde{\Sigma}\, \sigma \mid \xi_i \sum_{n=1}^{N} x_n \xi_n \mid$$

2. - A crucial point for our study[4] is that

$$f_A - f_{\bar{A}} = \sigma\,(X_A - X_{\bar{A}})$$

where A and $\bar{A}$ are the subsets of any arbitrary partition of the set $\mathcal{N}$ of the N variables $\xi_1,\dots,\xi_N$: $A + A = \mathcal{N}$.

By definition

$$f_A = \sum_{\alpha' \varepsilon A} f_{\alpha'}$$

$$f_{\bar{A}} = \sum_{\alpha'' \varepsilon \bar{A}} f_{\alpha''}$$

$$X_A = \sum_{n' \varepsilon A} x_{n'}$$

$$X_{\bar{A}} = \sum_{n'' \varepsilon \bar{A}} x_{n''}$$

The use of these facts will enable us to understand better the nonlinearity of the relations between f_α and f_i.

First, we establish a canonical ordering of the x's, i.e.: $x_1 \geq x_2 \geq \dots \geq x_n \geq 0$. This is accomplished by permutations of indices and changes of signs which do not affect the function

$\sigma \left| \sum_{n=1}^{N} x_n \xi_n \right|$. This ordering may be shown to imply that $f_1 \geq f_2 \geq \ldots \geq f_N \geq 0$, which is clearly a very powerful restriction.

If we now stay in this region of the tensor space, we show here that all f_α's may be expressed as linear combinations of the f_i's for $N \leq 5$. A complete discussion will be the object of forthcoming papers.

The crucial point, as mentioned, is

$$f_A - f_{\bar{A}} = \sigma(X_A - X_{\bar{A}}).$$

Let us look at the following example

$$Y_1 = -x_1 + x_2 + x_3 + \ldots + x_N$$

$$Y_2 = x_1 - x_2 + x_3 + \ldots + x_N$$

$$Y_1 + Y_2 = 2(x_3 + \ldots + x_N) > 0 \text{ since } x_3 \geq \ldots \geq x_N \quad 0.$$

$$-Y_1 + Y_2 = 2(x_1 - x_2) \geq 0 \text{ since } x_1 \geq x_2 \quad \text{, hence } \sigma\,(Y_2) = 1.$$

Take $Y_2 = X_A - X_{\bar{A}}$: then $f_A - f_{\bar{A}} = 1$.

Such a method may be carried out in general; the results for the first simple cases are shown below, the situation for $N > 5$ becomes more involved and will be studied elsewhere. The table below gives the relations for all f_α for $N \leq 5$.

Note that N refers to the number of f_i different from zero. This is since a threshold function of N variables corresponding to a set of values of $f_1, \ldots f_N$ where $f_N = 0$ is actually the same as a function of N-1 variables in which the effect of x_N is negligible.

To obtain the f_α for $N < 5$, simply take only those equations which have all indices less than or equal to N, e.g. N = 4

take f_{123}, f_{124}, f_{134}, f_{234}

set all other $f_\alpha = 0$.

$f_{123} = 1 - f_1 - f_2 - f_3$ $\qquad N \leq 5$

$f_{124} = 1 - f_1 - f_2 - f_4$

$f_{125} = 1 - f_1 - f_2 - f_5$

$f_{134} = 1 - f_1 - f_3 - f_4$

$f_{135} = 1 - f_1 - f_3 - f_5$

$f_{145} = 3f_1 + 2f_2 + 2f_3 + f_4 + f_5 - \frac{7}{2}$

$f_{234} = 2f_1 + f_2 + f_3 + f_4 - 2$

$f_{235} = 2f_1 + f_2 + f_3 + f_5 - 2$

$f_{245} = \frac{5}{2} - 2f_1 - f_2 - 2f_3 - f_4 - f_5$

$f_{345} = \frac{5}{2} - 2f_1 - 2f_2 - f_3 - f_4 - f_5$

$f_{12345} = f_1 + f_2 + f_3 + f_4 + f_5 - \frac{3}{2}$

We know that in the expansions for $f_1,\ldots,f_N$, i.e.

$$f_i = \frac{1}{2^N} \tilde{\Sigma}\, \xi_i\, \sigma\left|\sum_{n=1}^{N} x_n\, \xi_n\right| ,$$

there are 2^N signum terms which we characterized by $\sigma(Y_\alpha)$.

It turns out that only half of these are distinct since, for example:

$Y_1 = -x_1 + x_2 + \ldots + x_N = -(x_1) - (-x_2) - \ldots - (-x_N) =$

$= -(x_1 - x_2 - \ldots - x_N) = -Y_{23\ldots N}$

So each f_α is expressed as a combination of 2^{N-1} signum terms. However, as was shown before, many of these are constants. As well, there exists a hierarchy among these Y_α.
For example, let $\sigma(Y_1)=-1$.

Then $x_1 > x_2 + x_3 + \ldots + x_N$

then $x_1 + x_N > x_2 + x_3 + \ldots + x_{N-1}$

So $\sigma(Y_{1N}) = \sigma(-x_1 + x_2 + \ldots + x_{N-1} - x_N) = -1$, etc.

Hence, these signum functions are not independent.
All these restrictions have a very powerful effect on the actual values of $f_1, \ldots, f_N$. Some results are shown below.

N	f_1	f_2	f_3	f_4	f_5	f_6
1	1	0	0	0	0	0
2	1	0	0	0	0	0
3	$\frac{2}{2}$	0	0	0	0	0
	$\frac{1}{2}$	$\frac{1}{2}$	$\frac{1}{2}$	0	0	0
4	$\frac{4}{4}$	0	0	0	0	0
	$\frac{3}{4}$	$\frac{1}{4}$	$\frac{1}{4}$	$\frac{1}{4}$	0	0
	$\frac{2}{4}$	$\frac{2}{4}$	$\frac{2}{4}$	0	0	0

N	f_1	f_2	f_3	f_4	f_5	f_6
5	$\frac{8}{8}$	0	0	0	0	0
	$\frac{7}{8}$	$\frac{1}{8}$	$\frac{1}{8}$	$\frac{1}{8}$	$\frac{1}{8}$	0
	$\frac{6}{8}$	$\frac{2}{8}$	$\frac{2}{8}$	$\frac{2}{8}$	0	0
	$\frac{5}{8}$	$\frac{3}{8}$	$\frac{3}{8}$	$\frac{1}{8}$	$\frac{1}{8}$	0
	$\frac{4}{8}$	$\frac{4}{8}$	$\frac{4}{8}$	0	0	0
	$\frac{4}{8}$	$\frac{4}{8}$	$\frac{2}{8}$	$\frac{2}{8}$	$\frac{2}{8}$	0
	$\frac{3}{8}$	$\frac{3}{8}$	$\frac{3}{8}$	$\frac{3}{8}$	$\frac{3}{8}$	0
6	$\frac{16}{16}$	0	0	0	0	0
	$\frac{15}{16}$	$\frac{1}{16}$	$\frac{1}{16}$	$\frac{1}{16}$	$\frac{1}{16}$	$\frac{1}{16}$
	$\frac{14}{16}$	$\frac{2}{16}$	$\frac{2}{16}$	$\frac{2}{16}$	$\frac{2}{16}$	0
	$\frac{13}{16}$	$\frac{3}{16}$	$\frac{3}{16}$	$\frac{3}{16}$	$\frac{1}{16}$	$\frac{1}{16}$
	$\frac{12}{16}$	$\frac{4}{16}$	$\frac{4}{16}$	$\frac{2}{16}$	$\frac{2}{16}$	$\frac{2}{16}$
	$\frac{12}{16}$	$\frac{4}{16}$	$\frac{4}{16}$	$\frac{4}{16}$	0	0
	$\frac{11}{16}$	$\frac{5}{16}$	$\frac{5}{16}$	$\frac{3}{16}$	$\frac{1}{16}$	$\frac{1}{16}$
	$\frac{11}{16}$	$\frac{5}{16}$	$\frac{3}{16}$	$\frac{3}{16}$	$\frac{3}{16}$	$\frac{3}{16}$
	$\frac{10}{16}$	$\frac{6}{16}$	$\frac{6}{16}$	$\frac{2}{16}$	$\frac{2}{16}$	0
	$\frac{10}{16}$	$\frac{6}{16}$	$\frac{4}{16}$	$\frac{4}{16}$	$\frac{2}{16}$	$\frac{2}{16}$

f_1	f_2	f_3	f_4	f_5	f_6
$\frac{10}{16}$	$\frac{4}{16}$	$\frac{4}{16}$	$\frac{4}{16}$	$\frac{4}{16}$	$\frac{4}{16}$
$\frac{9}{16}$	$\frac{7}{16}$	$\frac{7}{16}$	$\frac{1}{16}$	$\frac{1}{16}$	$\frac{1}{16}$
$\frac{9}{16}$	$\frac{7}{16}$	$\frac{5}{16}$	$\frac{3}{16}$	$\frac{3}{16}$	$\frac{1}{16}$
$\frac{9}{16}$	$\frac{5}{16}$	$\frac{5}{16}$	$\frac{5}{16}$	$\frac{3}{16}$	$\frac{3}{16}$
$\frac{8}{16}$	$\frac{8}{16}$	$\frac{8}{16}$	0	0	0
$\frac{8}{16}$	$\frac{8}{16}$	$\frac{6}{16}$	$\frac{2}{16}$	$\frac{2}{16}$	$\frac{2}{16}$
$\frac{8}{16}$	$\frac{8}{16}$	$\frac{4}{16}$	$\frac{4}{16}$	$\frac{4}{16}$	0
$\frac{8}{16}$	$\frac{6}{16}$	$\frac{6}{16}$	$\frac{4}{16}$	$\frac{4}{16}$	$\frac{2}{16}$
$\frac{7}{16}$	$\frac{7}{16}$	$\frac{7}{16}$	$\frac{3}{16}$	$\frac{3}{16}$	$\frac{3}{16}$
$\frac{7}{16}$	$\frac{7}{16}$	$\frac{5}{16}$	$\frac{5}{16}$	$\frac{5}{16}$	$\frac{1}{16}$
$\frac{6}{16}$	$\frac{6}{16}$	$\frac{6}{16}$	$\frac{6}{16}$	$\frac{6}{16}$	0

3. - Since each set of values of $f_1,\ldots,f_N$ is characterized by the values of 2^{N-1} distinct signum functions, we may represent threshold functions in another way, that of plane geometric polygons; this is instructive, at least for small N, and we discuss some instances next.

Let N = 1. In this case we have only x_1, so the only polygon (which is not closed in this case) we can construct is a straight line

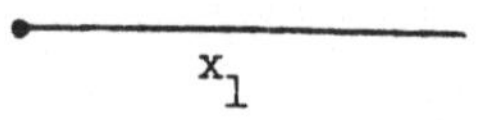

Note that for N = 1, there is only one set of values of $f_1,\ldots,f_N$, i.e. $f_i=1$.

Let N=2. We may form a closed polygon only by

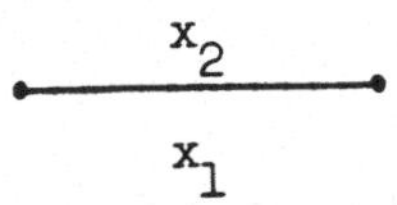

which is forbidden since $\sum_{n=1}^{N} x_n \xi_n \neq 0$ and clearly $x_1=x_2$, N=2 contradicts this. The only other possible form would be when x_2 is negligible with respect to x_1, and here we cannot construct a polygon. This, of course, is similar to the case N=1 where we could not construct a polygon; we point out that the only set of values of $f_1,\ldots f_N$ for N=2 is $f_1= 1, f_2 = 0$, which is the same as taking N=1 and ignoring x_2 and f_2.

Now take N = 3. The general method of constructing a closed polygon such that the sides are ordered, i.e. x_1 joins x_2 joins x_3 joins x_1, (omitting redundant forms due to rotations, etc.) is:

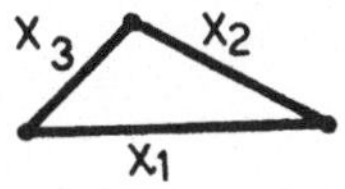

Remember that $x_1 \geq x_2 \geq x_3 > 0$.

Clearly then, the case $f_1 = f_2 = f_3 = \frac{1}{2}$ corresponds to any choice of sides x_1,x_2,x_3 which permits to construct a triangle with non-vanishing area; if this is impossible, $f_1=1$, $f_2=f_3= 0$ necessarily.

With N = 4 the case becomes somewhat less trivial. The general

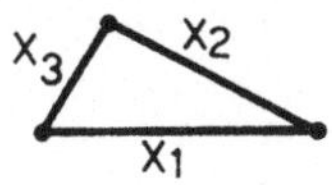

form of the polygon is

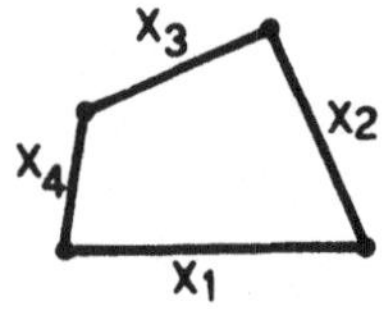

It can be shown for any N that a necessary condition for having $f_N > 0$ is that such a closed polygon with sides $x_1, \ldots x_N$ be construable.

It may be not without interest to verify that the polygonal representation contains information about the values of all the signum terms of the expansion of the f_i

This may be accomplished in the following way. To determine, for example, $\sigma\ (Y_{12})$, consider the diagonal separating x_1 and x_2 from the rest of the polygon

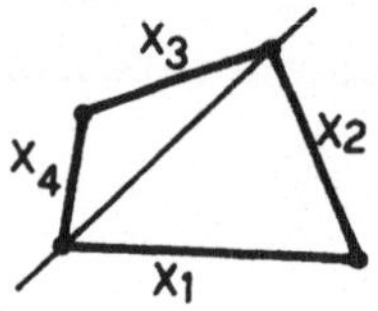

If we seize the vertices of the polygon lying on the diagonal and pull outward along the diagonal, the figure will collapse into a polygon with fewer sides.

In this case, since $x_1 \geq x_2 \geq x_3 \geq x_4 > 0$, it will become

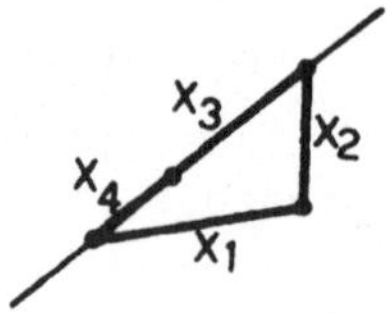

This is always the case, since $x_1 + x_2 > x_3 + x_4$. But, for example, we may not know the value of $\sigma (Y^1_{14})^2$.

Thus, depending on the actual values of x_1, x_2, x_3, x_4, the figure may collapse in two ways, illustrated below.

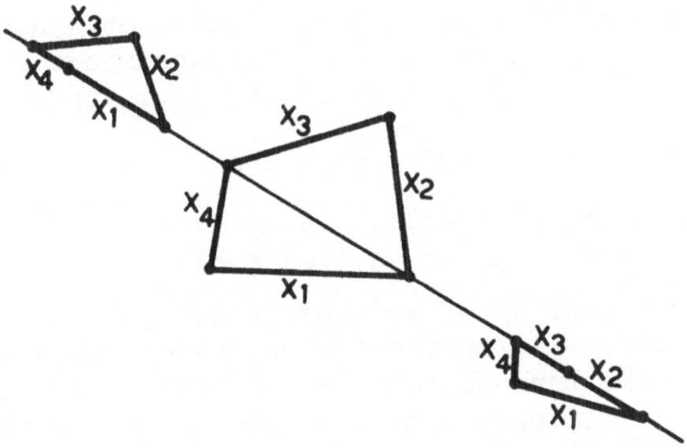

Figure (1) corresponds to $x_2 + x_3 > x_1 + x_4$ or $\sigma(Y_{14}) = 1$

Conversely, figure (2) corresponds to $x_1 + x_4 > x_2 + x_3$ or $\sigma(Y_{14}) = -1$

It turns out that for a function where $f_1, \ldots, f_4 \neq 0$ the polygon will always collapse to

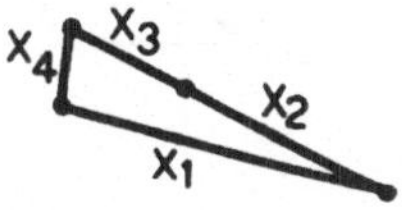

Otherwise, x_4 is negligible compared to the others and we have essentially a function of $N = 3$ variables.

If we continue this to $N = 5$, we find that, based on the manner in which the polygon collapses, there are precisely six different sets, or classes, of polygons. There is also the case where x_1 is so great that we cannot construct a polygon. Note that there also are only seven different sets of values of $f_1, \ldots, f_5$.

The cases are indicated below. Since most of the signum functions are constants, i.e. certain stretchings of the polygon will always result in the same collapsed form, we include only those stretchings which may vary. We leave it to the reader to verify that for each polygon, substitution of the values obtained by stretching for each signum function into the expansions for $f_1,\ldots,f_5$ does result in the set of values of $f_1,\ldots,f_5$ listed. The cases are:

Case 1 . Cannot construct a polygon, i.e. $x_1 > x_2 + x_3 + x_4 + x_5$ so $\sigma(Y_1) = -1$. Due to the hierarchy of dependence among the signum terms, $\sigma(Y_1) = -1$ implies the values of all other $\sigma(Y_\alpha)$. (Any term where $1 \in \alpha$ will be -1, any where $1 \notin \alpha$ will be 1; provided this does not contradict $x_1 \geq \ldots \geq x_5 > 0$).

Case 2 .

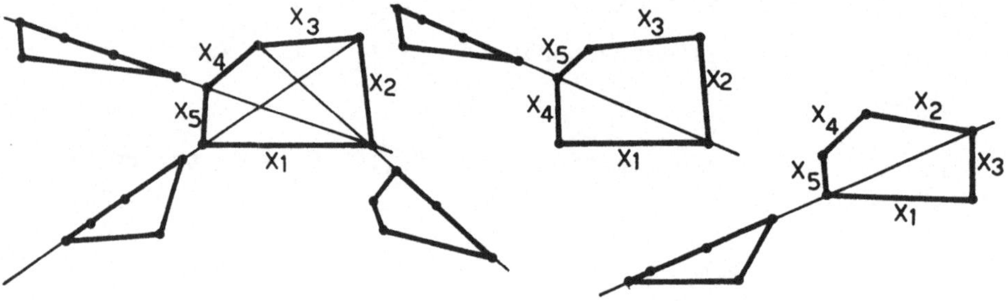

This corresponds to $f_1 = \frac{7}{8}$; $f_2 = f_3 = f_4 = f_5 = \frac{1}{8}$.

Case 3.

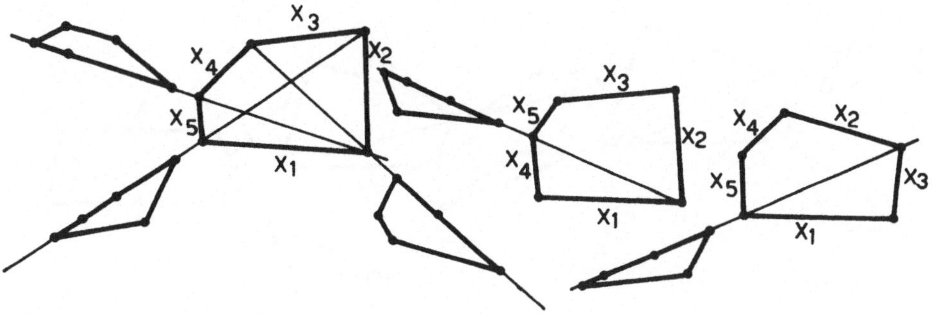

This corresponds to $f_1 = \frac{3}{4}$; $f_2 = f_3 = f_4 = \frac{1}{4}$; $f_5 = 0$ and is indistinguishable from

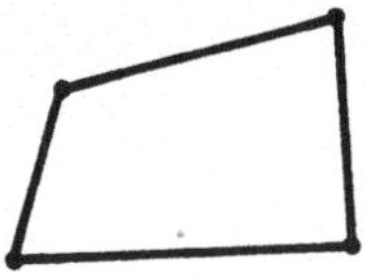

Case 4.

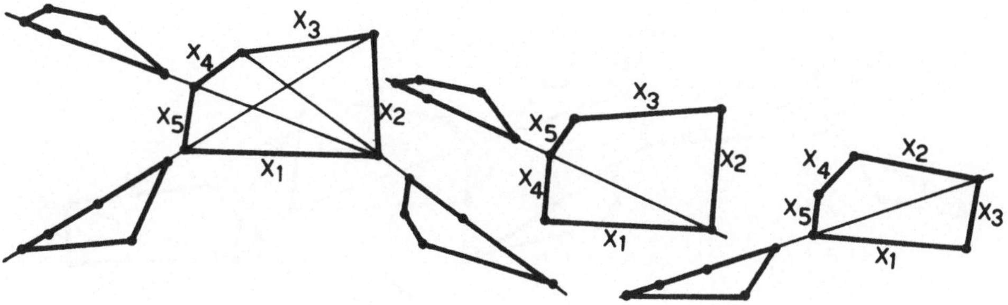

This corresponds to $f_1 = \frac{6}{8}$; $f_2 = f_3 = \frac{3}{8}$; $f_4 = f_5 = \frac{1}{8}$.

Case 5.

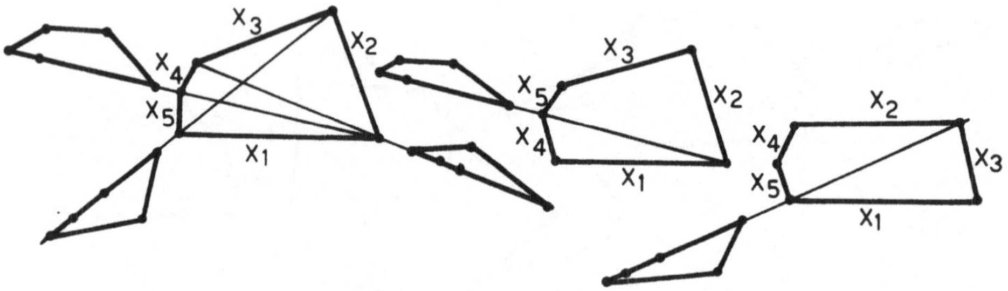

This corresponds to $f_1 = f_2 = f_3 = \frac{1}{2}$, $f_4 = f_5 = 0$ and is actually indistinguishable from

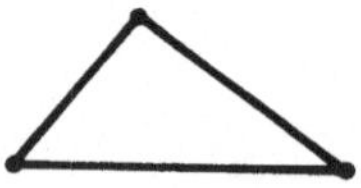

Case 6.

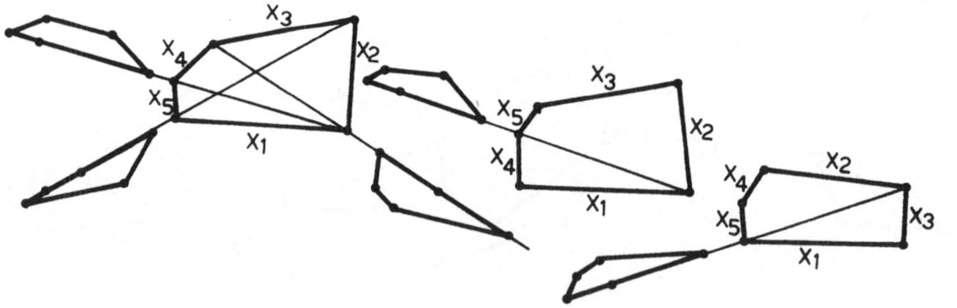

This corresponds to $f_1 = f_2 = \frac{1}{2}$; $f_3 = f_4 = f_5 = \frac{1}{4}$.

Case 7.

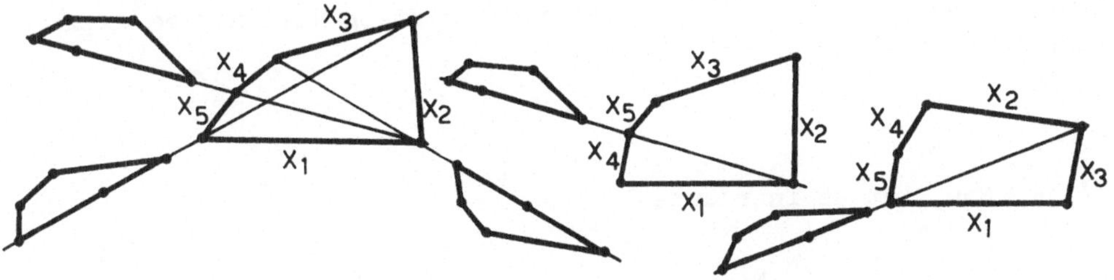

This corresponds to $f_1 = f_2 = f_3 = f_4 = f_5 = \frac{3}{8}$

Note that to distinguish some terms, i.e. for some stretchings, it was necessary to break the polygon and rotate one piece before stretching
e.g. for $\sigma(Y_{13})$

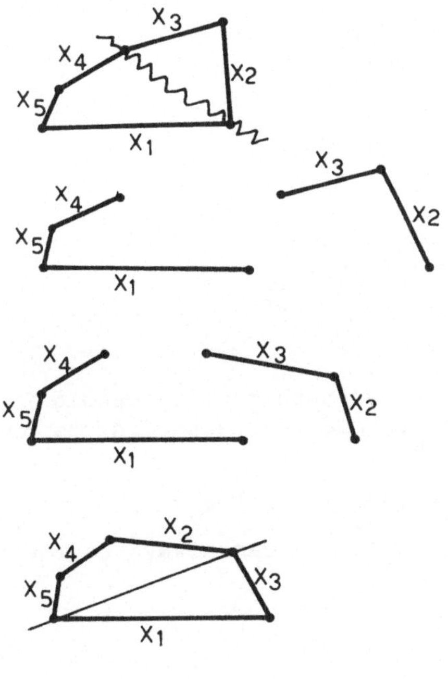

This form is general for any N. That is, each set of values of $f_1,\ldots,f_N$ is characterized by a class of polygons distinguished by a series of stretchings and the resultant degenerate polygons. This stems from the fact that for each set of values of the signum terms, there corresponds one set of values of $f_1,\ldots,f_N$. The properties of the polygonal representation will not be examined further here.

4. - Now, let us introduce

$$\Omega_N = \frac{2^N N!}{2^z \Pi s_i!}$$

which numbers all distinct sequences that can be obtained from $f_1 > f_2 > \ldots > f_N > 0$, by permutations and changes of signs where z is the number of $f_i = 0$ and s_i is the number of signs of f_i having a same value $\sum_{i=1}^{N} s_i = N$.

Using this, we have, for N = 4 for example:

f_1	f_2	f_3	f_4	Ω	
1	0	0	0	$\frac{2^4 \cdot 4!}{2^3 \cdot 1! \cdot 3!} =$	8
$\frac{3}{4}$	$\frac{1}{4}$	$\frac{1}{4}$	$\frac{1}{4}$	$\frac{2^4 \cdot 4!}{1! \cdot 3!} =$	64
$\frac{1}{2}$	$\frac{1}{2}$	$\frac{1}{2}$	0	$\frac{2^4 \; 4!}{2 \cdot 1! \cdot 3!} =$	32
					104

We note that 104 is the known number of threshold functions of three variables and a threshold (corresponding to our N = 4 as was shown earlier).

In conclusion, we remark that given the values of $f_1,\ldots,f_N$, a possible function $\sigma\left|\sum_{n=1}^{N} x_n \xi_n\right|$ can always be found.

Although this point is rather trivial, an illustration of it may serve as a useful exercise.

Take N = 5 and $f_1 = \frac{5}{8}$; $f_2 = f_3 = \frac{3}{8}$; $f_4 = f_5 = \frac{1}{8}$

We may use several methods to find the values of the signum terms involved; the following is perhaps the simplest. The expansions for f_i are

$$f_1 = \frac{1}{16}\{\sigma(Y_o) - \sigma(Y_1) + \sigma(Y_2) + \sigma(Y_3) + \sigma(Y_4) + \sigma(Y_5) - \sigma(Y_{12}) - \sigma(Y_{13})$$

$$- \sigma(Y_{14}) - \sigma(Y_{15}) + \sigma(Y_{23}) + \sigma(Y_{24}) + \sigma(Y_{25}) + \sigma(Y_{34}) + \sigma(Y_{35}) + \sigma(Y_{45})\}$$

$$f_2 = \frac{1}{16}\{\sigma(Y_o) + \sigma(Y_1) - \sigma(Y_2) + \sigma(Y_3) + \sigma(Y_4) + \sigma(Y_5) - \sigma(Y_{12}) + \sigma(Y_{13})$$

$$+ \sigma(Y_{14}) + \sigma(Y_{15}) - \sigma(Y_{23}) - \sigma(Y_{24}) - \sigma(Y_{25}) + \sigma(Y_{34}) + \sigma(Y_{35}) + \sigma(Y_{45})\}$$

$$f_3 = \frac{1}{16}\{\sigma(Y_o) + \sigma(Y_1) + \sigma(Y_2) - \sigma(Y_3) + \sigma(Y_4) + \sigma(Y_5) + \sigma(Y_{12}) - \sigma(Y_{13}) +$$

$$+ \sigma(Y_{14}) + \sigma(Y_{15}) - \sigma(Y_{23}) + \sigma(Y_{24}) + \sigma(Y_{25}) - \sigma(Y_{34}) - \sigma(Y_{35}) + \sigma(Y_{45})\}$$

$$f_4 = \frac{1}{16}\{\sigma(Y_o) + \sigma(Y_1) + \sigma(Y_2) + \sigma(Y_3) - \sigma(Y_4) + \sigma(Y_5) + \sigma(Y_{12}) + \sigma(Y_{13})$$

$$- \sigma(Y_{14}) + \sigma(Y_{15}) + \sigma(Y_{23}) - \sigma(Y_{24}) + \sigma(Y_{25}) - \sigma(Y_{34}) + \sigma(Y_{35}) - \sigma(Y_{45})\}$$

$$f_5 = \frac{1}{16}\{\sigma(Y_o) + \sigma(Y_1) + \sigma(Y_2) + \sigma(Y_3) + \sigma(Y_4) - \sigma(Y_5) + \sigma(Y_{12}) + \sigma(Y_{13})$$

$$+ \sigma(Y_{14}) - \sigma(Y_{15}) + \sigma(Y_{23}) + \sigma(Y_{24}) - \sigma(Y_{25}) + \sigma(Y_{34}) - \sigma(Y_{35}) - \sigma(Y_{45})\}$$

If we collect the constant terms we get

$$f_1 = \frac{1}{16}\{10 + \sigma(Y_1) - \sigma(Y_{12}) - \sigma(Y_{13}) - \sigma(Y_{14}) - \sigma(Y_{15}) + \sigma(Y_{23})\} = \frac{5}{8}$$

$$f_2 = \frac{1}{16}\{4 + \sigma(Y_1) - \sigma(Y_{12}) + \sigma(Y_{13}) + \sigma(Y_{14}) + \sigma(Y_{15}) - \sigma(Y_{23})\} = \frac{3}{8}$$

$$f_3 = \frac{1}{16}\{4 + \sigma(Y_1) + \sigma(Y_{12}) - \sigma(Y_{13}) + \sigma(Y_{14}) + \sigma(Y_{15}) - \sigma(Y_{23})\} = \frac{3}{8}$$

$$f_4 = \frac{1}{16}\{2 + \sigma(Y_1) + \sigma(Y_{12}) + \sigma(Y_{13}) - \sigma(Y_{14}) + \sigma(Y_{15}) + \sigma(Y_{23})\} = \frac{1}{8}$$

$$f_5 = \frac{1}{16}\{2 + \sigma(Y_1) + \sigma(Y_{12}) + \sigma(Y_{13}) + \sigma(Y_{14}) - \sigma(Y_{15}) + \sigma(Y_{23})\} = \frac{1}{8}$$

This yields

$$\sigma(Y_1) = \sigma(Y_{23}) = \sigma(Y_{14}) = \sigma(Y_{15}) = 1$$
$$\sigma(Y_{12}) = \sigma(Y_{13}) = -1.$$

One finds thus the following inequalities:

$$x_2 + x_3 + x_4 + x_5 > x_1 \; ; \quad x_1 + x_4 + x_5 > x_2 + x_3 \; ;$$

$$x_2 + x_3 + x_5 > x_1 + x_4 \; ; \quad x_2 + x_3 + x_4 > x_1 + x_5 \; ;$$

$$x_1 + x_2 > x_3 + x_4 + x_5 \; ; \quad x_1 + x_3 > x_2 + x_4 + x_5 ;$$

which are easily seen to have solutions such as

$$\sigma| 5\xi_1 + 4\xi_2 + 2\xi_3 + \xi_4 + \xi_5 |.$$

and all the others of its equivalence class, such as

$$\sigma| 5\xi_1 + 3\xi_2 + 3\xi_3 + \xi_4 + \xi_5 | .$$

Acknowledgment.

One of the authors (W. E. L. Grimson) is indebted to Prof. G. Papini and the University of Saskatchewan, Regina, Canada, for making it financially possible to spend these months in Naples.

REFERENCES

1. Caianiello, E. R.: Outline of a Theory of Thought Processes and Thinking Machines. J.Theoret. Biol. 2, 204 (1961)

2. Caianiello, E. R.; De Luca, A.; Ricciardi, L.M.: Reverberations and Control of Neural Networks. Kybernetik 4, 10 (1967)

3. De Luca, A.: On Some Representation of Boolean Functions Kybernetik 9, 1 (1971)

4. Caianiello, E.R.: Some Remarks on the Tensorial Linearization of General and Linearly Separable Boolean Functions. Kybernetik 12, 90 (1973).

AUTOMULTIPROCESSING (AMP): A NEW CAPABILITY FOR LSI/MOS COMPUTERS WITH APPLICATIONS TO HYPERNUMBER ALGEBRAS AND ARITHMETICS

C. Musès

Research Centre for Mathematics and Morphology
Editorial Offices, Santa Barbara, California

Editorial Board, Journal of Bio-Medical Computing, England

INTRODUCTORY REMARKS

There are several different usages of the term "parallel processing" but they all overlap in the essential concept of multiple, independent operations. Thus either hologram image realization or nonlinear selective picture scanning and description could be considered as parallel processing. In this paper we shall use the term in the sense of two or more independent programs operated together by an electronic computer over a wide spectrum of purposes.

AMP (INCLUDING A BRIEF HISTORY OF PARALLEL PROCESSING)

By 1962 G.W. Brown[1] — anticipating multi- or parallel processors — had realized the need for a processing method that would not be subject to the "kind of overspecification of sequence constraints that characterize present programming." Nothing really practical was accomplished in this problem area for some years, however. Only in 1968 was the idea of a new program language voiced[2] that might enable the determination of program sequence and concurrence without explicit indication. By 1968 there was also an awareness[2] of the need for parallel computation of arithmetic expressions and two years before, in 1966, D.Martin[3] had worked out some methods of automatic sequencing and assignment of computations in parallel processor systems. By 1969 there was a growing interest in automatic program analysis.[4]

This key trend is due to climax in maximally (optimally) unspecified automatic programs resulting in solutions of problems otherwise very difficult or impossible to gain. Since "automatic" here actually means self-steering or self-guiding, it becomes inevitable that cybernetic program components arise, and the approach and announcements of this paper explicitize this statement.

Although with large scale integrated circuits (LSI) parallel processing can and has taken large strides, little or no advance has been made in the region of minimally specified automatic programs for equation solving, even in real equations — not to mention those involving complex numbers or even more sophisticated hypernumbers.[5] This chapter, based on research[6] since 1962, is designed to begin to fill that hiatus, so that — unlike methods now in use — a human operator need not constantly intervene before the solution is gained to desired accuracy. We call this method AMP.

Self-guided automatic equation-solving by the parallel processing power of AMP (Automatic Multi-Processing with self-guidance, especially over a broad spectrum of types) (cf. Table I), has a future of indefinitely high potential. Equation-solving capability without intermittent and repeated human intervention during the solution can not beoffered at present even by the largest computer manufacturers. On the other hand, the AMP approach, which may also be firm-wired, has successfully solved equations automatically which would be very troublesome or even next to impossible by the means in ordinary use involving recurring human intervention; for example, equations with complex coefficients (2nd equation,item 7, Table I) as well as those containing irrational powers of the unknown (item 6), mixed transcendental forms (item 2, Table I, and item 6, 2nd

TABLE I: TYPE EXAMPLES OF THE CAPABILITY OF AMP (AUTOMULTIPROCESSING)

1. $x^{1/2} + x^{1/3} + x^{1/4} + x^{1/5} + x^{1/6} + x^{1/11} - 17 = 0$ yields $x = 35.9914857695$

2. $e^{x} + e^{2x} + e^{3x} + e^{\cos x} - 1001 = 0$ yields $x = 2.2666356880$

3. $\sin x + 2 \sin 3x + 5 \sin 7x - 1 = 0$ yields $x = 0.0239153458$ radians $= 1° 22' 13''$ to nearest sec.

4. $7x^5 - 3x^2 + x + x^{\frac{1}{2}} - 1 = 0$ yields $x = 0.6497655959$

5. $x^{3103} - x^{2218} - 10^{99} = 0$ yields $x = 1.07622879705$

   Above solution times: less than 13 seconds to about 1+ min.

6. Solve jointly: $6x^{13.7621} - 8x^{9.5142} - 15x - x^{2^{\frac{1}{2}}} - 17 = 0$

   and $e^{\cos x} + \ln x + x - 1 = 0$

   Yielded in order are the solutions $x = 1.2312478512$ and $x = -0.7863899677$ radians

   Solution time for both roots processed together: approx. 1½ min.

7. Solve jointly: $x^{79} - x^{67} - 7 = 0$; $x^{17} + 7x + 5 - 3i = 0$ $(i = \sqrt{-1})$; and $x^5 + 9x^{2/27} - 13x^{5/187} - 27x + 1 = 0$

   Yielded in order are the solutions $x = 1.0381963861$; $x = 1.1450884225 \pm 0.2057505002i$, where the absolute value is 1.1634263232; and $x = 2.3040203866$.

   Solution time for the three roots processed jointly: approx. 2 min.

equation), high powers (item 5), or fractional powers of high denominator (item 7, 3rd equation).

PARALLEL PROCESSING AND HYPER-ARITHMETICS

In the context of hypernumbers (of which complex numbers are the simplest) and their applications, parallel and de facto parallel processing have immense relevance, because each number may have several components each of which must be kept track of in the computation. The writer has had to work with hypernumbers involving 24 or more components, several sets of which are governed by quite distinct arithmetics (hence distinct algebras and function theories). The AMP approach can successfully program hyperarithmetics and their corresponding algebras, handling them with ease and elegance in either a soft or firmware program.

As an example, hitherto unpublished, of a hypernumber operation we shall consider a ring of 24 operators that embeds Cayley (and hence quaternion and complex) algebra. It also, incidentally, embeds the algebra of the Pauli spinors used in quantum physics. Eight of these twenty-four hypernumber operators are matrices, and form a sub-ring of their own, defined as follows:

$$i_1 = i = \begin{pmatrix} 0 & -1 \\ 1 & 0 \end{pmatrix} \qquad \epsilon_1 = \quad = \begin{pmatrix} 0 & i \\ -i & 0 \end{pmatrix}$$

$$i_2 = \begin{pmatrix} i & 0 \\ 0 & -i \end{pmatrix} \qquad \epsilon_2 = \begin{pmatrix} 1 & 0 \\ 0 & -1 \end{pmatrix}$$

$$i_3 = \begin{pmatrix} 0 & i \\ i & 0 \end{pmatrix} \qquad \epsilon_3 = \begin{pmatrix} 0 & 1 \\ 1 & 0 \end{pmatrix}$$

$$i_o = \begin{pmatrix} i & 0 \\ 0 & i \end{pmatrix} \qquad \epsilon_o = 1 = \begin{pmatrix} 1 & 0 \\ 0 & 1 \end{pmatrix}$$

Note that $i_n^2 = -1$, $\epsilon_n^2 = +1$, and $i_n \epsilon_n = \epsilon_n i_n = i_o$. The lack of notice of the important Jordan matrix i_o has confused and often marred the work of many writers in this field, including even E. Cartan and P.Dirac. We now continue with the other 16 elements.

The optimal way to obtain these is as follows. The basic property

$$i_n \epsilon_n = \epsilon_n i_n = i_o \tag{1}$$

implies

$$i_o \epsilon_n = \epsilon_n i_o = i_n \tag{2}$$

and also

$$i_o i_n = i_n i_o = -\epsilon_n \tag{3}$$

From (1), (2), (3) and the two defining equations

$$\epsilon_n^2 = -i_n^2 = 1 \qquad (4),(5)$$

together with the basic matrices just given, the simplest consistent set of rules for the hypernumber arithmetic of the ring of 24 elements can be derived. To do this we first define a bimatrix.

The term bimatrix denotes an operator composed of two matrices separated by a slant stroke, the second matrix "behind" the stroke being in a separate domain (the stroke domain) which is normally inaccessible but nevertheless connected to the ordinary domain preceding the stroke by certain operations to be specified. Thus A/B is a bimatrix where A and B are matrices. If B represents a real number, then A/B = AB, and if A is a real number then A(/B) = A/B where /B = 1/B and A/1 = A(1) = A. Instead of a stroke, a square bracket opening on the right may be used if there is any need for it occasioned by the use of the slant stroke for fractional notation at the same time. Normally no confusion will arise.

In this arithmetic and its algebra the only matrices that can enter the stroke domain are ϵ_o or 1, ϵ_1 or ϵ, i_1 or i, and i_o. Otherwise, as will be easily evident, there would occur a violation of the fundamental theorem first proved by A. Hurwitz that if the product of a sum of n squares by another sum of n squares can still be represented as the sum of n squares, then n cannot exceed 8.

NUMBERS WITH 24 COMPONENTS: THE OPERATOR RING OF M-ALGEBRA

By M-algebra we refer to the Meta-cayley algebra described in the second paragraph of the previous section. The 24-dimensional hypernumber arithmetic at the heart of this algebra (for hypernumber units generate algebras and not the other way around) is closely connected with M.J.E. Golay's 1954 discovery of a lossless or perfect triple-error correcting code, in turn intimately connected with the closest packing of equal spheres in 24 dimensions. C.A. Rogers' density criterion shows that the 24th dimension (also the key dimension in the theory of elliptic functions, so deeply related to number theory) is probably unique in packing capacity, and that hence the Golay code is likewise unique and that no perfect or close-packed code exists capable of automatic detection and correction of more than 3 errors, that is, 3 per code word.

We are now in a position to write down the entire ring of 24 elements:

$i_o (= \epsilon_n i_n = i_n \epsilon_n)$	$\epsilon_o (= \epsilon_n \epsilon_n = -i_n i_n)$
$i_1 \ (\equiv i)$	$\epsilon_1 \ (\equiv \epsilon)$
i_2	ϵ_2
i_3	ϵ_3
$i_4 = /i$	$\epsilon_4 = /\epsilon$
$i_5 = i/i$	$\epsilon_5 = i/\epsilon$
$i_6 = -i_2/i$	$\epsilon_6 = -i_2/\epsilon$
$i_7 = i_3/i$	$\epsilon_7 = i_3/\epsilon$
$i_8 = i_o/\epsilon$	$\epsilon_8 = -i_o/i$
$i_9 = \epsilon/\epsilon$	$\epsilon_9 = -\epsilon/i$
$i_{10} = \epsilon_2/\epsilon$	$\epsilon_{10} = -\epsilon_2/i$
$i_{11} = \epsilon_3/\epsilon$	$\epsilon_{11} = -\epsilon_3/i$

It was the writer's compact formula for sphere-packing in dimensions 1 through 8 (cited by H.S.M. Coxeter in the third volume of "Lectures in Modern Mathematics edited by T.L. Saaty in 1966) that he used, after Coxeter introduced us, to bring to J. Leech's attention that the latter's conception of the packing capacity of 24-dimensional space was undoubtedly far too small. This evaluation spurred Leech to try to contradict it, which he could not do. But his feverish search bore serendipitous fruit in a far denser 24-dimensional packing than he had ever imagined. Naturally, a finite group is also associated with that packing, because the packing reflects M-algebraic structure. In fact, I had explained to Leech that my work on characterizing the generalization of Cayley algebra indicated beyond reasonable doubt that his original value for the number of equal hyperspheres that would fit around a central one was far too small.

The phenomena of the highly dense 24-dimensional sphere packing, the Golay close-packed error-correcting code, and the related finite group are all rooted in M-algebra as Euler's product-of-sums-of-four-squares theorem is rooted in quaternion algebra. The algebras in turn are rooted in hypernumber units (which also generate distinct spaces and function theories) and it is in this sense among others that hypernumbers can be called the core of mathematics.

Turning to the list on the previous page, the absolute value and also the vector length of each of these 24 elements is unity. We distinguish between the two concepts of absolute value or modulus as distinct from vector length because in epsilon-type numbers the two are no longer necessarily identical. For example, the absolute value of the number $1+\epsilon_n$ is zero, but its length is the square root of two or approximately 1.4142, considerably more than zero. Also unlike imaginary numbers, the equation $e^{\theta x} = x^k$ no longer holds if x is changed from i_n to ϵ_n. In that case, the left-hand side of the equation remains in a 2-space, but the right-hand expression now requires a 4-space, and in general the equality does not hold. Thus $e^{\theta\epsilon} = \cosh\theta + \epsilon \sinh\theta$, whereas ϵ^k is given by the expression $\cos^2 \pi k/2 + \epsilon \sin^2 \pi k/2 - (i/2)(1-\epsilon) \sin \pi k$, where k may be real, complex, or countercomplex i.e. of the form $a+b\epsilon_n$, where a and b are real. Note, incidentally, that the absolute value of this number is the positive square root of $|a^2 - b^2|$ whereas the absolute value of $a+bi_n$ is the positive square root of $a^2 + b^2$. Also, the natural logarithm of ϵ is given by $(-\pi i/2)(1-\epsilon)$.

Great care must be exercised, when composing programs involving hypernumbers, not to fall into pre-conditioned familiar assumptions which no longer hold. Thus the logarithms even of such comparatively lowly hypernumbers as quaternions begin to abrogate the laws of ordinary logarithms. Calling the i-operators of the first three subscripts, that is i_1, i_2, and i_3, "i, j, and k" respectively, we have $\ln i = \pi i/2$ and $\ln j = \pi j/2$. But their sum is not the natural logarithm of k, which is $\pi k/2$ and not $(\pi/2)(i+j)$. Yet $i_1 i_2 = i_3$.

The essence of hypernumbers is that they in some measure abrogate one or more rules of the arithmetic of ordinary numbers. Since the number of those rules is finite, the number of their abrogations is also finite, irrelevancies of course not being included. Thus the number of species of hypernumbers is finite also.

In conclusion, it will be seen that the above specification of M-algebra fully obeys all previously known findings in quaternion and Cayley algebra. To make this fact clear we briefly state the rules for bimatrix operation, completing the preliminaries previously given.

BIMATRIX ARITHMETIC

The addition rule is simple:

$$A/B + C/D = (A+C)/(B+D) \tag{6}$$

The multiplication rules are based on anticommutation; except in the case of zero-subscript operators, which are commutative:

$$i_0(A/B) = i_0 A/B = A i_0/B \tag{7}$$

$$/i_o = i_o;\ /(-i_o) = /-i_o = -i_o \tag{8}$$

$$(/A)^2 = /(A^2) \equiv /A^2 \tag{9}$$

$$(A/B)^2 = -A^2/B^2;\ A \neq i_o,\ \epsilon_o \tag{10}$$

$$(i_o/B)^2 = -/B^2 \tag{11}$$

$$(A/B)(C/D) = -CA/BD;\ A,C \neq i_o,\ \epsilon_o \tag{12}$$

The operation of these rules is illustrated in the examples that follow. Thus $i_o(i_o/i) = i_o^2/i = -/i$; $(/i)^2 = -1$; again, $(i/\epsilon)^2 = -i^2/\epsilon^2 = 1$; $(i/i)^2 = -i^2/i^2 = -1$; $(\epsilon/\epsilon)^2 = -\epsilon^2/\epsilon^2 = -1$; $(i_o/\epsilon)^2 = -/\epsilon^2 = -1$; $i_o(/\epsilon) = i_o/\epsilon$; $(/\epsilon)(/i) = /i_o = i_o$; and finally, $(i/i)(i_3/i) = -i_3 i/i^2 = -i_2/-1 = i_2$; $(i/i)(-i_2/i) = i_2 i/i^2 = -i_3/-1 = i_3$, and so on.

Reference to our table of hypernumber operators three pages back, in conjunction with the above rules and examples, will show that the usual Cayley algebra relations hold, indeed follow from the bimatrix definitions of the elemental Cayley operators. Thus $i_3 i_5 = i_4 i_2 = i_6$; $i_1 i_6 = i_2 i_5 = i_3 i_4 = i_7$; and $i_1 i_4 = i_5$. It is worth noting that the subscript formation by modulo addition is important in the analysis of these algebraic-arithmetic structures.

All these operations may be computer-processed by the AMP capability. Note that $\epsilon_{1,2,3}$ are the Pauli spinors. As the writer pointed out in an invited lecture at the NASA Ames Research Center in 1970, there is a basic set of thirty-two matrices implied by the Pauli and Dirac spinor matrices, and including them. This set of 32 matrices is classifiable into four structure-types, which we call $S, \overline{Z}, \overline{S}$, and Z or 0,1,2, and 3 respectively, the particular type depending on whether the principal or secondary diagonal is tenanted with like members of like or unlike sign. Thus $\begin{pmatrix}\epsilon & 0\\ 0 & \epsilon\end{pmatrix} = \epsilon_{01}$, which is to say, an epsilon operator with structure 0 and element 1; also $\begin{pmatrix}0 & \epsilon_2\\ \epsilon_2 & 0\end{pmatrix} = \epsilon_{32}$, and so on. Sr. Research Physicist for Hofstra University Gary R. Gruber has just joined the writer in a collaborative project aimed at specifying implications of hypernumbers in quantum physical applications. With the necessary time and facilities made possible by funding now in process of negotiation, new and firm results are projected for within a year of the formal beginning of the project.

Some Hints in Programming Epsilon Arithmetic and Algebra

Countercomplex arithmetic, hence algebra, is not wholly associative when interaction between its zero divisors and infinity divisors (reciprocals of zero divisors) is considered. The only othrecent writer to discuss countercomplex numbers[7] (he uses the more opaque term "double numbers") has asserted that epsilon, which he calls "e", has no even roots and hence no square root. However, in fact epsilon possesses not only a square root and even roots but also all real roots, as explicitly displayed in the formula for ϵ^k with k real, given two pages back, second line from top. That author is not to be blamed, for hypernumber research often takes difficult and unexpected turns, the more so since it is such a new and relatively unconsidered field. Rather, that author is to be given credit for the courage to publish and so stimulate others. He also missed finding the important expression for the epsilon exponential (two pages back, top line) because he limited himself to hyperbolic geometrical considerations and did not explore epsilon algebra per se, which is the quickest and most natural route.

For the same reason the same author did not notice the anomalous nonassociation in epsilon algebra where interactions between zero divisors and their reciprocals are concerned. Thus this lack renders his discussion of numbers like $(1+\epsilon)/(1-\epsilon)$ unsafe or misleading. But again he deserves credit for bringing up the subject.

Epsilon algebra and arithmetic may be termed "almost associative" in the sense that, although the combinations causing nonassociation form an infinite set, the ratio of that set to the total infinite set of the entire arithmetic or algebra is zero. The relation of these two infinite sets can quite exactly be stated as that of the set of points lying on the asymptotes of an equilateral hyperbola to the entire set of points of the plane.

Epsilon numbers of the form $k(1 \pm \epsilon)$, where k is real and not 0, are all zero divisors. Although neither factor is zero, yet the product $(1+\epsilon)(1-\epsilon) = 0$. We have proved that the arithmetic and hence algebra of zero divisors is not associative.[8] Thus we have

$$\epsilon\left[(1+\epsilon)(1/1+\epsilon)\right] = \epsilon \tag{13}$$

but

$$\left[\epsilon(1+\epsilon)\right](1/1+\epsilon) = 1 \tag{14}$$

Moreover[8] the zeroth power of a zero divisor is not unity, and hence the first negative power of such a number does not equal its reciprocal. Thus

$$(1+\epsilon)^{0} = (1+\epsilon)/2 \tag{15}$$

which, like all zeroth powers, is an idempotent as the reader can easily check. But

$$(1+\epsilon)^{-1} = (1+\epsilon)/4 \tag{16}$$

which does not equal the reciprocal, $1/(1+\epsilon)$.

THE FUTURE POTENTIAL OF AMP

In view of growing demands for solving increasingly complicated or sophisticated equations by modern physics, meteorology and space exploration, the AMP capability in ordinary or hypernumbers has a vital and unique role to fulfill. Because of basic negotiations now in progress, specific details of method cannot be discussed explicitly at this time. However, within few years, judging from results so far achieved, this capability can be expected to be firmwired in even minicomputers and advanced programmable calculators.

Among the feasible candidates for AMP that could gain large increases of power and use by incoporating it, may be cited the outstanding 230-25 medium scale computer made by Fujitsu Ltd (not to mention their larger 230-60), IBM's 360 or 370 series, and the excellent Burroughs, Honeywell, and Hewlett-Packard machines. The interesting teamwork in both support and development between the government and information industry in Japan that was begun by special law enacted early in 1970, offers a very viable alternative in the computer design field and one also realistic and flexible enough to be able to adapt to new ideas and needs as they arise.[9]

Some of the forthcoming fourth generation computers now on the drawing boards or still being designed may profit from the opportunity of further modification enabling inclusion of AMP capability. Full AMP capability inclusion would mean not only automatic equation solving and hypernumber algebras, but also linguistically natural, highly efficient machine response in the areas of game theory and strategy as well as in the area of psychological analysis and therapy — important applications of the new approach for which there was insufficient space on this occasion.

Another significant feature of the AMP capability lies in its offsetting the present awkward exponential trend toward more and more cumbersome program libraries. Many formerly separate programs would be replaced by a single new operation, preferably firm-wired, thus providing de facto automatic program generation as well as automatic operation. This feature possesses current practical value in view of the shortage of able software composers and technicians.

Computers embodying the full spectrum of AMP capabilities are an assured part of the future of computer design, program realization and concurrent processing throughout the information sciences.

REFERENCES

1. Brown, G.W., A new concept in programming, in The Computer and the World of the Future, Greenberger, M., Ed., MIT Press, Cambridge (Mass.), 1969.
2. Tesler, L.G., and Enea, H.J., A language design for concurrent processes, Spring Joint Computer Conference Proceedings, 32, 403, 1968.
3. Martin, D., Automatic assignment and sequencing of computations on parallel processor systems, Reports of the University of California (Los Angeles) Department of Engineering, no.4, 1966.
4. Russell, E.C., Jr., Automatic program analysis. Ibid., no.12, 1969.
5. Musès, C., Hypernumbers and their spaces, Journal for the Study of Consciousness, 5, 251, 1972-3.
6. Musès, C., Automatic multi-processing in LSI computer programming of equation solutions, real or complex, RCM Progress Reports of Classified Distribution, no. 12, 1972. Research Centre for Mathematics and Morphology, Ed. Offices, Santa Barbara, California.
7. Yaglom, I.M., Complex Numbers in Geometry, Academic Press, New York, 1968.
8. Musès, C., Fractional dimensions and their experiential meaning, with some new theorems proved on divisors of zero, Journal for the Study of Consciousness, 5, 315, 1972-3.
9. Fujii, A., Software in Japan, Datamation, February 15, 1971.

Acknowledgement

The Monroe-Litton Company kindly lent their model 1880 for certain demonstrations during the June 1973 session of the NATO Advanced Study Institute at Capri.

Addendum. Trying to program Bernoulli numbers B(s) led to a long needed generalization of them for any real index: Where s is real and ζ is the Riemann zeta function, $B(s) = -s!2\cos(\pi s/2)\,\zeta(s)/(2\pi)^s$. Thus, recalling that $\zeta(0) = -\frac{1}{2}$ we have correctly $B(0) = 1$ and $B(2) = 1/6$. Also $B(-\frac{1}{2}) = -2\pi\,\zeta(-\frac{1}{2})$ and $B(\frac{1}{2}) = -\frac{1}{2}\,\zeta(\frac{1}{2})$, etc. And s may be complex.

ON A FAMILY OF RATIONAL LANGUAGES

A. Restivo and S. Termini

Laboratorio di Cibernetica
Arco Felice - Napoli

"Time, easily confutable on the sensitive level is not equally so on the intellectual one, whose essence seems inseparable from the concept of sequentiality" (*)
J. L. Borges

1. Motivations and introductory remarks

It is well known that every parallel process can be simulated by a sequential one. The converse, however, is not true: it is not possible to simulate every sequential process by means of a parallel one (1). One is generally interested in "parallel processing" for, clearly, a computation is speeded-up by using more than one processor at the same time (2). In this paper, however, we do not deal with this problem but want to give a brief survey of some concepts arising in the field of finite automata which are related to the notion of parallel analysis. The process with which an automaton "recognizes" if a string belongs to a certain set is typically sequential:

(*) "Sentirse en muerte" in Historia de la eternidad, Buenos Aires, 1936; perhaps it is not trivial to stress that Borges has been very often strongly inclined to deny reality to time.

the automaton scans the string symbol by symbol from the left to the right changing its internal state after reading each symbol. Is it possible to simulate every finite automaton by means of a parallel device? The answer, as in the case of a general computation, is again negative. In these last years, however, more and more attention has been payed to particular families of rational languages (i.e. the languages recognized by finite automata). These ones have been widely investigated both for many very deep and elegant mathematical properties characterizing them, and to understand what a finite device is able to do when other restrictions are added. One of these subclasses of rational languages, namely locally testable ones, is conceptually very strongly related with parallel analysis. In fact an automaton recognizing a locally testable language analyzes every string by only independently looking at its segments of a fixed length. The order in which the various segments are looked at does not matter. This kind of analysis, then, does not evidently require a sequential way of operating of the automaton. We do not present here the theory of these languages, for which the reader is referred to the monograph of McNaughton and Papert[2], but shall only summarize the basic definitions (section 3) needed to present some recent results (section 4).

In order to make the paper self-contained some classical results of the theory of free semigroups are reported in the next section. The style is colloquial and for the proofs of the quoted theorems the reader is referred to the original papers. We want finally remark that the choice of the arguments presented here is strongly biased by our personal preference and work of the last period and does not, then, pretend to be a general survey of the actual problems of locally testable languages.

2. Free Semigroups

In this section we report a classical theorem of the theory of free semigroups and show its relationship with the notion of very pure subsemigroups. (3)

Given a free semigroup S and subsemigroup S' of S it is important to find the conditions under which S' is also free. The answer to this important question is given by the following theorem due to M. P. Schützenberger (4):

Theorem 2.1 Let S' be a subsemigroup of a free semigroup S. S' is free iff for any s'εS' and s ε S, ss' ε S' and s's ε S' implies s ε S'.

Let us now give the definition of an algebraic property of a subsemigroup of a free semigroup that arose in connection with the notion of conjugacy. A purely algebraic definition will here given. The reader wanting to see its relationships with the notion of conjugacy is referred to Restivo[7] and Lentin and Schützenberger[8].

Definition 2.2 (5) Let S' be a subsemigroup of a free semigroup S. S' is very pure iff for any s ε S and s' ε S, ss'ε S' and s's ε S' imply s ε S' and s' ε S'.

The following corollary of theorem 2.1 holds:

Corollary 2.3 Let S' be a subsemigroup of a free semigroup S. If S' is very pure then S' is free.

3. Rational and locally testable languages

Let X be a finite set and X^* the free monoid (i.e. semigroup with an identity element) on X. The elements of X will be called letters, the elements of X^* words and the subsets of X^* languages. A language is rational iff belongs to the class Rat defined in the following way:

1) All the singletons {x}, xεX, the identity element λ of X^* and the empty set ϕ belong to Rat.
2) If A and B belong to Rat then $A \cup B$ belongs to Rat.
3) If A and B belong to Rat then $A \cdot B = \{a \cdot b \mid a \in A, b \in B\}$ belongs to Rat.
4) If A belongs to Rat then A^*, the submonoid generated by A, belongs to Rat.

The classical, important, result of Kleene[9] asserts that the class Rat of the rational subsets of X^* coincides with the class Rec of the subsets of X^* recognizable by a finite automaton (6). We shall now take into account subclasses of Rat; this means that the languages under study are recognized by particular finite automata (7).

Kleene proved also that the class Rat is closed under the boolean operations, that is

5) If A belongs to Rat then the complement $\bar{A} = X^{*} \setminus A$ of A belong to Rat.

6) If A and B belong to Rat then $A \cap B$ belongs to Rat.

An interesting subclass of Rat, the aperiodic (or star-free) languages, is obtained by dropping axiom 4 and adding to the list of axioms the properties 5 and 6.
Locally testable languages are a subclass of the aperiodic ones. We shall now give the definitions of strictly k - testable and strictly locally testable languages.

Definition 3.1 A language L on X is strictly k - testable iff exist three subset P,S and I of X^k such that for all $f \varepsilon X^*$, $|f| \geq k$, $f \varepsilon L$ iff $P_k(f) \varepsilon P$, $S_k(f) \varepsilon S$, $I_k(f) \subset I$, where $P_k(f)$ is the k-length prefix of f, $S_k(f)$ the k-length suffix of f and $I_k(f)$ the set of k-length internal segments of f.

Definition 3.2 A language L is strictly locally - testable iff exists an integer k such that L is strictly k-testable.

Strictly locally testable languages are not closed under boolean operations. Consider, for istance, the following example. Let L_x, $x \varepsilon X$, the language formed by all and only the words on X that begin and end with the same letter x of X. L_x is strictly locally testable. Let us now consider the language $L = L_a \cup L_b$, where a and b are two different letters of X. L is formed by all and only the words that either begin and end with a or begin and end with b; L, clearly, is not strictly locally testable.
The term locally testable language is used to indicate any element of the boolean closure of the family of strictly locally testable ones.

4. Some results

In this last section some recent results concerning very pure and free submonoids and locally testable languages will be reported. The notion of local testability is a formalization of an intuitive idea of the operations of a machine while the property of a submonoid to be "very pure" is a purely algebraic concept. They are, however, related; the two notions are even equivalent under suitable hypothesis, as the following theorem shows.

Theorem 4.1 (8) Let X be a finite set, X^* the free monoid generated by X and M a submonoid of X^*. If M is free and finitely generated (i.e. generated by a finite subset of X^*) then M is very pure iff M is strictly locally testable.

The two hypothesis of the freedom of M and of the finiteness of its basis are essential. The following two examples show that without one of them the equivalence between the two notions does not still hold. It is noteworthy, however, that if only one of them is dropped one of the two implications is still true, while if both are dropped neither of the two notions implies the other.

In particular if M is free but not finitely generated it is no more true that M very pure implies M strictly locally testable but it is still true that M strictly locally testable implies M very pure.

If M is finitely generated but not free it is no more true that M strictly locally testable implies M very pure. However the implication M very pure implies M strictly locally testable, if M is finitely generated, is true without explicitely requiring its freedom since, by corollary 2.3, the freedom of M is contained in the hypothsis that M is very pure.

We now give the example of a free but not finitely generated submonoid which is very pure but not locally testable.

Example 1. Let be X = {a,b} and consider the submonoid M of X^* generated by the infinite set $(a^2)^*b$. M is very pure; indeed all its words are characterized by the following properties:

1) Either they begin with the letter b or there is an even number of letters a before the first letter b.

2) Between two letters b either there is no letter a or there is an even number of them.

3) they end with the letter b.

If the previous properties are satisfied both by the words ss' and s's (that are cyclic trasformations each of the other) then they are satisfied also by s and s' separately. M is then very pure. However it is not locally testable since the number of letters a between two consecutive b (and of initial a) must be even but does not admit an upper bound.

The following example shows a finitely generated but not free submonoid which is strictly locally testable.

<u>Example 2.</u> Let M be the submonoid of X^* generated by the set $\{v^2, v^3, w^2, w^3\}$ where v and w belong to X^*.
Clearly M is not free and then, by corollary 2.3, is neither very pure. It is nevertheless strictly locally testable since all and only the words of X^* of the form vwp, wvp, pvw, pwv, pwvwq, pvwvq, where p and q are arbitrary words of X^*, are excluded.

Footnotes

(1) A still open problem is to find necessary and sufficient conditions for an algorithm to be parallelizable. A condition of this type has been suggested by C. Bohm in this same school. (see page this volume)

(2) For a detailed account of the problems arised by the parallel use of more processors and for the solution of many of them we refer to the works of S. Winograd and in particular to his contribution to this same school. (see page this volume)

(3) For the definitions of semigroup, free semigroup and free subsemigroup see a standard reference text as Clifford and Preston[3].

(4) See Schützenberger[4]; the same condition was found independently one year later by Sçevrin[5]. An equivalent condition has been given by P.M. Cohn[6].

(5) Let us note a linguistic discrepancy between the way in which the word "very pure" is used here and in Restivo[7]; in the present paper the word is referred to the subsemigroup while in Restivo[7], where only free submonoids are considered, it is an attribute of its basis.

(6) For a more precise discussion on rational languages and automata see a standard reference text as Eilenberg[10]

(7) A detailed discussion on locally testable languages is found in McNaughton and Papert[2];see also McNaughton[11] for more recent developements.

(8) The proof of the theorem is in Restivo[7], where the two notions considered in the theorem are also related to the codes with finite synchronization delay.

REFERENCES

1. Borges, J.L., Nueva Confutacion del tiempo, Otras Inquisiciones, Buenos Aires, 1960.

2. Mc Naughton, R. and Papert, S.
Counter Free Automata.
MIT Press, Cambridge Mass., 1971.

3. Clifford A.H. and Preston, G.B.
The Algebraic Theory of Semigroups, Math. Surv. 7, Vol. I and II, A.M.S., Providence R.I., 1961 and 1967.

4. Schützenberger, M.P.
Sur certain sous-demi-groups qui interviennent dans un problem de mathematiques appliquee,
Publ. Sci. Univ. d'Alger, Sez. A 6, 85, 1959.

5. Scevrin, L.N.
On subsemigroups of free semigroups.
Dokl. Akad. Nank SSSR, 133, 537, 1960 (russian); english translation, Soviet Math. Dokl. 1, 892, 1960.

6. Cohn, P.M.
On subsemigroups of free semigroups.
Proc. A.M.S., 347, 1962.

7. Restivo, A.
On a question of Mc Naughton and Papert
(to appear in Information and Control).

8. Lentin, A. and Schützenberger, M.P.
A Combinatorial problem in the theory of free monoids in Combinatorial Mathematics and its Applications, Bose and Dowlings Eds.
University of North Carolina Press, 128, 1969.

9. Kleene, S.C.
Representation of Events in Nerve Nets and Finite Automata in Automata Theory, Shannon and McCarthy Eds.
Princeton, 1956.

10. Eilenberg, S.
Book in print

11. McNaughton, R.
Algebraic decision procedures for local testability.
(preprint)

OUTLINE OF A NEW LOGICAL APPROACH TO INFORMATION THEORY

A. De Luca and E. Fischetti

Laboratorio di Cibernetica del CNR
Arco Felice, Naples, Italy.

0. Introduction.

The most usual approach to information theory (1) is the "probabilistic" one based on the notion of "entropy" of a random variable (Shannon[1]). However, Shannon's information measure and the fundamental code-theorems on the transmission of information are meaningfully applicable only in a narrow context related to the statistical communication theory. To quote Kolmogorov[2]:"the probabilistic approach is natural in the theory of information transmission over communications channels carrying "bulk" information consisting of a large number of unrelated or weakly related messages obeying definite probabilistic laws".

Since the ordinary definition of information uses only probabilistic concepts, questions such as "what is the content of information in a mathematical theorem (or physical law) about another theorem (or law) ?" do not make any sense. On the contrary, as we shall see in the following, these questions become natural in a "logical" approach to information theory. As pointed out by von Neumann[3] the two existing theories on which it seems that an information theory has to be founded are "formal logics" and "thermodynamics" (2).

For the foregoing reasons the intuitive notion of information is only partially formalized in Shannon's theory. Apart from this, a proper use of probabilistic information requires that one knows what "probability" means exactly and moreover when it can be applied to the descriptions of physical phenomena.

It is well known that the "probability theory" has been formalized in 1933 by Kolmogorov,[4] who gave a set of axioms from which it is possible to derive the probability calculus. Such an axiomatization, that essentially reduces the probability theory to measure theory, appeared to solve most of the diatribes among frequentistic, logicistic and subjectivistic schools about the interpretation to be given to the term "probability". In fact, except for some cases (3), any "explicatum" of the intuitive notion of probability seems to obey Kolmogorov's axioms. However, as stressed by Kolmogorov[5] himself, the problem of finding the bases of real applications of the theory of probability is outside the theory itself.

Historically this latter problem was initially posed in 1919 by von Mises[6] in the setting of a frequentistic interpretation of probability (Kollektiv theory). Important contributions to this theory have been given later by Wald,[7] Ville[8] and Church.[9] Church was the first who recognized as essential, in order to give an adequate "explicatum" of the informal notion of infinite random sequence, the use of the formal concept of "algorithm". The major difficulty which arises in a theory of infinite random sequences is that the limit frequencies of the considered events cannot be calculated neither a-priori, since one cannot construct a random sequence, nor a-posteriori, since any run of real statistical experiments is of finite length.

Kolmogorov[2,5] recently reproposed the same problematics but for finite sequences (that one can always construct by means of suitable algorithms) using for his treatment the notion of "complexity" of the algorithms. By means of a suitable definition of it, Kolmogorov was able to furnish an algorithmic approach for information theory as well as probability theory.

We briefly outline Kolmogorov's approach. Let X and Y be two sequences in a certain alphabet. The complexity $K_A(Y/X)$ of Y given X with respect to the algorithm A is the minimum length of a "program" p such that the algorithm A is able to produce Y starting from X and the program-string p (that means $A(p,X) = Y$). If there is no such program $K_A(Y/X) = +\infty$.

This theory has been developed by Kolmogorov[2,10] and Martin-Löf[11] and beautiful results have been obtained. On one hand, by means of universal Turing machines, it is possible to introduce a suitable class of algorithms, called asymptotically optimal, with respect to which the complexity measures for high values of them do not depend on the particular algorithm of the class. On the other, Martin-Löf can characterize as random those sequences whose complexity is approximately equal to their length: such strings possess all conceivable statistical properties of randomness.

The Kolmogorov algorithmic approach to information theory appears, however, to be not very satisfactory since the complexity measure used by him is not a very "good" explicatum for the informal notion of complexity of an algorithm. In fact, such a quantity seems to be related more to a "dynamic" measure of the "resource" needed in a computation (as number of steps, length of tape, etc.) than to the program or "input data". A theory of computational complexity based on "dynamic" measures has been developed by Rabin[12] and Blum.[13]

In this paper a new logical approach to information theory is outlined. To this end we start by considering the notion of "formal system" or "logic" in its most general formulation. In these systems the concept of "complexity-proof" measures is axiomatized in a manner such as to preserve the most general aspects of the intuitive notion, and at the same time obtain a definition of "minimal complexity" satisfying requirements of effective computability. The latter requirement for certain complexity measures can be fulfilled only by suitable logics. We then arrive at a definition of "quantity of information", conveyed by one object about another, that is purely logical in its formulation. This quantity shows properties similar to mutual probabilistic information, except for the commutative property, not valid for our measure. The problem of varying the information changing the complexity measure and the logic is also investigated.

In conclusion, we stress that we have not yet strong indications showing that our results can be also utilized, as in Kolmogorov's theory, for a logical approach to the problem of randomness.

1. Mathematical preliminaries.

Let A be a countable (non-empty) set, called alphabet, and A^* the free monoid generated by A, i.e. the set of all finite sequences or words X of symbols of A including the empty sequence Λ. The monoid operation in A^* is the juxtaposition XY of a pair of words X and Y; the length of a word X will be denoted by $L(X)$. Any subset of A^* is called event or language. Let $(A^*)^k$ be the cartesian product $A^* \times \ldots \times A^*$ k-times that we call k-th power of A^*. In the following we consider word predicates $\mathcal{R}$ defined in suitable powers of A^*.

Definition 1.1 - A finite set of recursive word predicates, none of which being singulary, is a set of rules of inference (or, for short, simply "rules"). If $\mathcal{I} \equiv \{\mathcal{R}_s\}_{s=1}^{p}$ $(p \geq 1)$ is a set of rules and the predicate $\mathcal{R}(X_1,\ldots,X_k,Y)$ belongs to $\mathcal{I}$ we say that Y

is a logical consequence of $X_1,\ldots,X_k$ by the rule $\mathcal{R}$, if $\mathcal{R}(X_1,\ldots,X_k,Y)$ is true.

Definition 1.2 - A logic $\mathfrak{L}$ is a pair $\mathfrak{L} \equiv (\mathcal{A},\mathcal{I})$ where $\mathcal{A}$ is a recursive set of words called axioms and $\mathcal{I}$ a set of rules.

Definition 1.3 - A proof in $\mathfrak{L}$ is any finite sequence of words $X_0,X_1,\ldots,X_n$ such that for any X_i $(i=0,1,\ldots,n)$ either

i. $X_i \in \mathcal{A}$, or

ii. $X_i = X_j$ with $j<i$ (4), or

iii. there exists a $\mathcal{R} \in \mathcal{I}$ and a set of integers $j_1,\ldots,j_s$ with $j_s<i$ such that $\mathcal{R}(X_{j_1},\ldots,X_{j_s},X_i)$ is true.

Definition 1.4 - A word W is called a theorem of $\mathfrak{L}$, and one writes $\vdash_{\mathfrak{L}} W$, if and only if there exists a proof whose last word is W.

We shall denote by $\mathcal{P}(\mathfrak{L})$ and $\mathcal{T}(\mathfrak{L})$ the set of all the proofs and theorems of $\mathfrak{L}$, respectively. It is well known that $\mathcal{P}(\mathfrak{L})$ is recursive whereas $\mathcal{T}(\mathfrak{L})$ is only recursively enumerable (see, for instance, Davis[14] and Rogers[15]). By decision problem for the logic $\mathfrak{L}$ one means the problem of deciding whether or not an arbitrary word W is a theorem of $\mathfrak{L}$. By definition the decision problem for a logic $\mathfrak{L}$ is recursively solvable if $\mathcal{T}(\mathfrak{L})$ is recursive; otherwise it is recursively unsolvable.

Definition 1.5 - A logic $\mathfrak{L} \equiv (\mathcal{A},\mathcal{I})$ is called combinatorial if and only if each rule $\mathcal{R}_s \in \mathcal{I}$ $(s=1,\ldots,p)$ depends only on two word variables. In this case the predicate "Y is a direct consequence of X" can be written as

$$C(X,Y) \quad \leftrightarrow \quad \bigvee_{s=1}^{p} \mathcal{R}_s(X,Y)$$

Definition 1.6 - A combinatorial logic is called finite if and only if $\mathcal{A}$ is a finite set and for any predecessor word X there exists only a finite (possibly empty) set of words such that $C(X,Y)$ is true.

Definition 1.7 - A finite combinatorial logic is called strictly finite if and only if the function $N(X) := \#\{Y \mid C(X,Y)\}$ is recursive.

Typical examples of latter systems are the 1-antecedent Post's[16] canonical systems. We stress that in these cases there exists an effective procedure by which for any X one can produce in a finite number of steps all the direct consequences of X.

2. Complexity measures on proofs.

In this section we introduce for any proof in a logic $\mathfrak{L}$ a complexity measure. We start by giving the following

Definition 2.1 - Two logics $\mathfrak{L} \equiv (\mathcal{A},\mathcal{T})$, $\mathfrak{L}' \equiv (\mathcal{A}',\mathcal{T})$ are similar if and only if the set of rules of inference is the same for both, whereas the set of the axioms can be different.

Let $\mathfrak{L} \equiv (\mathcal{A},\mathcal{T})$ be a logic and $\mathcal{B}$ a recursive set of words. We can then consider the similar logic $\mathfrak{L}_{\mathcal{B}} \equiv (\mathcal{A} \cup \mathcal{B}, \mathcal{T})$ obtained from $\mathfrak{L}$ adding the words of $\mathcal{B}$ to the set $\mathcal{A}$ of the axioms. One has $\mathfrak{L}_{\emptyset} = \mathfrak{L}$, $\emptyset$ denoting the empty set, and, from definition 1.3

$$\vdash_{\mathfrak{L}} W \quad \rightarrow \quad \vdash_{\mathfrak{L}_{\mathcal{B}}} W \tag{2.1}$$

$$\vdash_{\mathfrak{L}} W \ \wedge \ \vdash_{\mathfrak{L}_W} Z \quad \rightarrow \quad \vdash_{\mathfrak{L}} Z \ , \tag{2.2}$$

where $\mathfrak{L}_W \equiv \mathfrak{L}_{\{W\}}$ is the similar logic obtained from $\mathfrak{L}$ adding the singleton W to the set of the axioms. In the following, for short, we shall denote $\vdash_{\mathfrak{L}_X} W$ as $\vdash_X W$. From now on we assume the set $\mathcal{B}$ to be formed only by a single sequence even though what we say can be obviously generalized when $\mathcal{B}$ is any finite set of sequences.

Let a logic $\mathfrak{L}$ be given, we denote, for any pair X,Y of words of A^*, by $\mathcal{P}_Y(\mathfrak{L}_X)$ the recursive set (possibly empty) of all the proofs of Y in $\mathfrak{L}_X$.

Definition 2.2 - A complexity measure on proofs is a partial recursive function Φ on the set of all finite sequences of words, assuming integral values, that satisfies the following axioms (5)

A1. $\quad \mathrm{Dom}\ \Phi \supseteq \mathcal{P}(\mathfrak{L}_X)$, for all $X \in A^*$.

A2. $\quad$ The function $M_{\mathfrak{L}}^{\Phi}$ defined as

$$M_{\mathfrak{L}}^{\Phi}(X,Y,k) = \begin{cases} 1 & \text{if } \bigvee\limits_{P} P \in \mathcal{P}_Y(\mathfrak{L}_X) \wedge \Phi(P) < k, \\ 0 & \text{otherwise,} \end{cases}$$

is (total) recursive.

The meaning of A2 is that, although one generally is not able to decide whether a word Y is provable or not in $\mathfrak{L}_X$, one can, however, decide, when Φ is a complexity measure on proofs, whether does or does not exist a proof of Y in $\mathfrak{L}_X$, whose complexity does not exceed a fixed amount k.

For a given logic $\mathfrak{L}$ and complexity measure on proofs Φ we define the following quantity

$$K_{\mathfrak{L}}^{\Phi}(Y/X) = \begin{cases} \text{minimum value of the complexity } \Phi \\ \text{of a proof of } Y \text{ in } \mathfrak{L}_X \quad \text{if } \vdash_X Y, \\ +\infty \quad \text{otherwise.} \end{cases} \tag{2.3}$$

We shall call $K_{\mathfrak{L}}^{\Phi}(Y/X)$ <u>conditional complexity of Y given X with respect to the complexity measure on proofs Φ and to the logic $\mathfrak{L}$</u>. When $X \in \mathscr{A}$ we denote $K_{\mathfrak{L}}^{\Phi}(Y/X)$ simply by $K_{\mathfrak{L}}^{\Phi}(Y)$, and call it <u>absolute complexity</u> of Y, with respect to Φ and $\mathfrak{L}$.

We emphasize that the quantity $K_{\mathfrak{L}}^{\Phi}$ even though similar in the definition to Kolmogorov's notion of minimal program complexity, is conceptually very different from it. In fact this latter quantity is a "static" measure related to input data and program complexity, whereas (2.3) is a "dynamic" measure related to the used resource (number of steps, length of tape, etc.). From A2 and definition (2.3) one has

$$\begin{aligned} &\text{i.} \quad \text{Dom } K_{\mathfrak{L}}^{\Phi} \equiv \{(X,Y) \mid \vdash_X Y\} \\ &\text{ii.} \quad K_{\mathfrak{L}}^{\Phi}(Y/X) = \mu\, k(M_{\mathfrak{L}}^{\Phi}(X,Y,k)=1), \end{aligned} \tag{2.4}$$

so that $K_{\mathfrak{L}}^{\Phi}$ is a partially computable function, that is not generally true in Kolmogorov's approach. Furthermore we note the similarity between (2.4) and Blum's[13] axioms of computational complexity.

We observe that axiom A2 is automatically verified if one considers a logic $\mathfrak{L}$ and a function Φ satisfying A1, such that

A3. The set $\{P \in \mathscr{P}(\mathfrak{L}_X) \mid \Phi(P) \leq k\}$ is finite and the number of elements of it $\nu_{\mathfrak{L}}^{\Phi}(X,k)$ is a recursive function.

In fact in such a case there exists an effective procedure by which, for any X and k, one can yield all the proofs P in $\mathfrak{L}_X$ of complexity less than or equal to k. From this finite set one can then extract the subset formed by the proofs whose terminal is Y. If this subset is empty M=0, otherwise M=1. The function M is therefore a recursive function.

Let us consider the following examples. We assume first that the function Φ satisfying A1 is equal to the number of steps or length of a proof. In this case, $K_{\ell}^{\Phi}(Y/X)$ is the minimum number of steps required to prove Y in the logic ℓ_X. However, in the case of a general logic the number of proofs of a fixed length can be infinite so that generally $K_{\ell}^{\Phi}(Y/X)$ is not partial recursive. A typical feature of strictly finite combinatorial logics is, on the contrary, the fact that the length of a proof satisfies the requirement A3, so that K_{ℓ}^{Φ} is partially computable.

In general logics, when the alphabet A is finite, one can, however, consider, for instance, as complexity measure of a proof $P \equiv \xi_1, \ldots, \xi_{n-1}, \xi_n$ the length of the word $\xi_1 \xi_2 \ldots \xi_{n-1}$ that is $\Phi(P) = \sum_{i=1}^{n-1} L(\xi_i)$. In fact the number of proofs in ℓ_X with complexity equal to n is less than or equal to k^n, with $k = \# A$, and one can effectively generate them for all X and n. A3 is then satisfied. We call this complexity measure word-length of a proof.

3. Properties of the function K_{ℓ}^{Φ}.

In this section we shall analyze some general properties of the function $K_{\ell}^{\Phi}(Y/X)$ defined in (2.3).

Property 3.1 - For any pair X,Y of words of A^*

$$K_{\ell}^{\Phi}(Y/X) \leq K_{\ell}^{\Phi}(Y). \qquad (3.1)$$

This property is due to the fact that any proof in ℓ is a proof in ℓ_X too, whereas the contrary is not generally true.

Property 3.2 - Let $\ell \equiv (\mathscr{A}, \mathscr{T})$ be a logic. If $\ell^* \equiv (\mathscr{A}, \mathscr{T}^*)$ denotes another logic with $\mathscr{T}^* \supseteq \mathscr{T}$, then, when $\mathrm{Dom}\ \Phi \supseteq \mathscr{P}(\ell_X) \cup \mathscr{P}(\ell_X^*)$, one has

$$K_{\ell^*}^{\Phi}(Y/X) \leq K_{\ell}^{\Phi}(Y/X). \qquad (3.2)$$

In fact any proof in $\mathfrak{L}_X$ is also a proof in $\mathfrak{L}_X^*$.

In order to get a property of $K_{\mathfrak{L}}^{\Phi}$ which is very meaningful for the informational interpretation of it, that we shall give in the next section, let us make on the complexity measures on the proofs Φ the following hypothesis

H1. $Y \in \mathscr{A} \cup \{X\}$ if and only if there exists a proof P of Y in $\mathfrak{L}_X$ such that $\Phi(P)=0$, that is
$M_{\mathfrak{L}}^{\Phi}(X,Y,0) = 1 \quad \leftrightarrow \quad Y \in \mathscr{A} \cup \{X\}$.

Such kind of measures Φ will be called "natural". Examples of natural measures are the <u>number of steps of a proof minus one</u> and the <u>word-length of a proof</u>, that we shall denote in the next by Φ° and Φ^*, respectively. With respect to natural measures $K_{\mathfrak{L}}^{\Phi}$ satisfies the following

<u>Property 3.3</u> - $\qquad K_{\mathfrak{L}}^{\Phi}(Y/X) = 0 \quad \leftrightarrow \quad Y \in \mathscr{A} \cup \{X\}$
<u>for any pair X,Y of words of A^*.</u>

Let us now introduce the quantity $I_{\mathfrak{L}}^{\Phi}$, defined as

$$I_{\mathfrak{L}}^{\Phi}(Y/X) := K_{\mathfrak{L}}^{\Phi}(Y) - K_{\mathfrak{L}}^{\Phi}(Y/X) \tag{3.3}$$

$I_{\mathfrak{L}}^{\Phi}$ is a partial recursive function that is not defined when $K_{\mathfrak{L}}^{\Phi}(Y) = +\infty$; in this case if $K_{\mathfrak{L}}^{\Phi}(Y/X) < +\infty$ we set, as natural, $I_{\mathfrak{L}}^{\Phi}(Y/X) = +\infty$ (6). From definition (3.3) $I_{\mathfrak{L}}^{\Phi}$ satisfies the following properties, for any $X \in A^*$ and $Y \in \mathscr{T}(\mathfrak{L}_X)$

$$\begin{aligned} &\text{i.} && 0 \leq I_{\mathfrak{L}}^{\Phi}(Y/X) \leq K_{\mathfrak{L}}^{\Phi}(Y) \\ &\text{ii.} && I_{\mathfrak{L}}^{\Phi}(X/X) = K_{\mathfrak{L}}^{\Phi}(X) \\ &\text{iii.} && I_{\mathfrak{L}}^{\Phi}(Y/X) = K_{\mathfrak{L}}^{\Phi}(Y) \leftrightarrow Y \in \mathscr{A} \cup \{X\} \end{aligned} \tag{3.4}$$

i. is a consequence of property 3.1 and ii.,iii. derive from property 3.3. If $Y \in \mathscr{A}$ then $K_{\mathfrak{L}}^{\Phi}(Y)=0$ so that from iii. $I_{\mathfrak{L}}^{\Phi}(Y/X)=0$.

A further hypothesis that one can make on the complexity measures on proofs Φ is

H2. If P is a proof of Z in $\mathfrak{L}_X$ and Q is a proof of Y in $\mathfrak{L}_Z$, then there exists a proof Π of Y in $\mathfrak{L}_X$

such that

$$\Phi(\Pi) \leqslant \Phi(P) + \Phi(Q) \qquad (3.5)$$

H2 is satisfied by the measures Φ^{o} and Φ^{*}. In fact if $P \equiv X_1,\ldots,X_n=Z$ is a proof of Z in $\mathfrak{L}_X$ and $Q \equiv Y_1,\ldots,Y_m=Y$ is a proof of Y in $\mathfrak{L}_Z$, then the sequence of words $\Pi \equiv X_1,\ldots,X_{n-1},Y_1,\ldots,Y_m=Y$ is a proof of Z in $\mathfrak{L}_X$ for which (3.5) is satisfied by Φ^{o} and Φ^{*} with the equal signum.

If Φ satisfies H2 one has

Property 3.4 - For any triplet X,Y,Z of words of A^{*}

$$K^{\phi}_{\mathfrak{L}}(Y/X) \leqslant K^{\phi}_{\mathfrak{L}}(Y/Z) + K^{\phi}_{\mathfrak{L}}(Z/X). \qquad (3.6)$$

This triangular inequality derives from the fact that certainly a proof of Y starting from the set of axioms $\mathscr{A} \cup \{X\}$ can be obtained proving first a theorem Z and afterwards proving Y starting from the set of axioms $\mathscr{A} \cup \{Z\}$. Therefore, making use of definition (2.3) and the hypothesis H2, (3.6) follows. A corollary of this property is the inequality

$$K^{\phi}_{\mathfrak{L}}(Y) \leqslant K^{\phi}_{\mathfrak{L}}(X) + K^{\phi}_{\mathfrak{L}}(Y/X). \qquad (3.7)$$

Property 3.5 - If Z and Z' are two words of A^{*} such that $\vdash_Z Z' \leftrightarrow \vdash_{Z'} Z$, then

$$|K^{\phi}_{\mathfrak{L}}(Y/Z) - K^{\phi}_{\mathfrak{L}}(Y/Z')| \leqslant C_{ZZ'}, \qquad (3.8)$$

where $C_{ZZ'}$ is a constant whose value, for any $\mathfrak{L}$ and Φ, depends on Z and Z' only.

In fact from property 3.4 one has

$$K^{\phi}_{\mathfrak{L}}(Y/Z) \leqslant K^{\phi}_{\mathfrak{L}}(Y/Z') + K^{\phi}_{\mathfrak{L}}(Z'/Z)$$

$$K^{\phi}_{\mathfrak{L}}(Y/Z') \leqslant K^{\phi}_{\mathfrak{L}}(Y/Z) + K^{\phi}_{\mathfrak{L}}(Z/Z'),$$

so that (3.8) follows.

From (3.8) one has that for large values of the complexity if Z is derivable from Z' and viceversa then $K^{\phi}_{\mathfrak{L}}(Y/Z) \approx K^{\phi}_{\mathfrak{L}}(Y/Z')$.

Making use of the triangular inequality one gets also the following property for $I^{\phi}_{\mathfrak{L}}(Y/X)$

iv. $$I^{\phi}_{\mathfrak{L}}(Y/X) \leqslant K^{\phi}_{\mathfrak{L}}(X), \qquad (3.9)$$

for $X \in A^{*}$ and $Y \in \mathscr{T}(\mathfrak{L}_X)$.

4. Interpretation of $I_{\mathfrak{L}}^{\Phi}$ and $K_{\mathfrak{L}}^{\Phi}$.

We shall give now an interpretation of the quantities $I_{\mathfrak{L}}^{\Phi}(Y/X)$ and $K_{\mathfrak{L}}^{\Phi}(Y/X)$, formally introduced in the previous sections, in "information terms". As we stressed in the introduction, in a probabilistic information theory the question "if X and Y are two objects, for instance two sequences, what is the content of information in X about Y ?", is meaningless. In our logical approach to information theory, the previous question is, on the contrary, the basic one. The intuitive idea is that the intrinsic content of information in X about Y can be obtained by measuring the reduction of the minimal complexity of a proof of Y when the word X is added to the set $\mathscr{A}$ of the axioms of some logical system $\mathfrak{L}$. In our framework this quantity can be just measured by $I_{\mathfrak{L}}^{\Phi}(Y/X)$ defined by (3.3). The numerical valuation of the foregoing content of information depends, of course, on the complexity measure on proofs Φ and on the logical system $\mathfrak{L}$, that is the set of initial data (axioms) and the rules of inference at our disposal. It is natural that $I_{\mathfrak{L}}^{\Phi}(Y/X)$ assumes the value $+\infty$ when $K_{\mathfrak{L}}^{\Phi}(Y/X) < +\infty$ and $K_{\mathfrak{L}}^{\Phi}(Y) = +\infty$, since Y cannot be proved by using the axioms of the set $\mathscr{A}$ only, while can be proved by adding X to $\mathscr{A}$.

The properties (3.4) are the most natural that an "information measure" of Y given X has to satisfy. We remember here that the probabilistic average mutual information $\mathfrak{I}(Y:X)$ of two (discrete) random variables X and Y, defined as

$$\mathfrak{I}(Y:X) := H(Y) - H(Y/X) , \tag{4.1}$$

where

$$\begin{aligned} H(Y) &:= -\sum_{y\in Y} p(y)\ln p(y) \\ H(Y/X) &:= -\sum_{x\in X} p(x) \sum_{y\in Y} p(y/x)\ln p(y/x) , \end{aligned} \tag{4.2}$$

satisfies the properties

$$\begin{aligned} &\text{i.} \quad \mathfrak{I}(Y:X) \geq 0 \text{, where the equal signum holds if and only if X and Y are independent random variables.} \\ &\text{ii.} \quad \mathfrak{I}(X:X) = H(X) \\ &\text{iii.} \quad \mathfrak{I}(Y:X) \leq \begin{cases} H(X) \\ H(Y) \end{cases} \end{aligned} \tag{4.3}$$

iv. $\mathfrak{I}(Y:X) = \mathfrak{I}(X:Y)$. (4.3)

The formal analogy between (3.3) and (4.1) and the properties i.,ii.,iv. of $I_{\mathfrak{L}}^{\Phi}(Y/X)$ and i.,ii.,iii. of $\mathfrak{I}(Y:X)$ suggests giving to $K_{\mathfrak{L}}^{\Phi}(Y/X)$ the following interpretation as "quantity of information". Intuitively the larger is the complexity of a proof of a theorem in a logic, the larger is the amount of "data" that one needs to construct the proof or, that is the same, that comes out in performing the proof itself. $K_{\mathfrak{L}}^{\Phi}(Y/X)$ measures the minimal value, expressed in Φ-units, of the previous amount of data, that can be interpreted as a quantity of information needed to prove Y in $\mathfrak{L}_X$. If Y is not a theorem in $\mathfrak{L}_X$, then, since there is not a finite complexity proof of Y, we say that the quantity of information needed to prove Y is infinite. If, for instance, Φ is the word-length of a proof then $K_{\mathfrak{L}}^{\Phi}(Y/X)$, in the alphabet $A \equiv (0,1)$ is just the minimum number of binary digits that one has to write before arriving at the theorem Y of the logic $\mathfrak{L}_X$.

In our formalism the last of the properties (4.3) of the average mutual information is not true. By using triangular inequality one can, however, show that is always true that

$$|I_{\mathfrak{L}}^{\Phi}(Y/X) - I_{\mathfrak{L}}^{\Phi}(X/Y)| \leqslant \max\{K_{\mathfrak{L}}^{\Phi}(Y/X), K_{\mathfrak{L}}^{\Phi}(X/Y)\}. \quad (4.4)$$

The main difference between the interpretation here given to the quantities $I_{\mathfrak{L}}^{\Phi}(Y/X)$, $K_{\mathfrak{L}}^{\Phi}(Y/X)$ and the corresponding quantities in Kolmogorov's theory lies in the complexity measure. We emphasize that there exist cases in which to a very short program corresponds a very large quantity of the resource used in computation and viceversa.

In order to make more cogent the similarity between $I_{\mathfrak{L}}^{\Phi}(Y/X)$ and $\mathfrak{I}(Y:X)$ we shall give the following

Definition 4.1 - The word Y is said to be independent from the word X, with respect to the logic $\mathfrak{L}$ and to the measure Φ, if and only if there exists a proof of Y in $\mathfrak{L}_X$ of minimal complexity in which the word X does not appear.

In other words to add the word X to the set $\mathcal{A}$ of the axioms does not give any "gain", in the sense of reducing the minimal complexity of a proof of Y, so that the information furnished by X about Y has to be zero. One can state

<u>Proposition 4.1</u> - <u>A necessary and sufficient condition in order that Y be independent from X is that $I_{\mathfrak{L}}^{\phi}(Y/X) = 0$.</u>

In fact, if Y is independent from X, by definition 4.1, there is a proof of Y in $\mathfrak{L}_X$ of minimal complexity in which X does not appear. This is therefore also a proof of Y in $\mathfrak{L}$ so that $K_{\mathfrak{L}}^{\phi}(Y/X) \geqslant K_{\mathfrak{L}}^{\phi}(Y)$. Since in general $K_{\mathfrak{L}}^{\phi}(Y/X) \leqslant K_{\mathfrak{L}}^{\phi}(Y)$ it follows that $I_{\mathfrak{L}}^{\phi}(Y/X)=0$. Viceversa if $I_{\mathfrak{L}}^{\phi}(Y/X)=0$ the minimal complexity of a proof of Y in $\mathfrak{L}_X$ is equal to the minimal complexity of a proof of Y in $\mathfrak{L}$. In this way, since any proof of Y in $\mathfrak{L}$ is also a proof of Y in $\mathfrak{L}_X$, there exists a proof of Y of minimal complexity in which X is absent. This means that Y is independent from X.

We stress that, unlike what occurs for random variables, if Y is independent from X this <u>does not imply</u> that X is independent from Y (cfr. eq. (4.4)).

Let us now define the quantity $K_{\mathfrak{L}}^{\phi}(Y,Z)$ equal to the <u>minimal complexity of a proof of Y containing the word Z too, if there exists such a proof, equal to $+\infty$, otherwise.</u>

Extending to this case the triangular inequality one has

$$K_{\mathfrak{L}}^{\phi}(Y,Z) \leqslant K_{\mathfrak{L}}^{\phi}[(Y,Z)/Z] + K_{\mathfrak{L}}^{\phi}(Z) .$$

Now $K_{\mathfrak{L}}^{\phi}[(Y,Z)/Z] = K_{\mathfrak{L}}^{\phi}(Y/Z)$ except when Y is independent from Z. In fact, if $K_{\mathfrak{L}}^{\phi}(Y/Z) < K_{\mathfrak{L}}^{\phi}(Y)$ this implies that a proof of Y in $\mathfrak{L}_Z$ of minimal complexity has to contain Z, so that the above equality follows.

Therefore, when $I_{\mathfrak{L}}^{\phi}(Y/Z) \neq 0$, one can write

$$K_{\mathfrak{L}}^{\phi}(Y,Z) \leqslant K_{\mathfrak{L}}^{\phi}(Z) + K_{\mathfrak{L}}^{\phi}(Y/Z) = K_{\mathfrak{L}}^{\phi}(Z) + K_{\mathfrak{L}}^{\phi}(Y) - I_{\mathfrak{L}}^{\phi}(Y/Z). \quad (4.6)$$

The foregoing relation is similar to the classical formula of information theory

$$H(Y,Z) = H(Z) + H(Y/Z) = H(Z) + H(Y) - \mathfrak{I}(Y:Z) , \quad (4.7)$$

where $H(Y,Z) = H(Z,Y)$ denotes the entropy of the <u>joint</u> probability distribution of Y and Z. In our case generally $K_{\mathfrak{L}}^{\phi}(Y,Z) \neq K_{\mathfrak{L}}^{\phi}(Z,Y)$ and furthermore (4.6) holds with the signum

$\leqslant$ only when $I^{\phi}_{\mathfrak{L}}(Y/Z) \neq 0$.

It is easy to prove that, referring to the measure Φ°, it is always true that

$$K^{\phi^{\circ}}_{\mathfrak{L}}(Y,Z) \leqslant K^{\phi^{\circ}}_{\mathfrak{L}}(Z) + K^{\phi^{\circ}}_{\mathfrak{L}}(Y/Z) + 1 .$$

Moreover in the case of combinatorial logics, if $I^{\phi^{\circ}}_{\mathfrak{L}}(Y/Z) \neq 0$, then

$$K^{\phi^{\circ}}_{\mathfrak{L}}(Y,Z) = K^{\phi^{\circ}}_{\mathfrak{L}}(Z) + K^{\phi^{\circ}}_{\mathfrak{L}}(Y/Z) .$$

5. Complexity and length of the theorems.

The complexity of a theorem of a logic generally does not depend on its length. In fact the latter can be very short and the minimal complexity of a proof of it, very large. We stress that, generally, one cannot find a recursive upper bound for the complexity of a theorem, depending on the length of it. Indeed in the case of a logic $\mathfrak{L}$ whose decision problem is recursively unsolvable, no recursive function h can exist such that

$$K^{\phi}_{\mathfrak{L}}(Y/X) \leqslant h(X,Y,L(Y)) , \qquad \text{for all } Y \in \mathcal{T}(\mathfrak{L}_X).$$

Otherwise, since the function $M^{\phi}_{\mathfrak{L}}$ is recursive, one would be able to decide, by an effective procedure, whether an arbitrary word is, or not, a theorem of $\mathfrak{L}$.

It is straightforward to prove that the length of a theorem has a recursive upper bound depending on its complexity. In fact the function defined as

$$G(X,Y,m) = \begin{cases} \max \{L(Y), K^{\phi}_{\mathfrak{L}}(Y/X)\} & \text{if } K^{\phi}_{\mathfrak{L}}(Y/X) = m, \\ 0 & \text{otherwise.} \end{cases}$$

is recursive, so that

$$G(X,Y,K^{\phi}_{\mathfrak{L}}(Y/X)) \geqslant L(Y). \qquad (5.1)$$

To find a recursive relation more meaningful that (5.1), let us associate, in the case of a finite alphabet A, to any recursive function $F(X,Y,k)$ with $X,Y \in A^{*}$ and assuming integral values , the recursive function $F^*(Y,k)$ defined as

$$F^*(Y,k) := \max \{F(\xi,Y,k) \mid L(\xi) \leqslant L(Y)\} . \qquad (5.2)$$

One then gets, by (5.1)

$$G^*(Y, K_{\ell}^{\phi}(Y/X)) \geqslant L(Y) , \quad \text{for } L(Y) \geqslant L(X). \qquad (5.3)$$

We shall see now that, referring to a quite large class of logics, for particular complexity measures on proofs one can find a recursive lower bound for the complexity of a theorem, depending on its length. For the sake of simplicity we consider strictly finite combinatorial logics Γ whose rules of inference $\{\mathscr{R}_s\}$ satisfy, for all the words X and Y of A^*, the following requirement

$$\mathscr{R}_s(X,Y) \rightarrow L(Y) - L(X) \leqslant \beta_s , \qquad (5.4)$$

where β_s are suitable integers. Rules of this kind are, for instance, those of semi-Thue systems (see, Davis[14]) or of Post's[16] normal systems, that can be expressed respectively by productions as

$$\xi g_s \eta \rightarrow \xi \bar{g}_s \eta \qquad \text{and} \qquad g_s \xi \rightarrow \xi \bar{g}_s ,$$

where $g_s, \bar{g}_s$ denote fixed words, and ξ, η variable words of A^*. In these cases β_s can be taken equal to $L(\bar{g}_s) - L(g_s)$.

Let us now assume as complexity measure on proofs the length of a proof minus one. In this case, denoting simply by $K_\Gamma^o(Y/X)$ the conditional complexity of Y given X in Γ and setting

$$L_M := \max\{L(A_i) \mid A_i \varepsilon \mathscr{A}\} , \ \beta := \max \{\beta_s\} , \qquad (5.5)$$

it is easy to derive, when $\beta > 0$,

$$K_\Gamma^o(Y/X) \geqslant \left[\frac{L(Y) - \max\{L_M, L(X)\}}{\beta} \right] , \qquad (5.6)$$

where [ ] denote the minimum integer not less than the number inside. The foregoing inequality gives, for $L(Y) \geqslant \max\{L_M, L(X)\}$, a recursive lower bound for the complexity $K_\Gamma^o(Y/X)$, (linearly) depending on the length of the theorem Y.

When $X \varepsilon \mathscr{A}$, then eq. (5.6) reduces to

$$K^{\circ}_{\Gamma}(Y) \geq \left[\frac{L(Y) - L_M}{\beta}\right]. \qquad (5.7)$$

If one considers as complexity measure on proof the word-length of a proof and denote by $K^{*}_{\Gamma}(Y/X)$ the conditional complexity in Γ, one has

$$K^{\circ}_{\Gamma}(Y/X) \leq K^{*}_{\Gamma}(Y/X) + 1 , \qquad (5.8)$$

so that also in this case there exists a recursive lower bound for the complexity depending on the length of the theorem.

6. Recursive relations among the complexity measures.

The complexity of a theorem of a logic $\mathfrak{L}$ depends on the used complexity measure on proofs. However for any two measures the corresponding complexities are bounded to each other recursively. More precisely, if Φ and $\hat{\Phi}$ denote two complexity measures on proofs then, in the case of a finite alphabet, a recursive function h^{*} exists such that, for $L(Y) \geq L(X)$

$$\begin{aligned} h^{*}(Y, K^{\Phi}_{\mathfrak{L}}(Y/X)) &\geq K^{\hat{\Phi}}_{\mathfrak{L}}(Y/X) \\ h^{*}(Y, K^{\hat{\Phi}}_{\mathfrak{L}}(Y/X)) &\geq K^{\Phi}_{\mathfrak{L}}(Y/X) , \end{aligned} \qquad (6.1)$$

To show this we follow a typical procedure used in computational complexity theory (Blum[13]). One defines the function

$$h(X,Y,n) = \begin{cases} \max\{K^{\Phi}_{\mathfrak{L}}(Y/X), K^{\hat{\Phi}}_{\mathfrak{L}}(Y/X)\} & \text{if } K^{\Phi}_{\mathfrak{L}}(Y/X) \leq n \text{ or } K^{\hat{\Phi}}_{\mathfrak{L}}(Y/X) \leq n, \\ 0 & \text{otherwise} , \end{cases}$$

that is recursive, since one can effectively decide whether $K^{\Phi}_{\mathfrak{L}}(Y/X)$ or $K^{\hat{\Phi}}_{\mathfrak{L}}(Y/X)$ are equal to n, or not, and furthermore $K^{\Phi}_{\mathfrak{L}}$ and $K^{\hat{\Phi}}_{\mathfrak{L}}$ have the same domain. The recursive function h^{*}, obtained from h by using definition (5.2), is then such to satisfy (6.1), for $L(Y) \geq L(X)$.

From (6.1) one gets, however, no other information but the existence of a recursive relation between complexities. Therefore it is important, mainly in the applications, to find, for some classes of logics and particular measures on proofs, recursive relations between the corresponding complexities more meaningful than (6.1). If, for instance, one refers to combinatorial logics Γ satisfying the requirement (5.4) and to measures Φ° and Φ^{*}, then the respective complexities, that we simply denote by K_{Γ}° and K_{Γ}^{*}, are recursively related by (5.8) and it is easy to verify by

$$K_{\Gamma}^{*}(Y/X) \leq K_{\Gamma}^{\circ}(Y/X)\,(\max\{L_M,L(X)\} + (K_{\Gamma}^{\circ}(Y/X) - 1)\beta/2),\quad(6.2)$$

The minimal word-length of a proof of a theorem has therefore a recursive upper bound that depends quadratically on the length of a proof.

7. Dependence of complexity on the logic.

In the previous section we considered the problem of recursively relating the complexities of theorems of a logic evaluated with respect to two different complexity measures on proofs. In this section we shall analyze, on the contrary, the problem of relating the complexities of theorems of two logics but using a same complexity measure on proofs, as for instance, the length of a proof. To make this concept more clear, we consider the logics $\mathfrak{L}$ and $\hat{\mathfrak{L}}$ and a complexity measure on proofs Φ such that $\mathrm{Dom}\,\Phi \supseteq \mathscr{P}(\mathfrak{L}_X) \cup \mathscr{P}(\hat{\mathfrak{L}}_X)$, for all $X \in A^{*}$. In this way one can measure by means of Φ the complexity of proofs in $\mathfrak{L}_X$ and $\hat{\mathfrak{L}}_X$. Let us suppose now that, for all X of A^{*}, any theorem of $\mathfrak{L}_X$ is also a theorem of $\hat{\mathfrak{L}}_X$, that is $\mathscr{T}(\mathfrak{L}_X) \subseteq \mathscr{T}(\hat{\mathfrak{L}}_X)$. This relation implies $\mathrm{Dom}\,K_{\mathfrak{L}}^{\Phi} \subseteq \mathrm{Dom}\,K_{\hat{\mathfrak{L}}}^{\Phi}$, so that the function g defined as

$$g(X,Y,n) = \begin{cases} K_{\hat{\mathfrak{L}}}^{\Phi}(Y/X) & \text{if } K_{\mathfrak{L}}^{\Phi}(Y/X) = n\,, \\ 0 & \text{otherwise}\,, \end{cases}$$

is recursive. Therefore one has, for $(X,Y) \in \mathrm{Dom}\,K_{\mathfrak{L}}^{\Phi}$,

$$K_{\hat{\mathfrak{L}}}^{\Phi}(Y/X) = g(X,Y,K_{\mathfrak{L}}^{\Phi}(Y/X))$$

and, using definition (5.2)

$$K_{\hat{\mathfrak{L}}}^{\Phi}(Y/X) \leq g^{*}(Y,K_{\mathfrak{L}}^{\Phi}(Y/X))\,, \qquad \text{for } L(X) \leq L(Y).$$

In other words the theorems of $\hat{\mathfrak{L}}_X$ which are theorems of $\mathfrak{L}_X$ too, have a complexity that can be upper bounded by a recursive function of the complexity in $\mathfrak{L}_X$. When $\mathcal{T}(\mathfrak{L}_X) \equiv \mathcal{T}(\hat{\mathfrak{L}}_X)$, it is easy to show the existence of a unique recursive function r such that, for $L(X) \leqslant L(Y)$

$$K^{\Phi}_{\mathfrak{L}}(Y/X) \leqslant r(Y, K^{\Phi}_{\hat{\mathfrak{L}}}(Y/X))$$

$$K^{\Phi}_{\hat{\mathfrak{L}}}(Y/X) \leqslant r(Y, K^{\Phi}_{\mathfrak{L}}(Y/X)) .$$

The complexities are then bounded to each other recursively.

As we said in the foregoing section, to know that complexities are recursively related is not very meaningful, since the recursive relation, that depends on $\mathfrak{L}$, $\hat{\mathfrak{L}}$ and Φ, can be very large. For this reason we shall analyze how the above recursive relations may be specialized referring to class Σ of all combinatorial logics Γ satisfying the requirement (5.4) and assuming the complexity of a proof equal to the number of steps minus one. For any $\Gamma \in \Sigma$ we denote by β^{Γ} and L^{Γ}_M the values of β and L_M as defined in (5.5) and simply by $K_{\Gamma}(Y)$ the complexity of Y in Γ.

<u>Proposition 7.1 - For any positive integer β, there is a logic $U \in \Sigma$, with $\beta^U = \beta$, such that</u>

$$\frac{L(Y)}{\beta} \leqslant K_U(Y) < \frac{L(Y)}{\beta} + 1. \qquad (7.1)$$

To prove (7.1) let us consider, for any $\beta > 0$, the logic U whose set of axioms is formed by the empty word Λ and whose rules of inference are given by the set of <u>normal productions</u>

$$\xi \to \xi\, y_1 \dots y_k \quad , \quad \text{for all } y_1, \dots, y_k \in A \text{ and } 1 \leqslant k \leqslant \beta.$$

U is such that $\beta^U = \beta$; furthermore one can prove <u>any</u> word Y, starting from Λ , in a number of steps minus one that can be upper bounded by $\left\lceil \frac{L(Y)}{\beta} \right\rceil$. In this way $K_U(Y) \leqslant \left\lceil \frac{L(Y)}{\beta} \right\rceil < \frac{L(Y)}{\beta} + 1$. Furthermore from the inequality (5.7), eq. (7.1) follows.

A corollary of (7.1) is the relation

$$\frac{L(Y) - L(X)}{\beta} \leqslant K_U(Y/X) < \frac{L(Y)}{\beta} + 1. \qquad (7.2)$$

From (7.1) and (7.2) one derives that the information $I_U(Y/X)$ is upper bounded by $\left[\frac{L(X)}{\beta}\right]$. Moreover in the case of the alphabet $A \equiv \{0,1\}$, one has that for the $2^{L(Y)-L(X)}$ words of length $L(Y)$ having X as prefix, $K_U(Y/X) = \left[\frac{L(Y) - L(X)}{\beta}\right]$, whereas for all the others $K_U(Y/X) = K_U(Y) = \left[\frac{L(Y)}{\beta}\right]$, that is they are independent from X.

<u>Proposition 7.2</u> - <u>The logic U is such that for any other logic $\Gamma \in \Sigma$ one has</u>

$$\beta^U K_U(Y) < \beta^\Gamma K_\Gamma(Y) + (\beta^U + L_M^\Gamma). \qquad (7.3)$$

In fact when Y is a theorem of Γ then from (5.7) one has $L(Y) \leqslant \beta^\Gamma K_\Gamma(Y)+L_M^\Gamma$, so that by means of (7.1), eq. (7.3) follows. If Y is not a theorem $K_\Gamma(Y) = +\infty$ and (7.3) is still true.

Let us consider now the class of all the logics Ω of Σ satisfying, for all $\Gamma \in \Sigma$, the relation

$$\beta^\Omega K_\Omega(Y) \leqslant \beta^\Gamma K_\Gamma(Y) + C_{\Omega,\Gamma}, \qquad (7.4)$$

where $C_{\Omega,\Gamma}$ is a constant depending on Ω and Γ only. If Ω_1 and Ω_2 are two such logics one derives that, when $L(Y) \to +\infty$,

$$\beta^{\Omega_1} K_{\Omega_1}(Y) \approx \beta^{\Omega_2} K_{\Omega_2}(Y). \qquad (7.5)$$

Moreover from proposition 7.1 one has $\beta^U K_U(Y) \approx L(Y)$, so that for all the logics Ω one gets, when $L(Y) \to +\infty$,

$$\beta^\Omega K_\Omega(Y) \approx L(Y) \qquad (7.6)$$

or, when $L(Y)$ is very large with respect to β^Ω,

$$K_\Omega(Y) \approx \frac{L(Y)}{\beta^\Omega}. \qquad (7.7)$$

Eq. (7.7) shows that in systems like Ω the proof of almost all theorems can be speeded-up as one wishes, by taking a sufficiently large β^{Ω}.

8. Concluding remarks.

In the previous sections we exposed, even though in a incomplete and not exhaustive manner, a logical approach to the concept of information, based on a dynamic measure of computational complexity, that preserves some relevant properties of probabilistic information.

It would be desirable to evaluate the actual implications of our definition of information particularly with regard to the problem of "randomness". An infinite random sequence, according to the definition of von Mises-Church, is such that the set of all its initial segments cannot coincide with the set of the theorems of any logic. However, it is not presently clear whether, or in which way, it is possible, following our approach, to characterize "random" sequences of finite length, as done in Kolmogorov's theory.

Acknowledgements.

The authors are indebted to Dr. S. Termini and to Professor M.P. Schützenberger for their comments and suggestions.

Footnotes.

(1) Other approaches are the "combinatorial" (Kolmogorov[2]) and the "semantic" (Carnap and Bar-Hillel[17]).

(2) A thermodynamic-like approach to information theory outside a probabilistic (frequentistic) context is in Capocelli and De Luca.[18]

(3) The ordinary probability theory does not include, for example, the situations arising in quantum mechanics (see, for instance, Varadarajan[19]).

(4) For reasons of technical convenience, which will be clear in the following, we permit the repetition of a step of a proof any time one wishes (see, for example, Rosser,[20] Chap.IV).

(5) In the most general case, that we shall not consider here, the complexity measures on proofs may depend on the logics too.

(6) When $K^{\phi}_{\mathfrak{L}}(Y/X) = +\infty$ it could be convenient to define $I^{\phi}_{\mathfrak{L}}(Y/X) = 0$, since in this case, being Y not provable in $\mathfrak{L}_X$, to add X to the set of the axioms $\mathcal{A}$ does not give any "information" about Y. Being in this case Y intuitively independent from X the previous position is in agreement with definition of "independence" given in section 4, that implies $I^{\phi}_{\mathfrak{L}}(Y/X) = 0$. However, the function $I^{\phi}_{\mathfrak{L}}$ so defined would be generally not partial recursive.

REFERENCES

1. Shannon, C.E., A mathematical theory of communication, Bell Sys. Tech. J., 27, 379,623, 1948 - Reprinted in Shannon, C.E. and Weaver, W., The Mathematical Theory of Communication, University of Illinois Press, Urbana, 1962.

2. Kolmogorov, A.N., Three approaches to the quantitative definition of information, Problemy Pederachi Informatsii, 1, 3, 1965.

3. von Neumann, J., Theory and organization of complicated automata, in Theory of Self-Reproducing Automata, Burks, A.W. Ed., University of Illinois Press, Urbana and London, 1949.

4. Kolmogorov, A.N., Grundbegriffe der Wahrscheinlichkeitsrechnung, in Ergebnisse der Mathematik, Berlin, 1933 - Reprinted: Foundations of the Theory of Probability, 2nd ed., Chelsea, New York, 1956.

5. Kolmogorov, A.N., On tables of random numbers, Sankhya, 25, 369, 1963.

6. von Mises, R., Grundlagen der Wahrscheinlichkeitsrechnung, Mathematische Zeitschrift, 279, 1919.

7. Wald, A., Sur la notion de collectif dans le calcul des probabilités, Comptes Rendus, 202, 180, 1936.

8. Ville, J., Etude Critique de la Notion de Collectif, Gauthier-Villars, Paris, 1939.

9. Church, A., On the concept of a random sequence, Bull. Amer. Math. Soc., 46, 130, 1940.

10. Kolmogorov, A.N., Logical basis for information theory and probability theory, IEEE Trans. Information Theory, IT-14, 662, 1968.

11. Martin-Löf, P., The definition of random sequences, Information and Control, 9, 602, 1966.

12. Rabin, M.O., Degrees of difficulty of computing a function and a partial ordering of recursive sets, Tech. Rep. 2, Hebrew University, Jerusalem, 1960.

13. Blum, M., A machine independent theory of the complexity of recursive functions, J. ACM, 14, 322, 1967.

14. Davis, M., Computability and Unsolvability, McGraw Hill, New York, 1958.

15. Rogers, H.Jr., Theory of Recursive Functions and Effective Computability, McGraw Hill, New York, 1967.

16. Post, E.L., Formal reductions of the general combinatorial decision problem, Amer. J. Math., 65, 197, 1943.

17. Carnap, R. and Bar-Hillel, Y., An outline of a theory of semantic information, Tech. Rep. 247, M.I.T., Research Laboratory of Electronics, 1952 - Reprinted in Bar-Hillel, Y., Language and Information, Addison Wesley, Reading, Mass., 1962.

18. Capocelli, R.M. and De Luca, A., Fuzzy sets and decision theory, Information and Control, 23, 446-673, 1973.

19. Varadarajan, V.S., Probability in physics and a theorem of simultaneous observability, Comm. Pure Appl. Math., 15, 189, 1962.

20. Rosser, J.B., Logic for Mathematicians, McGraw Hill, New York, 1953.

SOLUTION NON COMMUTATIVE D'UNE EQUATION DIFFERENTIELLE CLASSIQUE

M.P. SCHÜTZENBERGER

(Fac. Sci. Paris et Laboratorio di Cibernetica del CNR, Arco Felice)

INTRODUCTION

Cette communication fait partie d'une série de recherches poursuivies depuis plusieurs années avec mon ami le Professeur D. Foata sur les nombres d'Euler, c'est à dire sur les coefficients de Hurwitz de la fonction $tgt + 1/cost$.

Les rapports que peuvent avoir de semblables questions arithmétiques avec certains aspects de la cybernétique ont été brillament illustrés par les deux conférenciers qui m'ont précédé et je me bornerai donc à discuter un problème de nature purement technique, à savoir la solution de l'équation différentielle classique $y'' = y\,y'$ dans le cas non commutatif, c'est à dire, par exemple, dans le cas où la fonction inconnue y et ses dérivées y' et y'' appartiennent à un anneau de matrices dont les entrées sont des fonctions de la variable indépendante t.

Dans le cas commutatif, la solution de cette équation qui satisfait les conditions initiales $y(0) = y'(0) = 1$ est précisément la fonction génératrice exponentielle $tgt + 1/cost$ des nombres d'Euler. (Cf (1), (4), (5)). Il me parait assez remarquable que l'on puisse en exprimer la solution dans le cas général au moyen d'une famille infinie de polynomes en les variables (non commutatives) $y(0)$ et $y'(0)$ dont les coefficients numériques dépendent assez simplement des nombres d'Euler.

Dans un premier chapitre nous rappelons quelques éléments du formalisme de la théorie des équations différentielles et, dans les deux suivants, nous appliquons ces notions au cas de $y'' = y\,y'$ Dans le dernier chapitre nous présentons une autre propriété remar quable de l'opérateur associé à cette équation.

I. GENERALITES

I.1. Pour simplifier au maximum ce rappel de notions connues nous nous bornons à considérer une équation différentielle du second ordre.

(1) $y'' = H\,(t, y, y')$

où H est un polynôme dont les coefficients appartiennent à un anneau donné $\mathcal{A}$ de caractéristique zéro. Formellement nous introduisons l'anneau $\bar{\mathcal{A}} = \mathcal{A}[y, y']$ des polynomes à coefficients dans $\mathcal{A}$ en les variables (non commutatives) y et y' et l'anneau $\bar{\mathcal{A}}(t)$ n'ayant qu'un nombre fini de termes non nuls. Par définition, une solution formelle de (1) est un élément

$$Y = \sum_{0 \le n} \frac{t^n}{n!} a_n \quad \text{de} \quad \bar{\mathcal{A}}(t)$$

satisfaisant les deux conditions suivantes :

(2) $a_0 = y$; $a_1 = y'$; $a_{n+2} \in \bar{\mathcal{A}}$ pour chaque $n \in \mathbb{N}$;

(3) $Y'' = H\,(t, Y, Y')$ où Y' et Y'' sont définies par

(4) $Y' = \sum_{0 \le n} \frac{t^n}{n!} a_{n+1}$; $Y'' = \sum_{0 \le n} \frac{t^n}{n!} a_{n+2}$.

Comme H est un polynome, le coefficient de t^n dans le membre de droite de (3) est pour chaque n un polynome en $a_{\grave{a}}$, a_1, ..., a_{n+1} cependant que le coefficient de t^n dans Y" est égal à $\underline{a_{n+2}}$ n . Il en résulte immédiatement que a_{n+2} est déterminé de façon unique par les a_i d'indices inférieurs, d'où par induction, que tous les a_n appartiennent à $\bar{\mathcal{A}}$. Autrement dit, l'équation (1) admet une et une seule solution formelle satisfaisant Y (0) = y , Y'(0) = y' .

Soit maintenant φ un morphisme de $\bar{\mathcal{A}}$ dans une algèbre normée. On peut montrer que pour chaque ε positif il existe un ε', positif lui aussi, tel que l'on ait identiquement
$||a_n \varphi|| \leq n!\ \varepsilon^n$ quand $||y\varphi||$ et $||y'\varphi||$ sont inférieurs à ε'. Ceci entraine que la série $Y\varphi = \Sigma \frac{t^n}{n!} (a_n\varphi)$ converge absolument au voisinage de l'origine et montre que la solution formelle est bien la solution de (1).

I.2. Nous en venons maintenant au calcul effectif des polynomes a_n et pour cela nous rappelons qu'une dérivation d'un anneau $\mathcal{B}$ quelconque est une application β de $\mathcal{B}$ dans lui même, satisfaisant l'identité :

(5) $(ab)\,\beta = (a\,\beta)\,.\,b + a\,.\,(b\,\beta)$ pour tout $a, b \in \mathcal{B}$.

$\mathcal{A}(t)$ des séries formelles en la variable t dont les coefficients sont dans $\mathcal{A}$. Par définition, une solution formelle de (1) est un élément

Nous considérons désormais le cas où $\mathcal{B} = \bar{\mathcal{A}}_{(t)}$ et nous notons ∂ la dérivation de noyau $\bar{\mathcal{A}}$ envoyant t sur 1. On a identiquement $t^n\partial = nt^{n-1}$.

Ceci entraine évidemment que β soit définie par la donnée de son action sur les générateurs de $\mathcal{B}$ et que $\beta\,\mu$ et $\mu\,\beta$ soient aussi des dérivations pour tout morphisme μ de $\mathcal{B}$ dans lui même.

Par conséquent si

$$X = \Sigma \frac{t^n}{n!} b_n$$

est une série formelle (dont les coefficients b_n sont dans $\bar{\mathcal{A}}$) nous aurons la formule habituelle

(6) $X \partial^p = \Sigma \frac{t^n}{n!} b_{n+p} \qquad (p \in \mathbb{N})$

d'où l'on déduit que $b_n = X \partial^n \theta$ où θ est le morphisme de $\bar{\mathcal{A}}(t)$ sur son sous-anneau $\bar{\mathcal{A}}$, laissant ce dernier invariant et envoyant t sur 0.

En particulier notre équation différentielle peut s'écrire sous la forme $Y \partial^2 = H (t, Y, Y \partial)$ et les calculs d'identification des coefficients a_n de $\frac{t^n}{n!}$ dans Y effectués plus haut sont équivalents à l'ensemble des équations :

(7) $Y \theta = y \; ; \; y \partial \theta = y' \quad ; \quad Y \partial^{n+2} \theta = H \partial^n \theta$.

Nous nous proposons de mettre ces relations sous une forme plus compacte. Pour cela nous considérons une suite de nouvelles variables non commutatives $\{x_j\}$ $(j \in \mathbb{N})$, l'algèbre large $\mathcal{A}(t, \{x_j\})$, une dérivation χ de cette algèbre définie par la condition que $a\chi = 0$ pour chaque $a \in \mathcal{A}$, $t\chi = 1$ et $x_j \chi = x_{j+1}$ pour chaque $j \in \mathbb{N}$ et enfin le morphisme ξ laissant $\mathcal{A}(t)$ invariant et envoyant chaque x_j sur $X\partial^j$ où $X = \Sigma \frac{t^n}{n!} b_n \quad (b_n \in \mathcal{A})$.

Comme on l'a mentionné plus haut, $\xi\partial$ et $\chi\partial$ sont deux dérivations et par définition elles coincident sur $\mathcal{A}(t)$. En outre pour chaque $j \in \mathbb{N}$ on a $x_j \xi \partial = x_{j+1} \xi = x_j \chi \xi$.

Par conséquent on a

(8) $K\xi\partial = \kappa\chi\xi$

quelque soit la série formelle K en t et les x_j .

Posons maintenant $x_0 = y$, $x_1 = y'$ et définissons une dérivation η de $\bar{\mathcal{A}}(t)$ par les conditions $a\eta = 0$ pour $a \in \mathcal{A}$, $t\,\eta = 1$; $y\,\eta = y'$, $y'\,\eta = H(t, y, y')$.

Si $X = Y$ est la solution formelle de notre équation, nous avons l'identité

$$Y\,\partial^{n+2} = H(t, x_0, x_1)\,\xi\,\partial^n = H(t, x_0, x_1)\,\chi^n\,\xi$$

d'où en particulier $x_2\,\xi = Y\,\partial^2 = H(t, x_0, x_1)\,\xi$ puis par induction sur n $x_{n+2}\,\xi = Y\,\partial^{n+2} = H\,\eta^n\,\xi = y\,\eta^{n+2}\,\xi$ ce qui donne enfin d'après (7) la formule cherchée :

$a_n = Y\,\partial^n\,\theta = y\,\eta^n\,\theta$ c'est à dire

$$(9)\quad Y = \sum_{o\leq n} \frac{t^n}{n!}\,(y\,\eta^n\,\theta)\,.$$

I.3. Considérons à titre d'exemple l'équation classique

$$y' = ay - yb + c \quad (a, b, c \in \mathcal{A})$$

dont la solution remonte à Wedderburn. En raison de sa forme particulière la dérivation η se réduit au morphisme de $\bar{\mathcal{A}}(t)$ laissant $\mathcal{A}(t)$ invariant et envoyant y sur $ay - yb + c$. En outre θ se réduit à l'identité. La solution Y est donc égale à

$$y + \frac{t}{1!}(ay - by + c) + \ldots + \frac{t^n}{n!}(y\,\eta^n) + \ldots$$

Par induction sur n on pourrait d'ailleurs vérifier que le coefficient $a_n = y\,\eta^n$ peut s'expliciter de la façon suivante :

$$a_n = \sum_{o\leq j\leq n} (-1)^j \begin{bmatrix} n \\ j \end{bmatrix} a^{n-j}\,y\,b^j + \sum_{o\leq j\leq n-1} (-1)^j \begin{bmatrix} n-1 \\ j \end{bmatrix} a^{n-1-j} c\, b^j$$

Un autre exemple est fourni par l'équation $y'' = y\,y'$ qui nous interessera plus particulièrement ici. La dérivation η laisse $\mathcal{A}(t)$ invariant et elle envoye y sur y' et y' sur $y\,y' = H$. Les premiers termes du développement de la solution Y sont $Y = y + \frac{t}{1!}\,y' + \frac{t^2}{2!}(y\,y') + \ldots$ et on calcule :

$$a_3 = (y\, y')\, \eta = y'\, y' + y\, y\, y'$$
$$a_4 = (y'^2 + y^2\, y')\, \eta = y\, y'\, y' + y'\, y\, y' + y'\, y\, y'$$
$$+ y\, y'\, y' + y^2\, y\, y' = (y^3 + 2\, y\, y' + 2\, y'\, y)\, y'$$
$$a_5 = a_4\, \eta = (y^4 + 3\, y^2\, y' + 5\, y\, y'\, y + 3\, y'\, y^2 + 4\, y'^2)\, y'$$
$$a_6 = a_5\, \eta = (y^5 + 4\, y^2\, y' + 9\, y^2\, y'\, y + 9\, y\, y'\, y^2 + 4\, y'\, y^2$$
$$+ 12\, y\, y'^2 + 10\, y'\, y\, y' + 12\, y'^2\, y)\, y'$$

....

L'expression des coefficients est passablement compliquée et sera donnée (de façon indirecte) dans la section suivante. Les polynomes a_n ont été définis dans un tout autre contexte (où ils sont dénotés par D_{n+1}). Leurs valeurs pour $y = y' = 1$ sont les nombres d'Euler.

II. L'EQUATION $y'' = y\, y'$

II.1. Nous gardons les notations utilisées dans l'exemple ci-dessus et nous nous proposons d'expliciter les polynomes $a_n = y\, \eta^n \in \overline{\mathcal{A}} = \mathcal{A}[y, y']$ au moyen d'un changement de variable. Pour cela nous considérons deux nouvelles variables x et z et l'anneau $\overline{\mathcal{B}} = \mathcal{A}[x, z]$ de leurs polynomes non commutatifs. Nous définissons le morphisme φ, l'anneau de polynomes et la dérivation σ de $\overline{\mathcal{B}}$ par les conditions suivantes :

(10.1) $x\varphi = y$; $z\varphi = -y^2 + 2\, y'$;

(10.2) $x\, \sigma = x^2 + z$; $z\, \sigma = 0$.

Nous établissons d'abord la remarque (11) :

Pour chaque $\eta \in \mathbb{N}$ on a $a_{n+1} = b_n\, \varphi\, y'\, 2^{-n}$ où les b_n sont des polynomes de $\overline{\mathcal{B}}$ définis par la récurrence

$$b_o = 1 \;;\; b_{n+1} = b_n\, \sigma + x\, b_n + b_n\, x \; (= b_n\, \lambda).$$

Preuve : On a $a_o = y$ et d'après $y\, \eta = y'$ $y'\, \eta = y\, y'$, il existe pour chaque $\eta \in \mathbb{N}$ un polynome a'_n tel que $a_{n+1} = a'_n y'$.

Comme $a'_{n+1}\, y' = a_{n+1} = a_n\, \eta = a'_n\, y'\, \eta = a'_n\, \eta\, y' + a'_n\, (y'\, \eta)$ $= a'_n\, \eta\, y' - a'_n\, y\, y'$, on a la récurrence $a'_{n+1} = a'_n\, \eta + a'_n\, y$ avec $a'_o = 1$.

Notons maintenant que φ est un isomorphisme dont l'inverse ψ est définie par $y\,\psi = x$ et $y'\,\psi = \frac{1}{2}(x^2 + z)$.
Posant $b_n = a'_n\, \psi \,.\, 2^n$ nous aurons donc $b_o = 1$ et $a_n = b_n\, \varphi \,.\, y'$ pour chaque n positif. De plus les b_n satisferont la relation de récurrence

$$b_{n+1} = a'_{n+1}\, \psi \,.\, 2^{n+1} = 2\,(a'_n\, \eta\, \psi \,.\, 2^n + a'_n \,.\, y\, \psi \,.\, 2^n)$$
$$= 2\,(b_n\, \varphi\, \eta\, \psi + b_n\, x) \,.$$

Maintenant, comme φ est un isomorphisme l'application linéaire $\bar{\eta} = \varphi\, \eta\, \psi$ est une dérivation de $\mathcal{B}$ dont l'action sur les générateurs x et z est donnée par le calcul

$$x\, \bar{\eta} = y\, \eta\, \psi = y'\, \psi = \frac{1}{2}(x^2 + z) = \frac{1}{2}\, x\, \sigma \;;$$
$$z\, \bar{\eta} = (2\, y' - y^2)\, \eta\, \psi = (2\, y\, y' - y\, y' - y'\, y)\, \psi$$
$$= (y\, y' - y'\, y)\, \psi = x\, \frac{1}{2}\,(x^2 + z) - \frac{1}{2}\,(x^2 + z)\, x$$
$$= \frac{1}{2}\,(x\, z - z\, x) \,.$$

Ceci permet d'écrire $\bar{\eta} = \frac{1}{2}(\sigma + \rho)$ où ρ est la dérivation envoyant x sur 0 et z sur $x\, z - z\, x$, et nous avons donc $b_{n+1} = b_n\, \sigma + b_n\, \rho + 2\, b_n\, x$.

Il suffit désormais de montrer que pour tout monôme u en x et z on a identiquement $u\, \rho + 2\, u\, x = x\, u + u\, x$. Or ceci est trivial quand $u = x^n$ puisque $x^n\, \rho = 0$ et $x^n\, x = \frac{1}{2}(x\, x^n + x^n\, x)$.

Supposons donc ce résultat établi pour le monome v et prenons $u = v\, z\, x^n$. On a :

$$u\,\rho + 2\,u\,x = v\,\rho\,.\,z\,x^{n} + v\,(x\,z - z\,x)\,x^{n} + 2\,v\,z\,x^{n+1}$$
$$= x\,v\,z\,x^{n} - v\,x\,z\,x^{n} + v\,x\,z\,x^{n} - v\,z\,x^{n+1} + 2\,v\,z\,x^{n+1}$$
$$= x\,u + u\,x \qquad \text{Q.E.D.}$$

Dans ce qui suit nous nous bornerons au calcul des polynômes $b_n = 1\,\lambda^n$ dont les premiers sont

$$b_o = 1 \;\; ; \;\; b_1 = b_o\,\lambda = x1 + 1\,x = 2\,x \;\; ;$$
$$b_2 = 2\,x\,\lambda = 2\,(x^2 + z) + x\,.\,2\,x + 2\,x\,.\,x = 6\,x^2 + 2\,z \;\; ;$$
$$b_3 = 24\,x^3 + 8\,(x\,z + z\,x) \;\; ;$$
$$b_4 = 120\,x^4 + 40\,(x^2\,z + x\,z + z\,x^2) + 16\,z^2 \;\; ;$$

Notre résultat principal est résumé par les formules 17 et 22 ci-dessous. Notant M le monoïde libre engendré par $\{x, z\}$, c'est à dire l'ensemble des monômes en x et z nous écrirons chaque polynôme b de B comme une somme finie $\Sigma\,\{\langle b, m\rangle\, m : m \in M\}$ dont les coefficients $\langle b, m\rangle$ appartiennent à $\mathcal{A}$. L'application $\langle\, ,\, \rangle$ sera étendue de façon naturelle à une application bilinéaire de $\mathcal{B} \times \mathcal{B}$ dans $\mathcal{A}$.

Le degré $|m|$ d'un mot $m \in M$ sera par définition $|m|_x + 2\,|m|_z$ où, comme d'usage, $|m|_x$ (resp. $|m|_z$) dénote le nombre d'occurence de x (resp. z) dans m.

On vérifie sans difficulté que chacun des termes d'un polynôme $m\,\lambda$ $(m \in M)$ est de degré $|m| + 1$.

Comme $b_o = 1$ est homogène de degré zéro, il en résulte que chaque $b_n = 1\,\lambda^n$ est homogène de degré n. Ceci entraine en particulier que chaque mot m ait le même coefficient dans $b_{|m|}$ et dans la série formelle $B = \sum_{o \le n} b_n$. Ceci nous permettra d'utiliser la notation $\langle B, m\rangle$ au lieu de $\langle b_{|m|}, m\rangle$.

III. CALCUL DES POLYNOMES b_n

III.1. Nous désignerons par μ la transposée de $\lambda\dagger$ de λ, c'est à dire l'application linéaire de $\mathcal{B}$ telle que pour chaque paire $m, m' \in M$ le coefficient de m dans $m'\mu$ soit égal au coefficient de m' dans $m\lambda$. On vérifie facilement que chaque $m'\mu$ est un polynôme. Par définition on a la formule

(12) $b\lambda = \Sigma \langle b, m\mu\rangle\, m$

pour tout polynôme (ou série formelle) b. En particulier

(13) $\langle B, m\rangle = \langle b, m\mu \rangle$ identiquement.

Comme λ est <u>rationnelle</u> au sens de Eilenberg, un théorème de cet auteur que nous nous bornerons à appliquer donne une technique générale pour calculer M. Pour cela nous écrivons $\sigma = \sigma_x + \sigma_z$ où σ_x est la dérivation envoyant x sur x^2 et σ_z la dérivation envoyant x sur z, les noyaux de ces deux dérivations étant les polynômes qui ne contiennent pas la variable x. La transposée μ de λ est la somme des transposées des applications linéaires σ_x, σ_z, $m \longrightarrow m\,x$ et $m \longrightarrow x\,m$.

En ce qui concerne ces deux dernières, ce sont les applications $m \longrightarrow m\,x^{-1}$ et $m \longrightarrow x^{-1}\,m$ où, comme d'usage, la notation $m\,x^{-1}$ (resp. $x^{-1}\,m$) désigne le monome m' tel que $m'\,x = m$ (resp. $x\,m' = m$) s'il en éxiste un et 0 autrement. En ce qui concerne $\sigma_z\dagger$ c'est l'application envoyant chaque $m \in M$ sur la somme des produits $m_1\,x\,m_2$ où $m_1, m_2 \in M$ satisfont $m_1\,z\,m_2 = m$.

Enfin $m\,\sigma_x\dagger$ est la somme des monômes $m_1\,x\,m_2$ étendue à toutes les <u>paires</u> $m_1, m_2 \in M$ telles que $m_1\,x\,x\,m_2 = m$.

On voit sans peine que $x^{-1}\,m + m\,x^{-1} + m\,\sigma_z\dagger$ a tous ses coefficients 0 ou 1 et qu'aucun mot du <u>support</u> de $m\,\sigma_z\dagger$ (c'est à dire de l'ensemble des mots ayant un coefficient non nul dans ce polynôme) n'appartient au support de m $m\,\sigma_x^{-1}$.

Par contre, si $m = m_1\, x^n\, m_2$ où $n \geq 2$ et où les mots m_1 et m_2 satisfont la condition $m_1 \notin M_x$, $m_2 \notin x\,M$ (c'est à dire, pour abréger si x^n est un x^*- facteur maximal de m), le mot $m' = m_1\, x^{n-1}\, m_2$ apparait dans $m\,\sigma_x\dagger$ avec la multiplicité n-2. Si en outre $m_1 = 1$, le même mot m' apparait dans $x^{-1}\,m$ ce qui fait que son coefficient dans $m\mu$ devient n-1. La même remarque vaut quand $m_2 = 1$.

A titre d'application nous calculons

(14) Pour chaque $n \in \mathbb{N}$, le coefficient $< B, x^n >$ est égal à $(n + 1)!$.

<u>Preuve</u> : Ceci est vrai pour $n = 0$, puisque $b_o = 1 = (0 + 1)!$. Soit donc n positif. On a $x^n\,\sigma_z\dagger = 0$ et par conséquent

$$x^n\,\mu = x^n\,\sigma_x\dagger + x^{-1}\,x^n + x^n\,x^{-1} = (n-1)\,x^{n-1} + x^{n-1} + x^{n-1}$$
$$= (n + 1)\,x^{n-1} .$$

Comme $< B, x^n > = < B, x^n\,\mu >$ d'après (13) le résultat en découle par induction. Q.E.D.

Nous établissons maintenant un énoncé technique essentiel pour la suite. Pour abréger nous notons $\mathcal{M}$ la relation dans $M \times M$ telle que $(m', m) \in \mathcal{M}$ ssi m appartient au support de $m'\,\sigma_x + x\,m' + m'\,x$ et nous étendons $\mathcal{M}$ à une relation $\overline{\mathcal{M}}$ dans $\mathcal{B} \times \mathcal{B}$ en posant $(b', b) \in \overline{\mathcal{M}}$ ssi il existe une bijection α entre les supports des polynômes b' et b telle que pour tout monôme m' ou ait d'une part $(m', m'\alpha) \in \mathcal{M}$ et d'autre part $< b', m' > = < b, m'\alpha >$.

<u>Propriété 15</u> : Soient $(m', m) \in \mathcal{M}$. On a $(m'\,\mu, m\,\mu - m') \in \overline{\mathcal{M}}$

<u>Preuve</u> : La vérification que $(m' \sigma_z\dagger, m \sigma_z\dagger) \in \overline{\mathcal{M}}$ est facile. Pour le reste, nous distinguons trois cas, et comme le résultat résulte immédiatement de (14) quand m et m' sont des puissances de x nous pouvons supposer que $|m|_z$ est positif.

<u>Cas i</u> : $m = x m'$.
Nous pouvons écrire $m' = x^n z m''$ $(n \in \mathbb{N}, m'' \in M)$ et comme $m' = x^{-1} m$ il nous suffira de vérifier d'une part $(m' x^{-1}, m x^{-1}) \in \overline{\mathcal{M}}$ ce qui est trivial et d'autre part $(m' \sigma_x\dagger + x^{-1} m', m \sigma_x\dagger) \in \mathcal{M}$.

Soit S l'ensemble des paires (m_1, m_2) telles que $m_1 x x m_2 = m$. Le sous ensemble S_1 de celles pour lesquelles $|m_1|_z$ est positif, est en correspondance biunivoque par $m_1 \longrightarrow x^{-1} m_1$ avec l'ensemble correspondant pour m'. En outre l'ensemble $S \setminus S_1$ est formé de n paires $(x^i, x^{n-i-1} z m'')$ $(o \leq i \leq n-1)$ ce qui fait que $x^n z m''$ a le coefficient n dans $m \sigma_x\dagger$.

On peut donc se limiter au cas de n positif et on vérifie de même que $x^{n-1} z m''$ a le coefficient $(n-1) + 1$ dans $m' \sigma_x\dagger + x^{-1} m'$ ce qui établit le résultat dans ce cas.

<u>Cas ii</u> : $m = m' x$.
Le même raisonnement s'applique par symétrie.

<u>Cas iii</u> : $m \neq m' x, x m'$.
On peut écrire $m' = m'' z x^n z m'''$ où z est positif et l'on a $m = m'' z \underline{x^{n+1}} z m'''$. Il est clair que $(x^{-1} m' + m' x^{-1}, x^{-1} m + m x^{-1})$ est dans $\overline{\mathcal{M}}$. Comme ci-dessus le sous ensemble $S_1 \subset S$ des paires (m_1, m_2) telles que $m_1 x x m_2 = m$ qui satisfont la condition supplémentaire que $|m_1|_z \neq |m''|_z$ est en correspondance biunivoque avec le sous ensemble correspondant pour m'. En ce

qui concerne les autres paires elles produisent le mot m' avec la multiplicité n et par conséquent le coefficient de m' dans $m\,\sigma_x\dagger - m'$ est $n-1$. Le mot correspondant $m''\, z\, x^{n-1}\, z\, m''$ a le coefficient $n-1$ dans $m'\,\sigma_x\dagger$ et le résultat est donc établi dans tous les cas. Q.E.D.

(16) Corollaire : Soit $(m', m) \in \mathcal{M}$. On a
$< B, m > \; = \; < B, m' > . (|m| + 1)$.

Preuve : Si $|m| = 1$, on a $m = x$ et il existe un seul m' , à savoir 1 tel que $(m', m) \in \mathcal{M}$. Comme $< B, x > = 2$ ainsi qu'on l'a vu dans (14), la formule vérifiée dans ce cas et nous pouvons procéder par induction sur le degré de m .
D'après (13) nous avons
$< B, m> = < B, m\mu >$ et $< B, m' > = < B, m'\,\mu >$.

De plus d'après (15) $m\,\mu$ est la somme de m' et d'un polynôme b tel que $(m'\,\mu, b) \in \mathcal{M}$. Comme b est homogène de degré $|m|-1$, il en résulte par l'hypothèse d'induction que
$< B, b > = < B, m'\mu > . |m|$. Par conséquent
$< B, m > = < B, m' > + < B, m'\,\mu > |m|$ où
$< B, m'> = < B, m'\mu >$ ce qui est le résultat cherché. Q.E.D.

Considérons maintenant un mot quelconque $m \in M$. On peut l'écrire sous la forme

$$m = x^{n_o}\, z^{p_1}\, x^{1+n_1}\, z^{p_2} \ldots x^{1+n_{k-1}}\, z^{p_k}\, x^{n_k}$$

où les p_i sont positifs et où k et les n_i sont non négatifs.

Désignons par $\hat{m}$ le mot $z^{p_1}\, x\, z^{p_2} \ldots x\, z^{p_k}$ (et par conséquent $\hat{m} = 1$ ssi $m = x^{n_o}$).

(17) Pour tout $m \in M$, on a $< B, m > = < B, \hat{m}> \dfrac{(1 + |m|)!}{(1 + |\hat{m}|)!}$

Preuve : Par induction sur $|m| - |\hat{m}|$ en observant que si $m \neq \hat{m}$, il existe au moins un m_1 tel que $(m_1, m) \in \mathcal{M}$ et que pour tout m_i satisfaisant cette dernière relation on a $\hat{m}_i = \hat{m}$.

Q.E.D.

(18) Quelques soient l'entier n et le polynome homogène a on a $< B, a x^n > = < B, a > \dfrac{(|a| + n + 1)!}{(|a| + 1)!} = < B, x^n a >$.

Preuve : Ceci résulte immédiatement de (17) et de ce que $\hat{m}_1 = \hat{m}$ pour chaque m_1 de la forme $m x^n$ ou $x^n m$. Q.E.D.

III.2. Le résultat principal

Définissons d'abord une suite infinie de nombres rationnels (d_k) par la récurrence

(20) $d_o = 1$; $d_{k+1} = (2k + 3)^{-1} \sum_{o \leq j \leq k} d_j d_{k-j}$.

Nous vérifions d'abord

(21) pour chaque entier positif k et chaque mot $g \in \{1\} \cup M x$ on a $< B, g z^k > = < B, g x^{2k} > d_k$.

Preuve : Le résultat est facilement vérifié pour $|g z^k| \leq 2$ et nous pouvons procéder par induction sur le degré.

Considérons d'abord le cas de $g = 1$. On a $z^{k+1} \mu = z^{k+1} \sigma_z \dagger = \sum_{o \leq j \leq k} z^{k-j} x z^j$.

D'après l'hypothèse d'induction et (17) on a

$$< B, z^{k-j} x z^j > = < B, z^{k-j} x^{2j+1} > d_j$$

$$= < B, z^{j-j} > d_j \frac{(2k + 2)!}{(2k - 2j + 1)!}$$

$$= < B, x^{2k-2j} > d_j \frac{(2k + 2)!}{(2k - 2j + 1)!}$$

$$= d_{k-j} \cdot d_j \,(2k+2) = (2k+3)^{-1} < B,\ x^{2k+2} > .$$

Par conséquent nous avons bien

$$< B,\ z^{k+1} > = <B,\ x^{2k+2}> . (2k+3)^{-1} \sum_{o \le j \le k} d_j\, d_{k-j} \ .$$

Supposons maintenant que g soit différent de 1, c'est à dire que $g = f\,x$ $(f \in M)$. On vérifie facilement que $g\, z^{k+1}\, \mu$ est la somme de $b = g\,(z^{k+1}\, \mu)$ et d'un autre polynôme c qui est égal à z^{k+1} si $f = 1$ et à $(f\,\mu)\, x\, z^{k+1}$ si $f \neq 1$.

D'après l'hypothèse d'induction, on trouve que

$$b = < B,\ g\, x^{2k+1} > \sum_{o \le j \le k} d_j\, d_{k-j}$$

$$= < B,\ g\, x^{2k+1} > (2k+3)\, d_{k+1} \quad ;$$

D'autre part $c = d_{k+1}$ si $f = 1$ et sinon, utilisant l'hypothèse d'induction et (17), on trouve

$$c = < B,\ g\, x^{2k+1} > (\,|f| + 1)\, d_{k+1} \ .$$

Par conséquent dans tous les cas

$$< B,\ g\, z^{k+1} > = < B,\ b > + < B,\ c >$$

$$= < B,\ g\, x^{2k+1} > \ (|g| + 2k + 3)\, d_{k+1}$$

ce qui est bien égal à $< B,\ g\, x^{2k+2} > d_{k+1}$. Q.E.D.

Nous sommes maintenant à même d'énoncer notre résultat principal :

(22) Si $m = x^{n_o}\, z^{p_1}\, x^{n+n_1} \ldots z^{p_k}\, x^{n_k}$

$(p_1, p_2, \ldots, p_k \ge 1)$ on a

$$< B,\ m > = (1 + |m|\,)!\, d_{p_1}\, , d_{p_2} \ldots d_{p_k}$$

Preuve : Pour $k = 0$ ceci est la formule (14). Le cas général s'en déduit par induction sur k au moyen de (22) et (17) . Q.E.D.

On notera que comme tous les coefficients intervenant dans λ sont des entiers non négatifs, il en est de même des coefficients $< B, m >$ des polynômes $b_n = 1\ \lambda^n$ $(n \in \mathbb{N})$ bien que les d_k soient des nombres fractionnaires.

III.3. Relations avec les nombres d'Euler.

Soit Y_o la solution de $y'' = y\ y'$ telle que $Y_o = 0$ pour $t = 0$. Elle est obtenue en faisant $y = 0$ dans les polynômes a_n décrits dans la première section de ce chapitre. Nous noterons a_{on} ces polynômes qui sont donc des polynômes en y'.

En particulier $a_{oo} = 0$ et $a_{o1} = y'$.

Rappelant la formule $a_{n+1} = b_n\ \varphi\ .\ y'\ 2^{-n}$ utilisée pour définir les polynômes b_n au moyen du morphisme φ envoyant x sur y et z sur $-y^2 + 2y'$, nous aurons
$a_{o,n+1} = b_{on}\ \varphi\ .\ y'\ 2^{-n}$ où $b_{o,n}$ est le polynôme obtenu en faisant $x = 0$ dans b_n. Comme b_n est homogène de degré n et que les degrés de x et z sont respectivement 1 et 2 nous savons que b_{on} est nul pour n impair et qu'il se réduit à $< B, z^k > z^k$ pour $n = 2k$ $(k \in \mathbb{N})$.

De plus, d'après (22) et (17) nous savons que $< B, z^k >$ est égal à $(2k + 1)!\ d_k$.

Il en résulte que $a_{o,n} = 0$ pour n pair et que pour $n = 2k + 1$ on a $a_{o,2k+1} = 2^{-k}\ (2k + 1)!\ d_k\ y'^{k+1}$ et

$$Y_o = \sum_{o \leq k} \frac{t^{2k+1}}{(2k+1)!}\ a_{o,2k+1} \quad .$$

Faisons maintenant $y' = 1$ dans Y_o. La série obtenue est la solution de l'équation différentielle (ordinaire) $y'' = y\ y'$

qui, pour $t = 0$, prend la valeur 0 cependant que sa dérivée prend la valeur 1, c'est à dire comme il est bien connu, la fonction tg (t). Les nombres $2^{-k}(2k+1)!\, d_k$ en sont les coefficients de Hurwitz; ce sont donc les nombres d'Euler d'indice impair.

IV. UNE AUTRE APPLICATION

IV.1. Nous commençons par un complément aux généralités de I. Considérons un anneau de polynomes (non commutatifs) $\mathcal{A}[u, v, w, \ldots]$ en les variables $u, v, w, \ldots$, son quotient $\mathcal{C}$ par la congruence $u\,v \equiv v\,u \equiv 1$ et une dérivation ζ telle que :

(23) $v\,\zeta = -\,v\,(u\,\zeta)\,v$.

Remarque : ζ est une dérivation de $\mathcal{C}$.

Preuve : Il suffit de montrer que pour tout $a, b \in \mathcal{A}[u, v, \ldots]$ on a $a\,u\,v\,b\,\zeta \equiv a\,v\,u\,b\,\zeta \equiv a\,b\,\zeta$.
Calculons $a\,u\,v\,b\,\zeta$. On a
$a\,u\,v\,b\,\zeta = (a\,\zeta)\,u\,v\,b + a\,u\,v\,(b\,\zeta) + a\,(u\,\zeta)\,v\,b + a\,u\,(v\,\zeta)\,b$.

La somme des deux premiers termes est congrue à $(a\,\zeta)\,b + a\,(b\zeta)$ c'est à dire à $a\,b\,\zeta$. La somme des deux derniers termes est congrue à 0 puisque
$a\,u\,(v\,\zeta)\,b = -\,a\,u\,v\,(u\,\zeta)\,v\,b \equiv -\,a\,(u\,\zeta)\,b\,v$.
Un calcul analogue vaut pour $a\,v\,u\,b\,\zeta$. Q.E.D.

Il résulte de ceci que $(u\,v)\,\zeta = (v\,u)\,\zeta = 1\,\zeta = 0$, donc que $(u\,v)\,\zeta^n = (v\,u)\,\zeta^n = 0$ pour tout n positif. Or, comme ζ est une dérivation, on a identiquement
$(a\,b)\,\zeta^n = \sum_{o \le j \le n} (a\,\zeta^j)\,(b\,\zeta^{n-j}) \begin{bmatrix} n \\ j \end{bmatrix}$ quelque soient a, b et $n \ge 0$.

Divisant par $n!$ on en déduit l'identité

$$(24) \quad \sum_{0\le j\le n} \frac{u\,\zeta^j}{j!} \cdot \frac{v\,\zeta^{n-j}}{(n-j)!} = 0 \quad \text{pour chaque } n \text{ positif.}$$

Introduisant une nouvelle variable t commutant avec u et v et posant

$$U = \sum_{0\le n} \frac{t^n}{n!} \cdot u\,\zeta^n \; ; \; V = \sum_{0\le n} \frac{t^n}{n!} \cdot v\,\zeta^n$$

il en résulte finalement la relation

(25) $U\,V = V\,U = 1$

par identification des coefficients de t^n et (24) .

IV.2. Nous revenons aux notations de II.1. et nous adjoignons à $\mathcal{A}(t)$ une nouvelle variable v satisfaisant $y'\,v = v\,y' = 1$ (ce qui entraine $v\,t = t\,v$ puisque $t\,y' = y'\,t$)

En conformité avec (23), nous étendons la dérivation η (définie par $y\,\eta = y'$, $y'\,\eta = y\,y'$) en posant $v\,\eta = -\,v\,(y'\,\eta)\,v$ ce qui donne $v\,\eta = -\,v\,y\,y'\,v = -\,v\,y$.

D'après (25) les deux séries formelles

$$Y' = \sum_{0\le n} \frac{t^n}{n!}\,(y'\,\eta^n) \quad \text{et} \quad V = \sum_{0\le n} \frac{t^n}{n!}\,(v\,\eta^n)$$

sont inverses l'une de l'autre et nous proposons de calculer directement les coefficients $v\,\eta^n$ de V .

<u>Remarque 26</u> : Pour chaque $n \in \mathbb{N}$ on a

$$2\,v\,\eta^n = v\,(C_n\,\varphi)$$

où les C_n sont des polynomes dans $\overline{\mathcal{B}}$ définis par la récurrence

$$C_o = 2 \; ; \; C_{n+1} = \frac{1}{2}\,(C_n\,\sigma - x\,C_n - C_n\,x) = C_n\,\nu.$$

<u>Preuve</u> : C'est essentiellement la même que celle de la remarque 11 dont nous utilisons les notations.

Nous avons $v\,\eta^0 = v$ et par conséquent nous pouvons prendre $C_0 = 2 \in \overline{\mathcal{B}}$

Supposons maintenant que $2\,v\,\eta^n = v\,(C_n\,\varphi)$ où $C_n \in \overline{\mathcal{B}}$.

Nous avons :

$$\begin{aligned} 2\,v\,\eta^{n+1} &= (v\,(C_n\,\varphi))\,\eta = v\,\eta\,.\,(C_n\,\varphi) + v\,(C_n\,\varphi\,\eta) \\ &= -\,v\,y\,(C_n\,\varphi) + v\,(C_n\,\varphi\,\psi\,\bar{\eta}\,\varphi\,) \\ &= -\,v\,x\,\varphi\,(C_n\,\varphi) + v\,(C_n\,\bar{\eta}\,\varphi)\ . \end{aligned}$$

Mettant v en facteur nous obtenons donc $2\,v\,\eta^{n+1} = v\,C_{n+1}$ où $C_{n+1} = -\,x\,C_n + C_n\,\bar{\eta}$ appartient à $\overline{\mathcal{B}}$ par induction.

Il ne reste plus qu'à vérifier que cette équation peut se mettre sous la forme plus simple

$$C_{n+1} = \frac{1}{2}\,(C_n\,\sigma\ -\,x\,C_n - C_n\,x) = C_n\,\nu$$

ce qui est facile compte tenu de $\bar{\eta} = \sigma + \rho$ et des propriétés de ρ établies dans la preuve de la remarque 11; Q.E.D.

On trouve pour les premiers termes

$C_0 = 2$; $C_1 = -\,2\,x$; $C_2 = x^2 - z$; $C_3 = x\,z + z\,x$;

On a les formules suivantes :

(27) $C_{2k} = (-1)^k\,(z^k - x\,z^{k-1}\,x)$ pour $k \geq 1$;

$C_{2k+1} = (-1)^{k+1}\,(x\,z^k + z^k\,x)$ pour $k \geq 0$.

<u>Preuve</u> : On vérifie directement que $C_1 = -\,2\,x$.

Supposant maintenant que C_{2k+1} a la forme (27), on trouve :

$$\begin{aligned} &2\,(-1)^{k+1}\,C_{2k+2} = 2\,(-1)^{k+1}\,C_{2k+1}\,\nu \\ &= (x^2 + z)\,z^k + z^k\,(x^2 + z) - x^2\,z^k - x\,z^k\,x - x\,z^k\,x - z^k\,x^2 \\ &= 2\,(z^{k+1} - x\,z^k\,x)\ . \end{aligned}$$

De même supposant que C_{2k} $(k \geq 1)$ a la forme (27) on obtient

$$2\,(-1)^k\ C_{2k+1} = 2\,(-1)^k\ C_{2k}\ \nu = -\,(x^2 + z)\ z^{k-1}\ x$$

$$-\ x\ z^{k-1}\ (x^2 + z) - x\ z^k + x^2\ z^{k-1}\ x - z^k x + x\ z^{k-1}\ x^2$$

$$= -\ 2\ (x\ z^k + z^k\ x)\ .$$

Q.E.D.

Posant $C = \sum_{o \leq n} t^n\ C_n$ les formules précédentes peuvent être résumées par la formule unique

(28) $C = 1 + (t - t\ x)\ (1 + t\ z)^{-1}\ (1 - t\ x)$

où, comme d'usage, $(1 + t\ z)^{-1}$ est une abréviation pour $\sum_{o \leq n} (-1)^n\ t^n\ z^n$ ce qui donnerait facilement l'expression de $\sum_{o \leq n} t^n\ (v\ \eta^n)$ en appliquant le morphisme φ . Il me parait très remarquable que cette dernière fonction se trouve ainsi être une <u>fonction rationnelle</u> en tous ses arguments.

IV.3. Nous terminons en montrant que la fonction V de t est la solution de l'équation différentielle

(29) $V'' = V'\ V^{-1}\ V' - 1$

où $V' = V\ \partial$, $V'' = V\ \partial$

et où ∂ est la dérivation de noyau $\overline{\mathcal{B}}$ envoyant t sur 1 qui a été définie et utilisée dans le chapitre I .

Tout d'abord, nous avons par hypothèse $Y'' = Y\ Y'$ d'où $Y''' = Y\ Y'' + Y'^2$ ce qui donne $Y = Y''\ Y'^{-1}$ et $Y''' = Y''\ Y'^{-1} Y'' + Y'^2$ ce qui peut se réécrire

$$Y'^{-1}\ Y''' Y'^{-1} = Y'^{-1}\ Y''\ Y'^{-1}\ Y''\ Y'^{-1} + 1\ .$$

Maintenant d'après (25) nous avons $V = Y'^{-1}$ d'où d'après (23)

$V' = -\ V\ Y''\ V$ puis $V'' = -\ V'\ Y''\ V - V\ Y'''\ V - V\ Y''\ V'$

$= 2\ V\ Y''\ V\ Y''\ V - V\ Y'''\ V$ ce qui est égal à $V\ Y''\ V\ Y''\ V - 1$

d'après l'expression trouvée plus haut pour

$$Y'^{-1} \, Y''' \, Y'^{-1} = V \, Y''' \, V$$

et le résultat s'en déduit en utilisant la relation

$$V \, Y'' \, V = - V' \quad .$$

Q.E.D.

REFERENCES

[1] D. André. Developpements de Sec x et tang x . C.R. Acad. Sc. Paris 88 (1879) - 965 - 967 .

[2] S. Eilenberg. Theory of Automata. (Sous presse)

[3] D. Foata et M. P. Schützenberger. Nombres d'Euler et permutations alternantes. In a Survey of Combinatorial Theory (J. N. Stivastaver Ed) Amsterdam 1973 .

[4] N. Nielsen. Traite élementaire des nombres de Bernouilli. Paris 1923 .

[5] I. Riordan. Combinatorial Identities. N. Y. 1970 .

[6] Wedderburn. Lectures on Matrices. Am. Math. Soc. Colloq. Publi. 17 1934 .